Homomorphic Encryption for Data Science (HE4DS)

Allon Adir • Ehud Aharoni • Nir Drucker •
Ronen Levy • Hayim Shaul • Omri Soceanu

Homomorphic Encryption for Data Science (HE4DS)

Allon Adir
IBM Research
Haifa University Campus
Mount Carmel Haifa, Israel

Ehud Aharoni
IBM Research
Haifa University Campus
Mount Carmel Haifa, Israel

Nir Drucker
IBM Research
Haifa University Campus
Mount Carmel Haifa, Israel

Ronen Levy
IBM Research
Haifa University Campus
Mount Carmel Haifa, Israel

Hayim Shaul
IBM Research
Haifa University Campus
Mount Carmel Haifa, Israel

Omri Soceanu
IBM Research
Haifa University Campus
Mount Carmel Haifa, Israel

ISBN 978-3-031-65493-0   ISBN 978-3-031-65494-7 (eBook)
https://doi.org/10.1007/978-3-031-65494-7

© The Editor(s) (if applicable) and The Author(s), under exclusive license to Springer Nature Switzerland AG 2024

This work is subject to copyright. All rights are solely and exclusively licensed by the Publisher, whether the whole or part of the material is concerned, specifically the rights of translation, reprinting, reuse of illustrations, recitation, broadcasting, reproduction on microfilms or in any other physical way, and transmission or information storage and retrieval, electronic adaptation, computer software, or by similar or dissimilar methodology now known or hereafter developed.
The use of general descriptive names, registered names, trademarks, service marks, etc. in this publication does not imply, even in the absence of a specific statement, that such names are exempt from the relevant protective laws and regulations and therefore free for general use.
The publisher, the authors and the editors are safe to assume that the advice and information in this book are believed to be true and accurate at the date of publication. Neither the publisher nor the authors or the editors give a warranty, expressed or implied, with respect to the material contained herein or for any errors or omissions that may have been made. The publisher remains neutral with regard to jurisdictional claims in published maps and institutional affiliations.

This Springer imprint is published by the registered company Springer Nature Switzerland AG
The registered company address is: Gewerbestrasse 11, 6330 Cham, Switzerland

If disposing of this product, please recycle the paper.

"If you want to keep a secret, you must also hide it from yourself." —George Orwell

Preface

IBM Research served as a development accelerator of the fully homomorphic encryption (FHE) technology since Craig Gentry laid the foundation for FHE in 2009. Since then researchers such as Gentry, Shai Halevi, and Victor Shoup continued developing, optimizing, and proposing new FHE schemes, such as the commonly used BGV scheme that is, at the time of writing, one of the schemes that are currently being standardized by ISO. IBM Research continued its contribution to the FHE domain by releasing the first open-source FHE library (HElib) with an optimized version of BGV and later on the CKKS FHE scheme as well.

Nevertheless, even with the major advancements made in the field of FHE, IBM identified that there is still a large gap between the expertise needed by end-users such as chief information security officers (CISOs), data scientists, and application developers and the low-level application interfaces (APIs) provided by FHE libraries. At the time, the use of these libraries was almost exclusively limited to applied cryptographers. FHE experts like these knew how to tune the FHE parameters for their specific needs and were able to construct mostly simple FHE-based applications such as linear or logistic regression.

It is for that reason that in 2019, IBM research has embarked on a journey to make FHE practical and viable for business usage providing a new software development kit (SDK) called HElayers. This SDK aims to close the gap mentioned above by providing a simple, easy-to-use, and accessible APIs that target non-cryptographers. By now, other companies, organizations, and academic institutions also identified the above gap and introduced other SDKs and compilers, which will be further described in Chap. 4.

When adapting a cleartext algorithm to FHE, it is often the case that just like with language translation, it is not enough to directly translate cleartext operations to FHE primitives. For example, the latency and memory costs, as well as other metrics, such as model accuracy or the F1 score for machine learning (ML) applications, may not stay the same under FHE. Minimizing this gap cannot be achieved solely by using the SDKs APIs, because some theoretical background that explains how to better utilize FHE-based applications is needed. Moreover, compilers such as HElayers may provide APIs for pre-defined applications that were custom-made to achieve the best performance. However, data scientists, application developers, and applied cryptographers may still need to collaborate when adapting or extending these out-of-the-box solutions.

This book aims to further narrow this gap. Unlike other books in the field, it does not aim to present the underlying primitives behind the FHE schemes but rather treats FHE as a black-box and builds layers of knowledge on top of it. The idea is to assume the existence of an efficient FHE implementation in software or hardware and to study how it can be utilized to build different AI and data science applications. We focus on this type of task due to the current proliferation of AI solutions and particularly the emergence of Generative AI, which rely on huge amounts of data. Training or evaluating such models at scale often requires using third-party untrusted environments. Maintaining privacy while using such environments is one of the main motivations for using FHE.

The book includes four parts with 11 chapters. In Part I we open the book with an introduction which includes a presentation of the basic FHE concepts used in the book. The book targets data science application developers and assumes familiarity with basic data science concepts; however, Chap. 1 also provides a very brief introduction to the data science and AI domains in order to level set. Readers who are familiar with these concepts can start their reading from Chap. 2. The introduction continues in Chap. 2, which presents mostly at a high level the basic concepts and properties of FHE schemes. Mastering these low-level concepts will enable readers to understand the motivation behind the higher-level concepts that will be discussed in other chapters of the book.

There are different ways of constructing privacy enhancing technology (PET) solutions, some without FHE, some combining FHE with other cryptographic primitives, and some relying exclusively on FHE. We focus only on FHE-based solutions because this topic already involves a large scope of accumulated knowledge for the reader. Still, we believe that readers who understand the eco-system of PETs and in particular FHE will be able to consider more alternatives and thus design and use better-optimized solutions that are stitched for their particular needs. For that purpose, Chap. 3 presents some basic security models and applications where FHE can be used. Interested readers may find in Sect. 3.1.4 some more discussion on multi-party computation (MPC) alternatives and how to combine these with FHE in their solutions.

The first part of the book concludes with Chap. 4 that extends the reader's toolbox with several commonly used techniques such as working with fixed point elements and polynomial evaluation under FHE. It also describes the eco-system of available FHE tools, their properties, and development directions.

Once the readers are familiar with the world of FHE, they are ready to learn about two principle aspects that are utilized by modern FHE schemes, namely polynomial approximations (Part II) and packing methods (Part III). The former part contains two chapters: Chap. 5 explains basic concepts of polynomial approximations, and Chap. 6 shows how to use the studied techniques to approximate standard functions such as comparisons, reciprocals, trigonometric functions, and activation functions. We note that approximation theory is a long-studied domain with a rich literature.

Our intention in providing this part is therefore not to bring the readers to the point where they know everything about approximation theory. Instead, we aim to cover the most useful techniques for implementing and evaluating an approximated primitive under FHE.

The part of the book dealing with packing methods contains three chapters. Chapter 7 reviews and explains the basic concepts of single instruction multiple data (SIMD) packing. Once these are made clear, Chap. 8 describes a data structure called tile tensor that offers a simple and versatile approach for handling sophisticated packing schemes, especially when working with tensors, as is common in AI. Finally, Chap. 9 provides advanced tile tensor techniques for more complicated computations, such as convolutional neural networks. The three chapters are designed to offer a comprehensive set of tools that can be applied for data science under FHE, and furthermore to lay out principles and design considerations with which these tools can be further extended and improved.

The last part, Part IV, demonstrates how to combine all the studied material to generate practical ML applications. Various ML applications are presented in Chaps. 10, and 11 specifically demonstrates how to perform inference operations over ResNet50 with ImageNet.

We hope that the availability of this book will increase the interest in FHE-based solutions by researchers as well as practitioners, and thus influence security and privacy decision makers of all kinds.

Mount Carmel Haifa, Israel	Allon Adir
Mount Carmel Haifa, Israel	Ehud Aharoni
Mount Carmel Haifa, Israel	Nir Drucker
Mount Carmel Haifa, Israel	Ronen Levy
Mount Carmel Haifa, Israel	Hayim Shaul
Mount Carmel Haifa, Israel	Omri Soceanu

Legal Note
This book aims to explain the FHE domain to a wide audience. However, it does not aim to replace any standard protocols or regulations. The authors strongly recommend that before using any cryptographic primitive, cryptographic protocol, cryptographic tool, cryptographic library, or cryptographic program, the reader should first consult with security or cryptography experts so as to be aware of and follow the relevant security standards and guidelines.

Acknowledgments

To our families and friends who supported us during the process of writing this book, and also to the group members from IBM's FHE and AI security teams with whom we had many interesting discussions that led to clearer descriptions of some of the concepts described in this book.

We want to thank several colleagues who graciously volunteered to read an earlier draft of this book and offered thoughtful comments and suggestions to improve technical content and readability. We gratefully acknowledge their extraordinarily helpful advice and their unusual generosity. While acknowledging that responsibility for the final text rests with the authors, we wish to extend our sincere thanks, using alphabetical order, to Elisheva Walzer, Eyal Kushnir, Greg Boland, Michael Mirkin, and Ramy Masalha.

Thanks and appreciation are also due to Springer for assistance and support with the editing and production of this book.

The work on this book has been partially supported by the European Union's Horizon 2020 research and innovation program under grant agreement No. 101021936.

Contents

Part I Introduction and Basic Homomorphic Encryption (HE) Concepts

1 Introduction to Data Science 3
   1.1 Data Science 3
         1.1.1 What is a Model? 6
         1.1.2 Training, Inference, and Other Machine Learning Activities (Pre-processing, Testing, and Evaluation) 7
   1.2 Privacy Concerns 8
         1.2.1 Description of the Data Security and Privacy Ecosystem 9
   References 10

2 Modern Homomorphic Encryption: Introduction 13
   2.1 Introduction 13
   2.2 History: From HE to FHE 15
   2.3 HE Spaces 17
         2.3.1 Cleartext Domains 18
         2.3.2 Cleartext Representation 19
   2.4 Noise 20
         2.4.1 Noise Sources: Homomorphic and Numeric 21
         2.4.2 Typical Operations: Multiplications, Additions, and Rotations 25
         2.4.3 Key Types 27
   2.5 Advanced: Intro to HE Math and Its Underlying Security Assumptions 29
         2.5.1 LWE and Ring-LWE 29
         2.5.2 Encryption-Decryption Example 31
   2.6 Lab Exercises: Using HE 32
   References 33

3 Modern HE: Security Models 37
3.1 Security Models and Assumptions When Using FHE.............. 37
3.1.1 Basic Cryptographic Definition 38
3.1.2 Privacy: Who Hides What from Whom? Data Owner Versus Function Owner and Client Versus Server 41
3.1.3 Threat Model and Adversary Capabilities................. 44
3.1.4 Multi-party FHE 45
3.1.5 Federated Learning with HE 51
3.1.6 Vanilla Federated Learning................................. 52
3.2 Security: Against Modification (Integrity, Malleability) 55
3.3 Client-Aided Designs 56
3.4 Other Privacy Risks 57
3.5 Business Use Cases with Privacy Requirements.................... 59
3.5.1 Fraud Detection... 59
3.5.2 Loan Approval 60
3.5.3 Medical: Image Classification and Anomaly Detection .. 61
3.5.4 Biometric: Authentication................................... 62
3.5.5 Client-Facing Service: ELECTRON....................... 64
References.. 65

4 Approaches for Writing HE Applications 69
4.1 Comparison Model Versus Circuit Model......................... 69
4.1.1 FHE Is a Circuit Model...................................... 70
4.1.2 A General Recipe 70
4.2 The Computation Depth Challenge................................. 73
4.2.1 Cost of Bootstrap 74
4.2.2 TFHE.. 74
4.3 Techniques for Addressing the Challenges of HE 75
4.3.1 Indicator Vectors.. 75
4.3.2 Comparing Bitwise Representations....................... 75
4.3.3 Traversing a Tree 76
4.3.4 Copy and Recurse 77
4.3.5 Communicating Indicator Vectors 79
4.4 Non-polynomial Computation 80
4.5 Masking and Two-Server Model 80
4.6 Loops 82
4.6.1 Loop Unrolling 82
4.6.2 Loops with Input-Dependent Conditions 82
4.7 Exploiting SIMD 83
4.7.1 How SIMD Improves Performance 84
4.7.2 The Effects of Packing 86
4.7.3 Comparing SIMD to Non-SIMD 87
4.8 Polynomial Evaluation Methods: Horner, Paterson-Stockmeyer, and Others 89

		4.8.1	The Naïve Method	90
		4.8.2	Horner's Method	90
		4.8.3	Exponentiation by Squaring	91
		4.8.4	Paterson and Stockmeyer Polynomial Evaluation Method	92
		4.8.5	Evaluation by the Polynomial Roots	93
		4.8.6	Considerations of Accuracy and Numerical Stability	95
	4.9	FHE Code as Circuits		96
		4.9.1	Circuit Representation	97
		4.9.2	Compiling HE Code into a Circuit	98
		4.9.3	Running a Circuit	99
	4.10	Lab Exercises		99
		4.10.1	Polynomial Evaluation Methods	99
	References			105

Part II Approximations

5 Approximation Methods Part I: A General Overview 111
  5.1 Approximating Functions with Finite Polynomials Under FHE ... 111
  5.2 Estimating a Function with a Taylor Polynomial 113
  5.3 Minimax Polynomial Approximation and the Remez Algorithm .. 115
  5.4 Estimating a Function by Composing Multiple Polynomials 119
  5.5 Lab Exercise: Implement and Test the Remez Algorithm 122
    5.5.1 Taylor Series 122
    5.5.2 Remez 122
    5.5.3 Polynomial Composition 122
  References 123

6 Approximation Methods Part II: Approximations of Standard Functions 125
  6.1 Approximating Comparisons: Equality and Inequality and General and Tailored Versions 125
    6.1.1 Estimating the Sign Function by Composing Polynomials 127
    6.1.2 Comparisons: Equality and Inequality of Integers 129
    6.1.3 Approximating Max 131
  6.2 ReLU and Other Activation Functions 133
  6.3 Computing Reciprocals: $\frac{1}{X}$ 134
    6.3.1 Computing Reciprocals When Working with Integers 134
    6.3.2 Estimating Reciprocals when Working with Approximate Arithmetic 135
  6.4 Chebyshev Polynomials and Approximation of Trigonometric Functions 136
    6.4.1 Chebyshev Polynomials 136
    6.4.2 Interpolation via Chebyshev Nodes 137
    6.4.3 Function Estimation with Chebyshev Expansions 139

		6.4.4	Estimating Sin and Cos Beyond the [−1, 1] Range	141
	6.5	Lab Exercises		144
		6.5.1	Estimate the Sign Function	144
		6.5.2	Compare Two Hybrid Bitwise Values with CKKS	145
		6.5.3	Computing Reciprocals Under FHE	146
	References			147

Part III Packing Methods

7	**SIMD Packing Part I: Basic Packing Techniques**		151
	7.1	Why Pack?	151
	7.2	The Rotate-and-Sum Algorithm	153
	7.3	Flattening	155
	7.4	Rotations and Pseudo-Rotations	158
	7.5	Simulating Small Ciphertexts	161
	7.6	Rotating Non-rotatable Dimensions	162
	7.7	Tensor Summation	163
	7.8	Complex Packing	164
	7.9	Toy Example: Matrix-Vector Multiplication	167
	7.10	Diagonal Packing Techniques	168
	7.11	Lab Exercise: Programming Encrypted Tensors and the Studied Primitives	171
	References		172

8	**SIMD Packing Part II—Tile Tensor Basics**		173
	8.1	What Are Tile Tensors?	173
	8.2	Elementwise Operators on Tensors	174
	8.3	Tile Tensor Overview	176
	8.4	Tile Tensor Basic Concepts	179
	8.5	Dimension Independence and Basic Tiling	180
	8.6	Used, Unused, and Unknown Slots	181
	8.7	Duplications	183
	8.8	Elementwise Operators	186
	8.9	Sum Over Dimension	188
	8.10	Linear Algebra Using Tile Tensors	191
	8.11	Performance Considerations	193
	8.12	Example: A Simple Neural Network	195
	8.13	Encrypted vs. Encoded Tile Tensors	196
	8.14	Lab Exercise: Program with Tile Tensors	197
	References		198

9	**SIMD Packing Part III: Advanced Tile Tensors**		201
	9.1	Interleaved Packing and Rotations	201
	9.2	Simulating Large Ciphertexts	206
	9.3	Slicing	208
	9.4	Convolution	210

	9.4.1	SISO	210
	9.4.2	Padding	213
	9.4.3	Strides	214
	9.4.4	Multiple Channels and Filters	216
	9.4.5	Image to Columns	219
9.5	Complex-Packed Tile Tensors		220
9.6	Complex Packing for Matrix-Matrix Multiplication		223
9.7	Diagonal Packing and Tile Tensors		223
	9.7.1	Diagonal Packing for Convolution	227
9.8	Matrix Multiplication Method Comparison and Summary		228
9.9	Summary		229
9.10	Lab Exercise: Program Advanced Circuits with Tile Tensors		230
References			231

Part IV Use Cases and Other Approaches

10 Privacy-Preserving Machine Learning with HE 235
   10.1 Training Under Encryption 235
      10.1.1 Resilience to Noise 236
      10.1.2 Training Approaches and Their Privacy Concerns 237
   10.2 Machine Learning Models 239
      10.2.1 Regressions (Linear and Logistic) 239
      10.2.2 Decision Trees 247
      10.2.3 Time-Series Analysis 250
      10.2.4 K-NNs 264
   10.3 Lab Exercises 266
      10.3.1 Statistical Metrics for Time-Series Analysis 266
   References 268

11 Case Study: Neural Networks 271
   11.1 Neural Networks 271
      11.1.1 Fully Connected Layers 272
      11.1.2 Activation Functions 273
      11.1.3 Convolution Layers 275
      11.1.4 Sum, Mean, and Max-Pooling 276
      11.1.5 Normalization 277
   11.2 Training FHE-Friendly Networks 278
      11.2.1 Training Methods 280
      11.2.2 A Two-Phase FHE-Friendly Training 280
      11.2.3 A Three-Phase FHE-Friendly Training 281
   11.3 The LeNet NN 282
   11.4 Obstacles with Implementing ResNet50 NNs and Even Deeper DNNs 286
   11.5 Reducing Computations Under Encryption 286
      11.5.1 Pruning 286
      11.5.2 Split Networks 288

11.5.3 Neural Architectural Search (NAS)	288
11.6 Lab Exercise: Train Large HE-Friendly NNs	289
References	290
Glossary	293
Index	299

Acronyms

AES	advanced encryption standard.
AI	artificial intelligence.
AKS	Ajtai, Komlós, and Szemerédi.
API	application interface.
AR	autoregressive.
ARIMA	autoregressive integrated moving average.
ARMA	autoregressive moving average.
AVX	advanced vector extensions.
B/FV	Brakerski / Fan and Vercauteren.
BDL	blind data linkage.
BFS	breadth-first-search.
BGV	Brakerski, Gentry, Vaikuntanathan.
CC	confidential computing.
CDF	cumulative density function.
CGGI	Chillotti-Gama-Georgieva-Izabachene.
CISO	chief information security officer.
CKKS	Cheon, Kim, Kim, and Song.
CNN	convolutional neural network.
CPU	central processing unit.
CRT	Chinese remainder theorem.
CVP	closest vector problem.
DAG	directed a-cyclic (hyper) graph.
DM	Ducas-Micciancio.
DNN	deep neural network.
DP	differential privacy.
DS	data science.
DT	decision tree.

EPES	electrical power and energy system.
ER	entity resolution.
FC	fully connected.
FFT	fast Fourier transform.
FHE	fully homomorphic encryption.
FL	federated learning.
GC	garbled circuit.
GDPR	general data protection regulation.
GPU	graphical processing unit.
HE	homomorphic encryption.
HIPAA	health insurance portability and accountability act.
IND-CCA2	Indistinguishability under adaptive chosen ciphertext attack.
IND-CCA1	Indistinguishability under non-adaptive chosen ciphertext attack.
IND-CPA	indistinguishability under chosen plaintext attack.
IP	intellectual property.
KD	knowledge distillation.
LDL	low-density lipoprotein.
LHE	levelled homomorphic encryption.
LoR	logistic regression.
LP	linear programming.
LR	linear regression.
LSB	least-significant bit.
LSH	local sensitive hash.
LSSS	linear secret sharing scheme.
LUT	lookup table.
LWE	learning with errors.
MA	moving average.
MIMO	multiple inputs multiple outputs.
MK-FHE	multi-key FHE.
ML	machine learning.
MP-FHE	multi-party FHE.
MPC	multi-party computation.
MSB	most-significant bit.
MTTD	mean time to detect.
NAS	neural architecture search.
NN	neural network.
NP	nondeterministic polynomial time.
NTT	number theoretic transform.
PBS	programmable bootstrapping.

PC	program counter.
PET	privacy enhancing technology.
PHE	partially homomorphic encryption.
PHI	personal health information.
PII	personally identifiable information.
PIR	private information retrieval.
PKE	public key encryption.
PPFL	privacy-preserving federated learning.
PPRL	privacy-preserving record linkage.
PPT	probabilistic polynomial time.
PSI	private set intersection.
RAM	random access memory.
RL	record linkage.
R-LWE	ring LWE.
RNS	residues number system.
RSA	Rivest–Shamir–Adleman.
SDK	software development kit.
SGD	stochastic gradient descent.
SHE	somewhat homomorphic encryption.
SIMD	single instruction multiple data.
SISO	single input, single output.
SOTA	state-of-the-art.
SPI	sensitive personal information.
SSE	streaming SIMD extensions.
SVP	shortest vector problem.
TEE	trusted execution environment.
TFHE	FHE over the torus.

Part I
Introduction and Basic Homomorphic Encryption (HE) Concepts

The proliferation of privacy enhancing technology (PET) solutions in the last decade has many advantages, but it can also make it hard for students, engineers, or decision-makers to select their starting point in studying the privacy domain. Questions like: What should be my starting point for studying? Which technology or primitive should I explore first? Where can I find the relevant materials? The answers to these questions are not always clear, and there are probably many starting points, many primitives to explore, and many books, tutorials, and manuals about every concept.

This part aims to be the starting point for a reader who would like to explore and learn more about the fully homomorphic encryption (FHE) cryptographic primitive, which is one of the most exciting PETs out there. This primitive has evolved so rapidly in the last decade that it now deserves its own book. Specifically, this part intends to educate the user about the world of FHE from a bird's-eye view. This includes several topics, such as the relevant terminology, the security models where FHE can be involved, and even the topic of how modern compilers can kick in for automating the process of writing FHE-based applications. A diligent reader who wants to dive deeper into the secrets and mathematical concepts of FHE can use one of the different prior manuals and books that have been published on the subject. A concrete list of references is given at the end of Chap. 2.

Chapter 1
Introduction to Data Science

Abstract Data science is the study of extrapolating insights out of data and information. It leverages tools and techniques from different academic fields such as mathematics, computer science, information science, and domain knowledge to analyze data and create data-driven observations, hypotheses, and conclusions. These hypotheses, or models, attempt to represent the underlying laws that govern the patterns we see. Using data science, we can now attempt to understand the data at a deeper level and even attempt to identify root causes that created the observed and collected data. Training and continuously updating the data models with different observations fine-tunes these models and can help to better approximate our understanding of the recorded phenomena. Thus, models can consequently be used to classify, infer, and predict new data points that were not part of the original training data. Since data is often representative of sensitive, private, or confidential information, its accumulation, processing, and insights are sometimes private as well. As such, data scientists are usually asked to adhere to various privacy regulations that restrict and regulate data science methodologies. This chapter provides a basic introduction to data science, data privacy concerns, and an overview of privacy-preserving techniques that attempt to address these concerns.

1.1 Data Science

Whether on clay tablets, papyri, or other means, records have been excavated depicting how ancient civilizations throughout the world have gathered and aggregated data. While these examples were major steps toward what we call data science (DS) today, they are not prime examples of modern day DS. The contributions of Pascal, Fermat, Bayes, and Galton are considered to be the ancestors of DS. While they were not formally labeled as data scientists in the seventeenth, eighteenth, and nineteenth centuries, their works represent the different steps that serve as the basis of modern DS methodologies.

One could indeed go back and point to different instances throughout history as early examples of DS, but it is only with the introduction of modern computing that DS as a field became clearly distinguishable from statistics and information science.

© The Author(s), under exclusive license to Springer Nature Switzerland AG 2024
A. Adir et al., *Homomorphic Encryption for Data Science (HE4DS)*,
https://doi.org/10.1007/978-3-031-65494-7_1

Computers have been instrumental to the analysis of large amounts of records and multidimensional data. With better computational power, scientists have been able to create more accurate hypotheses that are constructed with an increasing number of parameters.

It is common to think of the DS work as a cycle. As in other scientific fields, when examining the data, the scientists' understanding keeps on evolving. Hypotheses are formed and updated out of newly generated insights. The data life cycle [15] consists of eight steps that provide a basic structure and methodology to DS work. These eight steps start with data generation, followed by collection, processing, storage, management, analysis, visualization, and interpretation. Some of these steps can be mapped to the steps of the scientific method (observation, hypothesis, experiment, analysis, conclusions) that was developed hundreds of years ago to analyze observed phenomena. Just like with the scientific method [9], these steps can be considered as general principles and not a fixed sequence of steps.

Data generation—The multidimensional manifestations of an observed event, behavior, or phenomenon. Whether through sensors that measure natural phenomena, logs that record human or machine behaviors and activities, or videos, images, sounds, and texts that capture moments in time, data is continuously generated. The data generation might be happening irrespective to the start of a DS effort, e.g., data logs that were kept before any data scientist ever looked at them. The generated data might also contain data that would complicate the data scientists' job. For instance, a malfunctioning sensor might generate false information or a human being might lie when filling out a survey.

Collection—The curation (identifying relevant data to the DS process) and sampling of data is an important step in the data life cycle. Since data is generated in huge volumes at any point in time, it is crucial to identify what data would best serve the task at hand. One might say that "the flap of a butterfly's wings in Brazil" is a factor that should be taken into account when predicting weather patterns in Texas [10], but sampling too much might obscure more major factors that would better contribute to the understanding of such an event.

In addition, as a matter of practicality, a DS effort cannot be viewed irrespective of the resources at the disposal of the data scientists. Each data point has an inherent cost attached to it; this cost is embodied through acquisition, storage, or processing cost overheads. Answering the question of how much data is "too much" data is often a business decision, deciding the threshold when a certain amount of allocated resources and time will outweigh the benefits of a more accurate or performant model.

Processing—Once collected, data should be processed so that it can be used at later stages. This includes formatting the data, cleansing, and aggregating it, as well as compressing to reduce storage costs and encrypting it if required for compliance with rules or regulations. Encrypting the data at this stage is helpful for many reasons. For example, it may be dictated by different privacy regulations such as the general data protection regulation (GDPR) [8], or for business confidentiality reasons. Furthermore, encryption is important to ensure trust between the people whose data was recorded and the data collectors and processors. Without this trust, the earlier stage of data collection might prove to be more difficult as people and organizations might be hesitant to provide their personal data due to fear of it falling into the wrong hands.

Storage—After processing the data, it should be stored in a way that would facilitate quick and easy access to it. Storage should be reliable, robust, scalable, accessible, secure, and performant.

Management—Data management is the ongoing effort throughout the data life cycle that ensures the successful operations of the storage databases and optimizes the organization of the data within them to support the DS effort, for example, improving the response time for various SQL queries that are used for a specific task.

Analysis—One might say that this is the heart of the DS effort. At the analysis stage, the data scientist would extract structure and insights out of the stored data, hypothesize and conduct experiments based on the data, train a model that would best describe the data, and use it to infer insights about future data samples. Although machine learning (ML) can be used throughout the data life cycle, it is at this stage that it would be most extensively used. ML offers the necessary tools to use the data to draw insights and learn laws that might be too complicated to be described by a known algorithm or formula.

Visualization—When dealing with huge amounts of data and complex ML models, it is easy to forget that the consumer of the DS project is, more often than not, a human being. Visualizing the results and creating figures, plots, graphs, and diagrams to better convey the insights that are produced is sometimes no less important than the numbers themselves.

Example 1.1 (Florence Nightingale) Following the Crimean War, Florence Nightingale, a nurse that tended to British soldiers, drafted a report at the behest of Queen Victoria on the battlefield hospital's sanitary conditions. When the report did not generate the appropriate needed reform, Nightingale designed and created rose diagrams, visualizing the causes of mortality, as part of an annex to the report. Only after the visualization of the statistics was Nightingale able to persuade the British government to institute the reform [4]. Visualizations make the insights more consumable for the human reader and in Nightingale's case made the difference between a dusty report and lifesaving overhauls.

Interpretation—It is in this stage that the insights and visualizations are weaved into a story with pictures, conclusions are drawn, and one can decide whether to start another iteration of the data life cycle.

Examining these steps as a data life cycle, as well as a comprehensive story, allows us to better differentiate the field of DS from the fields of ML and artificial intelligence (AI). ML is the usage of data by computers to optimize a performance criterion [2]. It is a subfield of DS but also of AI, which is the study of methods designed to make computers intelligent. Specifically, AI tries to answer Alan

Turing's question "Can machines think?" [14] and provide tools for them to succeed in doing so. It is the study of building intelligent machines [12] that can at human intelligence level "operate successfully in the common sense informatic situation." [13].

1.1.1 What is a Model?

The core of every ML algorithm consists of an ML model. A model is a chosen hypothesis that converts input data to outputs using some mathematical multiparameter function. The parameters of the function are learned during a phased called the training phase using the training data. When the training is based on data that contains labels (ground-truth outputs to given inputs), the learning is considered to be supervised learning. Alternatively, when training data is not labeled, the learning process is referred to as unsupervised learning. As labeled data is often challenging to acquire, one might reluctantly be forced to choose unsupervised learning over the more accurate supervised learning. Nevertheless, a middle-ground third option exists. Semi-supervised learning describes learning methods that utilize both labeled and unlabeled data during the training process. In such cases, it is common that most of the data is unlabeled. Even if only a fraction is labeled, this part can serve as ground truth.

Some ML systems have output that illicit an action. In such cases, the result of the action (or more accurately the updated input, also known as a feature vector) can be fed back to the model and be used to change the model parameters to further optimize toward a successful criteria. This sort of learning is referred to as reinforcement learning. Formally, denoting a model as $g(\cdot)$, its parameters as θ, the input features or attributes vector as x, and the output or prediction of the model as y, [2] describes the model outcome as

$$y = g(x, \theta) \tag{1.1}$$

For supervised learning, it is a common practice to categorize models into two types according to their output types. Classification models are models whose output is a class or a category. Models of this type might output a binary classification, a category, or even in a word chosen from a million word dictionary. In contrast, Regression models usually output a floating-point number that represents a continuous value prediction.

The ML model can also be seen as an approximation of an underlying mathematical equation explaining some phenomenon, behavior, or event. As an approximation, it is not certain that all the factors effecting the phenomenon will be accounted for. The given input vector x might not include all the necessary features to provide an exact prediction of the outcome. For instance, one might try to model the speed of an object coming down a slope based solely on the slope's angle, the object's initial velocity, and the time that elapsed since the object was released. Although these parameters can provide a good approximation of the observed behavior, more features, such as, friction coefficients, shape, and wind velocity at different places and times, are necessary for a completely accurate model.

Some behaviors could not be easily modeled by the rules of physics and so one might never know which features are missing for an accurate model. For example, when trying to model the expected grade of a student in an upcoming exam, one might include in the input feature vector the student's previous grades, the time spent studying for the exam, amount of sleep the night before the exam, and IQ score of the student. These features are clearly insufficient, but what feature set could possibly be sufficient? Would including the student's age contribute to the accuracy of the model? Would the student's blood-alcohol level improve the accuracy? Similarly, the amount of parameters (learned and hyper-parameters) of the model (θ) might also be insufficient.

Since one cannot model every event using a simple linear equation, the model's equation and its number of parameters should be chosen to provide an accurate enough result given the input data and the resources at the modeler's disposal. However, choosing a model with more parameters than needed or a model that is not appropriate for the task at hand is also less ideal since it might unnecessarily prolong the training and overall latency of the model. However, over enough time (a model is often iteratively trained and each iteration is called a training epoch) and with sufficient amount of training data, the extra parameters might be nullified. Reaching an accurate model would require more resources and time than the approximated model, and it is unclear whether data scientists would ever reach the more accurate model. Thus, in some cases, choosing the right model, the right number of parameters, and the right feature set is key to obtaining an accurate model.

1.1.2 Training, Inference, and Other Machine Learning Activities (Pre-processing, Testing, and Evaluation)

During the training phase, a training algorithm optimizes the results of a goal/loss function over a training dataset, updating the model's parameters θ. In a supervised learning scenario, the training dataset would be a set of pairs of inputs and labels (ground truth outputs). The goal function should try to reward successful predictions of ground truth outputs given the training inputs. However, it should also avoid overfitting, training-set bias, or other forms of overreliance on unrepresentative training datasets.

As the quality of the trained model is inherently correlated with the quality of the training data, one should try to limit the amount of labeling mistakes in the training dataset. Nevertheless, it should be assumed that such mistakes might be present and mitigate the risk of them may contaminate the model.

One should make sure during the pre-processing phase that the data does not contain any invalid values such as $null$, inf, or values that do not match the expected type for a specific feature. When detecting errors within the data, one of the first questions to come up is: "How should these errors be fixed?". An easy solution might be to completely remove any record that contains missing or corrupted features. However, this might entail removing considerable amounts of data that might be usable following some small alterations. Different techniques are used to resolve such issues. Invalid values can be replaced by a minimum, maximum, or average value. Another solution is to train a smaller model to predict a certain feature using all other features and use that model to "fill-in" any missing or corrupted values. Some data scientists go even further and suggest to remove anomalous values, deeming them "probable mistakes." This however should be done quite carefully since anomalous values do not necessarily indicate a data-collection mistake.

Finally, when testing and evaluating the model, it is crucial to choose an appropriate evaluation metric that accurately represents the model's requirements. For example, if a bank would like to train a model to detect financial fraud, the evaluation metric might need to consider the cost of misclassifying a benign transaction, as well as the cost of missing a malicious transaction. Since one mistake might not have the same cost as another, a simple accuracy metric might not be sufficient and might misrepresent the underlying goal.

1.2 Privacy Concerns

It has been estimated that in 2024, on average 149 zetabytes (10^{21} bytes) of data would be consumed each day [1]. In spite of these high figures, readily available data, especially high-quality labeled data, for model training is still a valuable and scarce commodity, often unavailable to researchers. Data owners are reluctant to share their data due to privacy, confidentiality, and other concerns stemming from business, ethical, or legal/regulatory considerations.

Even after a model has been trained, data scientists may still be faced with different privacy concerns, since private and confidential data can be extracted from a trained model [5], and individual personal information can be extracted even from aggregated data [6]. Different anonymization techniques [11] can help mitigate some of the privacy concerns to a certain degree, but they often trade-off privacy with accuracy or with information availability.

The above concerns mainly relate to protection of personally identifiable information (PII), but one must also consider the need to protect against theft of intellectual property (IP) in the form of model weights. Attackers are able to either steal the whole model by gaining access to it or to employ query-based attack algorithms in black-box model usage scenarios.

Data being sent to models as query inputs can also leak to malicious actors through man-in-the-middle attacks. In such cases, a malicious hacker with access to the computing platform running the model would wait for the data to be decrypted prior to its processing by the analytical model and then steal the valuable cleartext information.

Finally, model outputs and analytical results might also be considered sensitive information, and one must secure the learning process so that results that relate to financial, healthcare, or other confidential information do not fall into the wrong hands.

1.2.1 Description of the Data Security and Privacy Ecosystem

Multiple approaches and technologies have been suggested to try and mitigate the privacy concerns described above, each with its corresponding trade-offs of benefits and drawbacks. The following is a brief discussion of some of the common approaches and technologies, besides FHE, which is the main topic of this book.

Multi-party computation (MPC) is a technology that provides strong cryptographic guarantees and protects the data without relying on dedicated hardware or software. MPC allows multiple parties to collaboratively compute the result of a function or algorithm, where the inputs provided by each party remain hidden from the other parties and only the result is revealed to either a designated party or to all the parties. However, MPC has a major limitation; the underlying algorithms implementing MPC may incur a significant network overhead caused by the large number of interactions between the multiple parties [3].

The concepts and ideas behind MPC were initially introduced for two parties by Yao and later generalized by Goldreich, Micali, and Wigderson for any number of parties. The paper [16] introduced the concept of garbled circuit (GC) which provided the basis for the theory around MPC. A GC is a protocol that allows performing a secure computation by two mistrusting parties without the need for a trusted third party.

The main motivation behind MPC is the need to generate insights from data shared across multiple entities and/or organizations while preserving the privacy of the data. There are endless numbers of scenarios and use cases in which the ability to safely share private data between organizations yields significant business benefits to all the parties concerned. Moreover, in some cases, the ability to share data in a privacy-preserving manner across multiple entities can result in the creation of completely new businesses.

Practical MPC protocols were developed for specific applications and use cases, e.g., private bidding and secure set intersection. Nevertheless, industry adoption was limited since for many use cases, the implementation proved to be challenging, and at the same time, computation and communication overheads were much higher than other privacy-preserving solutions.

Example 1.2 A major landmark for MPC was the Danish Sugar Beet Auction application in 2008 [7]. This was the first large-scale and practical application of

MPC in the business sector. A virtual auction was successfully run by a MPC protocol involving representatives of the sugar beet growers' association of Denmark, Danisco—the sugar beets' processor of Denmark and a team implementing and running the MPC protocol.

The use of MPC ensured that farmers' bids remain private from Danisco, the only sugar beets' processor in the Danish market. In addition, it also reduced the overall expenses of the auction process. Moreover, since farmers' bids may reveal their own economic positions and productivity, it was imperative that Danisco was not privy to this information and potentially use it for its advantage when selling contracts.

Confidential computing (CC) is a technology aiming to protect data in use by performing computations in a hardware-based Trusted Execution Environment (TEE), also known as secure enclaves. As part of deploying software onto a TEE, an attestation process is required to ensure that the software stack running in the TEE is authenticated and authorized. For most TEE implementations such as provided by IBM, Intel, and AMD, the data in memory is encrypted at all times and decrypted inside the central processing unit (CPU).

The main benefit of this approach versus other methods such as MPC and FHE comes in terms of a superior time performance. In contrast, its major limitation is its reliance on a limited security assumption: one must assume that the hardware and the software stack running in the secure enclave has not and will not be compromised. Such assumptions are inherently problematic, especially given the ever-growing number of software vulnerabilities and side-channel attacks. In addition, the secure enclave approach has an inherent limitation in that it does not allow for effective and secure collaboration on sensitive data between multiple parties.

Differential privacy (DP) is a privacy method that provides strong privacy guarantees by quantifying the amount of data that might leak. However, DP inherently reduces the utility and fidelity of the data and thus significantly limits its applicability for industrial use cases. DP preserves the privacy of individuals by adding random noise to private data attributes in a manner that still allows population trends to be accurately observable. The noise distribution is custom-tuned to hide the presence or absence of an individual in a dataset. Thus, DP ensures that an adversary cannot distinguish between the result of an analysis on a dataset with a specific individual and the same analysis conducted on a dataset, where that individual's data is replaced by that of another individual. The inability to detect whether a person's data is part of a dataset is often required by privacy regulations that demand that human participants would be unidentifiable.

References

1. 53 Important Statistics About How Much Data Is Created Every Day (2024). https://financesonline.com/how-much-data-is-created-every-day/. Last Accessed:04 Feb 2024
2. Alpaydin, E.: Introduction to Machine Learning. MIT Press, Cambridge (2020)

References

3. Berger, B., Cho, H.: Emerging technologies towards enhancing privacy in genomic data sharing. Genome Biol. **20**(1), 128 (2019). https://doi.org/10.1186/s13059-019-1741-0
4. Brasseur, L.: Florence Nightingale's visual rhetoric in the rose diagrams. Techn. Commun. Quarterly **14**(2), 161–182 (2005)
5. Carlini, N., Tramèr, F., Wallace, E., Jagielski, M., Herbert-Voss, A., Lee, K., Roberts, A., Brown, T., Song, D., Erlingsson, Ú., Oprea, A., Raffel, C.: Extracting training data from large language models. In: 30th USENIX Security Symposium (USENIX Security 21), pp. 2633–2650. USENIX Association, Berkeley (2021). https://www.usenix.org/conference/usenixsecurity21/presentation/carlini-extracting
6. Colbaugh, R., Glass, K.: Learning about individuals' health from aggregate data. In: 2017 39th Annual International Conference of the IEEE Engineering in Medicine and Biology Society (EMBC), pp. 3106–3109 (2017). https://doi.org/10.1109/EMBC.2017.8037514
7. Damgård, I., Toft, T.: Trading sugar beet quotas - secure multiparty computation in practice. ERCIM News **2008**(73), 32–33 (2008). http://ercim-news.ercim.eu/trading-sugar-beet-quotas-secure-multiparty-computation-in-practice
8. EU General Data Protection Regulation: Regulation (EU) 2016/679 of the European parliament and of the council of 27 April 2016 on the protection of natural persons with regard to the processing of personal data and on the free movement of such data, and repealing directive 95/46/EC (general data protection regulation). Official J. Europ. Union **119**, 1–88 (2016). http://data.europa.eu/eli/reg/2016/679/oj
9. Gauch Jr, H.G.: Scientific Method in Practice. Cambridge University Press, Cambridge (2002)
10. Lorenz, E.: Predictability: Does the flap of a butterfly's wing in Brazil set off a tornado in Texas? (1972). https://static.gymportalen.dk/sites/lru.dk/files/lru/132_kap6_lorenz_artikel_the_butterfly_effect.pdf. Last online Access 04 Feb 2024
11. Majeed, A., Lee, S.: Anonymization techniques for privacy preserving data publishing: a comprehensive survey. IEEE Access **9**, 8512–8545 (2021). https://doi.org/10.1109/ACCESS.2020.3045700
12. McCarthy, J.: Mathematical logic in artificial intelligence. Daedalus **117**(1), 297–311 (1988). http://www.jstor.org/stable/20025149
13. McCarthy, J.: From here to human-level ai. Artifi. Intell. **171**(18), 1174–1182 (2007). https://doi.org/10.1016/j.artint.2007.10.009. Special Review Issue
14. Turing, A.M.: Computing Machinery and Intelligence, pp. 23–65. Springer Netherlands, Dordrecht (2009). https://doi.org/10.1007/978-1-4020-6710-5_3
15. Wing, J.M.: The data life cycle. Harvard Data Sci. Rev. **1**(1) (2019). https://hdsr.mitpress.mit.edu/pub/577rq08d
16. Yao, A.C.C.: How to generate and exchange secrets. In: 27th Annual Symposium on Foundations of Computer Science (SFCS 1986), pp. 162–167 (1986). https://doi.org/10.1109/SFCS.1986.25

Chapter 2
Modern Homomorphic Encryption: Introduction

Abstract Homomorphic encryption (HE) is a cryptographic primitive that provides unique security guarantees in the privacy enhancing technologies (PETs) ecosystem. In Chap. 1, we discussed some privacy aspects that affect data science applications. The HE primitive can solve some of these privacy issues. This chapter dives into the properties and capabilities of modern HE schemes, how to use them, and their unique features and limitations. The goal of this chapter is mainly to introduce the HE terminology and the jargon, which are used throughout this book. In general, the book attempts to consider the HE primitive as a black box that provides a predefined set of APIs that application developers, applied cryptographers, and data scientists can use. For completeness, Sect. 2.5 includes some advanced HE topics that may help readers gain further understanding of the lower-level implementation details of some modern HE schemes.

2.1 Introduction

HE is a powerful cryptographic tool that may extend the toolbox of a PET designer with valuable capabilities. Thus, we start by providing the basic HE definitions required for understanding the HE concept. Some of these definitions may sound informal for an audience familiar with the formal cryptographic language. However, it is necessary to keep it this way so that readers from other domains, such as data scientists, will get a smooth introduction and a good understanding of the discussed topics.

Consider users who would like to use a cloud environment for their computations but are restricted from uploading sensitive data to the cloud, for example, by the GDPR regulations. The users can, of course, encrypt the data before uploading it to the cloud using symmetric encryption such as AES. However, by doing so, the users limit themselves to using the cloud only as a storage service, without the ability to perform further computations on the data. This ignores the main purpose of the cloud, which is to provide better computing capabilities for its users. HE comes to solve exactly this. It is a special type of encryption that allows computations to be performed on encrypted data. For example, it enables additions, multiplications, or both on ciphertexts, where the resulting ciphertext can be decrypted to have the same value as if the mathematical operations were performed directly on the encrypted data. This property is called homomorphism, hence the name of the HE primitive.

We start with an abstract definition of an HE scheme, which generally involves four methods:

1. **Key generation** ($k \leftarrow$ KeyGen()) that generates the required scheme keys. We further elaborate on the different key types used by different FHE schemes in Sect. 2.4.3.
2. **Encryption** ($c = \text{Enc}_k(m)$) that uses the generated set of keys k to transform an unencrypted input m to an encrypted ciphertext c.
3. **Decryption** ($m = \text{Dec}_k(C)$) that uses the generated set of keys k to extract the encrypted data m from an input ciphertext C.
4. **Evaluation** ($c' = \text{Eval}(f; (c_1, \ldots, c_n, m_1, \ldots, m_n))$) that receives multiple ciphertexts $c_1, \ldots, c_n$ and multiple plaintexts $m_1, \ldots, m_n$ as input together with a supported function f. Its output is a ciphertext c' that encrypts the evaluation of f on these inputs.

Remark 2.1 As we will see later on, not all HE schemes support all types of functions. For example, Paillier [32] supports only additions and thus cannot perform a complex operation such as square root inverse. Hence, in the above definition, the Eval method is defined only for functions f that are supported by the scheme operations.

For simplicity, we use small letters for both plaintext and ciphertext data. We perform a small abuse of notation where we leverage the homomorphic property, and when we perform operations on ciphertexts, we will refer to the data that they encrypt instead of the ciphertexts themselves. For example, consider the vector $m = (1, 2, 3, 4)$ and a ciphertext $c = \text{Enc}(m)$ that encrypts it; multiplying c by 2 results in a new ciphertext c' that encrypts the vector $(2, 4, 6, 8)$. For simplification, we may ignore the encryption notation and briefly say that 2 times c is the ciphertext $(2, 4, 6, 8)$.

Example 2.2 (Encrypted Homomorphic Computation) Figure 2.1 demonstrates the basic concept of applying an homomorphic computation to encrypted data. The diagram starts in the top left corner with three unencrypted data elements: $m_1, m_2,$ and m_3 that are encrypted to $c_1, c_2,$ and c_3, respectively, using a pre-computed key k.

Subsequently, we apply the function $f(x, y, z) = 2xy + z$ on both sides of the diagram. On the left side, we get $m_{res} = 2m_1 m_2 + m_3$, while on the right side, we get the encrypted value $c_{res} = \text{Eval}(f, (c_1, c_2, c_3)) = 2(c_1 \odot c_2) \oplus c_3$. Finally, we decrypt c_{res} and receive the expected value m_{res}.

Example 2.2 demonstrates the simplicity of using HE in a privacy-preserving design. Computing a function f on encrypted data is almost as simple as computing it on unencrypted data; just encrypt the data and perform the same computation on ciphertexts. It turns out, however, that there are some constraints and limitations

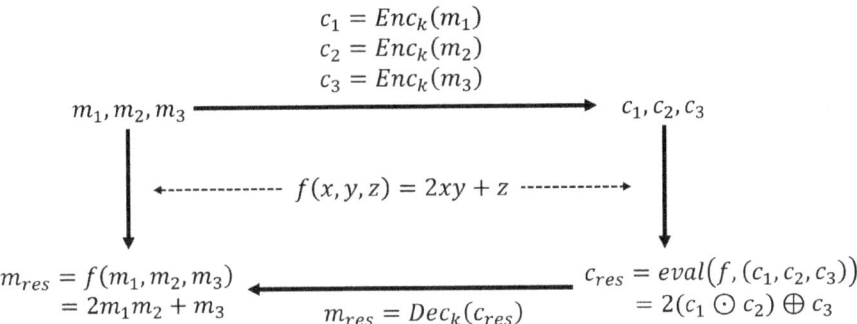

Fig. 2.1 An example of a HE computation

to the above claim, which should be considered. We explore some of them in later chapters.

2.2 History: From HE to FHE

The amazing concept of an encryption scheme that can perform operations on ciphertexts that can be magically translated to operations on the encrypted plaintext was introduced by Rivest et al. [34] in 1978. Clearly, this capability, when practical, opens the door to many technological opportunities. One such famous example that many companies and organizations consider is securely using cloud environments for performing computations over encrypted data. It seems that the north-star goal of HE is having an efficient cryptosystem that can encrypt *every* type of input data and can perform *every* type of computation on it. However, as we will observe, modern schemes still possess some limitations and gray areas that their users should familiarize themselves with.

Since 1978, different types of HE schemes were proposed, each one enabling a different set of computations over encrypted data. The first line of schemes is categorized as partially homomorphic encryption (PHE) schemes because they can only support functions with one type of operation such as addition or multiplication. Examples of such schemes include Rivest–Shamir–Adleman (RSA) [35] and ElGamal [18], which support an unbounded number of multiplications; the Goldwasser-Micali [21] scheme, which supports an unbounded number of exclusive-or operations; and Benaloh [15] and Paillier [32], which support an unbounded number of additions.

Examples of applications that use an additive PHE scheme include secure voting systems, simple federated learning (FL) protocols, and even some statistical computations such as average or standard deviation. In all cases, data is encrypted using a public key known to all participants, and uploaded to some agreed central location, which can be the voting committee servers, an untrusted cloud, or an aggregator in the FL case. Subsequently, the data is accumulated and the results are returned to the party who holds the decryption key. When an average computation is requested, this party can divide the summation results by the number of summands before using the data.

There are two main drawbacks to using the abovementioned schemes. First, the power provided by PHE schemes is limited to evaluating functions with only one type of operation. Second, they are considered to be non-post-quantum secure, which means that they are broken in the presence of quantum computers. The latter is critical because even though these schemes are considered secure against all known classic computer attacks, they allow adversaries to collect encrypted data today, store it, and break it in the future using quantum computers. Consequently, it is unclear yet whether users can rely on non-post-quantum secure protocols when data is required to remain secure for a decade or more. Note that not all PHE schemes are non-post-quantum secure. Just those that rely on the computational hardness of solving some modular arithmetic problems.

The next-generation scheme types are somewhat homomorphic encryption (SHE) and levelled homomorphic encryption (LHE), which allow performing both addition and multiplication over encrypted data. SHE and LHE are useful when the evaluated functions are small or limited either by the number of sequential multiplications or by some other properties of the evaluated functions. In a sense, SHE and LHE schemes suffer from flexibility issues as the users cannot just encrypt the data and evaluate every arbitrary function on the encrypted data. Instead, they have to either predefine the allowed number of serial multiplications of the evaluated function or limit the function to be from a set of predefined functions.

The last scheme type is called FHE schemes. Such schemes are based on the blueprint of Craig Gentry [20] and allow users to evaluate any polynomial function or binary circuit over encrypted data. It was only recently, in the last decade, that FHE schemes were shown to be practical. Schemes such as

- Brakerski, Gentry, Vaikuntanathan (BGV) [7],
- Brakerski / Fan and Vercauteren (B/FV) [6, 19],
- Cheon, Kim, Kim, and Song (CKKS) [10, 11],
- Ducas-Micciancio (DM) (a.k.a FHEW) [17], and
- Chillotti-Gama-Georgieva-Izabachene (CGGI) (a.k.a. FHE over the torus - TFHE) [14]

are now showing good performance results for many useful applications such as AI and data science applications. The main advantage of FHE schemes is that they allow the evaluation of arbitrary functions without artificial requirements such as setting a predefined multiplication level. This book only considers FHE schemes and thus by abuse of notation, it uses the terms HE and FHE interchangeably. The reader should note that some of the techniques described in this book that rely only on one operation may also apply to other HE schemes such as Paillier or RSA. Another advantage of modern FHE schemes is that they rely on the hardness of lattice-based problems that are considered post-quantum secure.

Approximate FHE Schemes The abstract FHE definition presented in Sect. 2.1 does not accurately capture all the available schemes. Specifically, it only refers to exact schemes, where the decryption of a ciphertext is always the expected value. In [10] a new family of FHE schemes was introduced, namely, approximated schemes, which allows the decryption results to include some noise. This is useful in many applications that do not require an exact answer. Consider, for example, a classification neural network (NN) that on an input image outputs an integer that represents one of ten possible categories. When the decrypted output is noisy, with noise less than 0.5, it can be cleaned while in plaintext to achieve the desired results. If the output is 3.4 or 3.7, it can be rounded to 3 or 4, respectively. Another example is a regression model that predicts the cost of a house based on parameters such as the city, street, and size of the house. In many cases, the users of this model can be satisfied with having a deviation of less than 0.1% which can be achieved with both exact and approximated schemes.

2.3 HE Spaces

Another way to categorize the different schemes is by considering the spaces on which they operate. An HE scheme usually involves three data spaces: (1) the cleartext space, where the data is represented in its native domain, e.g., floating-point numbers, Boolean data, or strings; (2) the plaintext space, where the cleartext is encoded to a specific format required for the encryption process but is still unencrypted; and (3) the ciphertext space, which includes encrypted plaintexts.

Example 2.3 (HE Spaces I) An HE application can consider five-letter word inputs. The application starts by pre-processing the words into the HE scheme's cleartext space. Here, every word is represented using 40 binary bits, where each letter is translated using its 8-bit ASCII encoding into its binary representation. Subsequently, every bit is encoded into an element of the plaintext space, for example, a polynomial space of polynomials with degree less than n. The encoding is performed by setting the first coefficient of some polynomial from that space to the cleartext value. This results in 40 (encoded) polynomials. The ciphertext space can also be a polynomial space, where each ciphertext polynomial encrypts one plaintext polynomial and hence it encrypts just one bit. To encrypt a five-letter word, 40 ciphertexts are required.

Example 2.4 (HE Spaces II) Alternatively, the input can be a digital image represented using 28×28 monochrome 8-bit pixels. Here, the input does not require pre-processing, and the cleartext space is matrices of 8-bit integers. To convert the cleartext to a plaintext, we encode it as a vector by placing all pixels in a row using row-major order, i.e., the plaintext is a vector of 784 8-bit elements and the ciphertext is its encryption.

Examples 2.3 and 2.4 demonstrate the principal concept of the three spaces. The cleartext is how HE users interpret their data; it has specific meaning and characterization. In contrast, plaintexts are intermediate representations, where the users encode their data in a way that makes it ready for encryption. While the plaintext format is dictated by the HE scheme, users still have plenty of options to optimize the process. Choosing the correct encoding method is critical because once the data is encrypted, many optimizations are no longer available or harder to achieve. That is one reason why modern compilers and software development kits (SDKs) provide unique application interfaces (APIs) for assisting the user in the process of selecting an optimal encoding. However, even with the presence of these compilers, when there is a requirement to program a new application, a manual process may still be required. Once the data is encoded, it is ready to be encrypted using the selected HE scheme.

2.3.1 Cleartext Domains

The choice of the plaintext and ciphertext domains dictates the capabilities that an HE scheme enables, such as whether we can operate over single, or vectors of, elements and with what precision level. However, application developers may find it easier to categorize the schemes by their cleartext domains instead. Specifically, consider the following possible domains:

- **A finite field** ($\mathbb{F}_q$), where the input elements belong to some predefined finite field ($\mathbb{F}_q$) with characteristic q and addition and multiplication operations are done in that field, e.g., using modular arithmetic. Two scheme examples that support operations on finite field elements are BGV and B/FV. For some use cases, working with finite fields enables an interesting set of operations beyond addition and multiplication in the field. For example, for a finite field of characteristic q, it is possible to compute the inverse a^{-1} of a number a by simply raising it to the power of $q - 1$, i.e., computing a^{q-1}. Another example is when q is a power of 2, addition operation translates to logical-xor operation on the binary representation of the inputs.
- **The real plane** ($\mathbb{R}$)—Here the input elements are real numbers represented using fixed-point integers. Schemes such as CKKS are a native fit for such inputs. In fact, CKKS can support an even wider field, namely, the complex plane ($\mathbb{C}$). Modern AI applications use floating-point computations, and thus schemes such as CKKS are often used for these tasks.
- **The integers** ($\mathbb{Z}$)—Another type of applications relies on integers; this can be represented either as fixed-point numbers as a subset of $\mathbb{R}$ and fed into a CKKS plaintext or as the representatives of a finite field elements, which can be consumed by BGV, B/FV, and TFHE. However, the computations when using integers must be bounded to avoid modular wraparound or fixed-point overflows. When using modern implementations of TFHE, for example, the size of the

integer must be bounded to a low number of bits, e.g., 16 or below, to maintain the computation accuracy.

- **Binary values {0,1}**—Interestingly, what brought FHE its reputation is its ability to evaluate **every function**. This can always be achieved when considering only binary (Boolean) inputs and using a Boolean circuit. These circuits are equivalent to Boolean circuits that are implemented in hardware with Boolean gates such as logical-xor and logical-and that translate to addition and multiplication under HE. However, one main difference is that the cost of performing multiplications in HE is often higher than the cost of additions. This is different from hardware, where multiplication is often costlier than addition, but the cost of logical-and and logical-xor is often similar. Boolean circuits can run easily on schemes such as TFHE and BGV and were until recently considered practical only for those schemes. Lately, a paper called BLEACH [16] showed that when considering a Boolean circuit that performs many parallel gates, e.g., many logical-and or many logical-xor in parallel, it is possible to achieve performant results also when using CKKS.

2.3.2 Cleartext Representation

After selecting the cleartext domain, the following task may be desired before encoding inputs into plaintexts. Specifically, we need to pre-process the input data into its representation in the cleartext domain. Some examples of data representations include bitwise, bytewise, and residues number system (RNS)-based representation (Example 2.6).

Example 2.5 (Bitwise Representation) Consider an input integer n of 8 bits. When the cleartext domain involves integer elements, then no pre-processing is needed. However, if the cleartext domain involves Boolean values, instead of passing the input as is to the desired plaintext HE encode operation, we first decompose it to its 8 binary bits $b_0, \ldots, b_7$. Subsequently, we encode every bit separately in a different plaintext or alternatively pack the 8 bits in eight slots of only one plaintext. In both cases, we use redundant representations because we occupy eight slots instead of just 1 and thus more space is required for this representation. Nevertheless, working with bits opens the door to more complex operations on n such as logical and/xor/or operations or working with a lower precision for accelerating some of the HE operations. On the other hand, implementing arithmetic operations such as integer multiplication becomes more complex. Some scheme examples where this representation is useful are TFHE, BGV with prime=2, and surprisingly even CKKS as we will see later.

Similarly to Example 2.5, one can work with bytewise representation by decomposing a number to bytes, or with RNS representation when the input belongs to some finite field. Upon decryption and decoding the data, an extra post-processing step is required to recompose the cleartext number to its original representation.

Example 2.6 (RNS Representation) Consider the mathematical ring $\mathbb{Z}_{70}$ with integer elements modulo 70. Mathematically, it is possible to represent every element x of $\mathbb{Z}_{70}$ using a unique combination of elements from the rings $\mathbb{Z}_3$, $\mathbb{Z}_5$, $\mathbb{Z}_7$ by reducing x modulo 3, 5, and 7, respectively. For example, the element $x = 25$ (mod 70) can be written as $(25 \pmod 3, 25 \pmod 5, 25 \pmod 7) = (1, 0, 4)$; to construct x back from $(1, 0, 4)$, the famous Chinese remainder theorem (CRT) should be used. In general, elements of $\mathbb{Z}_N$, $N = p_1 \cdots p_n$ for mutually co-primes $p_1, \ldots, p_n$ can be represented by the set of their residues modulo $p_1, \ldots, p_n$. This technique allows operating on smaller elements, which is useful when there are bound restrictions over the original data. Also here, pre-processing and post-processing of the data to move between the original and RNS representations are required.

Understanding the nature of the data is the first and most crucial step in choosing the right HE scheme. For example, CKKS will probably be selected for floating-point arithmetic that is used by AI developers; BGV or B/FV can be selected for computations involving modular arithmetic, which often appear in cryptography and security protocols; and lastly, TFHE will likely be the choice for working with quantized data such as quantized NNs, which involve reduced precision values, or by general-purpose compilers (see Chap. 4).

Interestingly, the boundaries dictating when to opt for a particular scheme are not always clear and may be refined with the development of the HE field. In practice, we anticipate that the complexity of selecting the right scheme could be eventually abstracted away. Compilers will most probably take on the responsibility of scheme selection, by analyzing the provided inputs and determining the most suitable scheme, freeing users from making this decision manually. But until we reach that point, it is better to further understand the HE domain and its limitations and capabilities.

2.4 Noise

One principal aspect that affects the operations supported by HE schemes is the noise added during the encryption process, also called the encryption error (see Sect. 2.5.1), which is crucial for enabling a valid and strong encryption. Managing its growth is not an easy task and is often left to experts or compilers. Still, when designing a new application, debugging the noise growth is a useful skill that can assist in the development process.

In fact, handling noise is not unique to HE and developers of applications that use floating-point arithmetic should already be familiar with ways to manage it. To see that, let us first revisit the structure of floating-point numbers. These are typically represented using three terms: mantissa, base, and exponent using the simplified equation: $mantissa \cdot base^{exponent}$. The base can be 2 for binary numbers and 10 for decimal numbers. The following example shows how rounding errors occur with floating-point numbers:

2.4 Noise

Example 2.7 (Floating Point Rounding Error) Consider the decimal number 0.3, which in binary is approximately 0.0100110011... and assume that we place this number in an 8-bit precision floating-point container that uses 3 bits for the exponent and 5 bits for the mantissa, while for simplicity we ignore the sign bit. We start by converting 0.3 to binary and use the 5 mantissa bits to get 0.01001, and then we normalize the number by shifting it by two digits to the left to get $1.0011 \cdot 2^{-2}$. To convert the number back to decimal, we use the formula above, which in decimal is 0.296875; we see that due to the precision limitations, just storing the data without even performing any other operations on it led to a rounding error.

The situation gets even more complex when considering operations such as multiplication, division, or even addition on rounded elements, as the error is accumulated with every computation. To reduce the effect of the error on the final results, developers of complex tasks tend to use high-precision elements, e.g., with single, double, or even higher precision. Nevertheless, increasing the precision comes at the cost of slower applications that may have high memory requirements. In HE, choosing the required precision is a parameter that a solution designer can control and should care about.

2.4.1 Noise Sources: Homomorphic and Numeric

Several error sources are considered when designing an FHE application.

- **Homomorphic noise.** This type of error is generated, for example, while encrypting a message, where an error term is added to the plaintext to cryptographically hide the plaintext content; see Sect. 2.5.1. It is getting larger, for example, when multiplying a ciphertext with another ciphertext or when having a computation that involves a ciphertext and an encrypted key such as the evaluation or rotation keys. Another source of error can be the encoding/decoding processes between cleartexts and plaintexts that may involve extra rounding steps.
  When using an approximate FHE scheme such as CKKS, it is often blended with the data. In contrast, in exact schemes, this error is cancelled out during decryption or reduced almost completely, during a process called bootstrapping, as long as it did not cross a predefined limit. Controlling the error is commonly automatically done through the compiler or by performing some manual analysis.
- **Numerical noise.** This type of noise occurs in programs due to unavoidable modifications to the computation process and data representation that FHE application developers must introduce to meet the limitations and requirements of the FHE-supported APIs. Two common examples are as follows:
  - **Polynomial approximation.** Some FHE schemes only support addition, subtraction, and multiplication and can thus only compute functions that can be represented as polynomials. In Chap. 5, we will see that one way to

evaluate a non-polynomial function is by approximating it with a polynomial. However, achieving an accurate approximation often requires using high-degree polynomials, which may harm the application's efficiency. Trading accuracy with efficiency is an important aspect that a solution designer should consider. When considering less accurate polynomial approximations, one must take into account that the error that is added to the program may be accumulated and eventually lead to inaccurate results.

- **Quantization.** Some schemes such as TFHE might be limited to operate over data with a word size of, e.g., 16 bits or less. Consider, for example, an NN trained with double precision weights of 64-bit size each. Directly reducing the precision of the weights to 16 bits may harm the accuracy of the network's classification results. One way to alleviate this issue is to train the network with 16-bit words in advance. Alternatively, there are other more sophisticated processes such as accounting for the introduced error in other intermediate computations such as polynomial approximations, which limit the noise added due to quantization.

  Luckily, several quantization techniques are already in common use by AI and ML applications even when used over unencrypted data. The motivation is however the same, since working with quantized networks of 16, 8, or even 4 bits over dedicated hardware increases the computation throughput and reduces latency of training and inferencing NNs. The availability of such techniques may make it easier for an encrypted ML designer to adapt applications that use quantization to run over encrypted data using HE schemes.

- **Numerical stability.** When dealing with approximate or rounded inputs, it becomes more critical to choose an algorithm that will reduce the overall error; otherwise, the computation may explode and hit an overflow. For example, Sect. 4.8 suggests different methods to evaluate polynomials with various latency-accuracy trade-offs.

2.4.1.1 Handling the Encryption Error

Different HE schemes place the encryption error (see Sect. 2.5.1) and the data in different positions inside ciphertexts. Figures 2.2, 2.3, and 2.4 illustrate some examples for the data and error placement, where each figure schematically presents the effect of performing addition or multiplication of two ciphertext containers of size q, i.e., containers that hold values that are smaller than q.

Both Figs. 2.2 and 2.3 show different ways to separate the plaintext data x and y from the corresponding encryption errors e_1 and e_2, during the encryption process. Figure 2.2 uses a BGV-style placement, where the data is placed in the lower area of the ciphertext, while the error is placed in its higher area. This is done by multiplying the error by a predefined factor t, i.e., $x+te_1$ and $y+te_2$. Upon decryption, the lower and higher parts are separated by reducing the ciphertext modulo t. This property

2.4 Noise

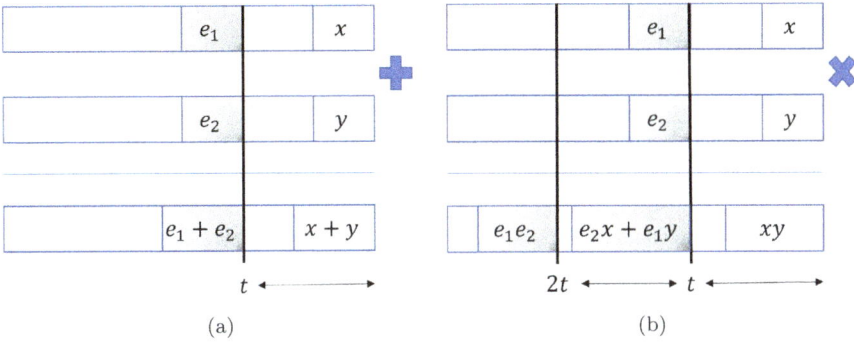

Fig. 2.2 BGV addition and multiplication. (**a**) Addition. (**b**) Multiplication

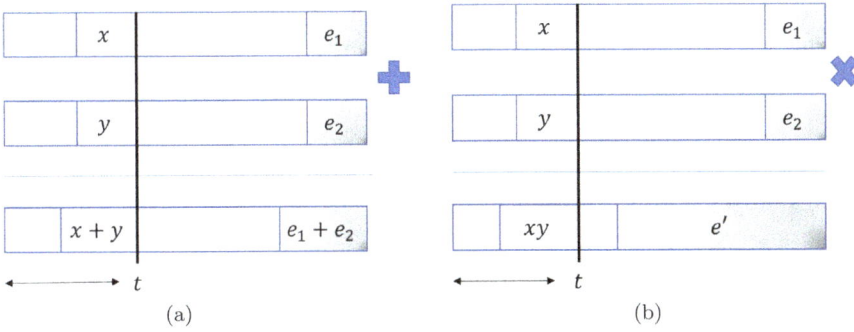

Fig. 2.3 A simplified illustration of the B/FV-style error packing when used for addition and multiplication. (**a**) Addition. (**b**) Multiplication

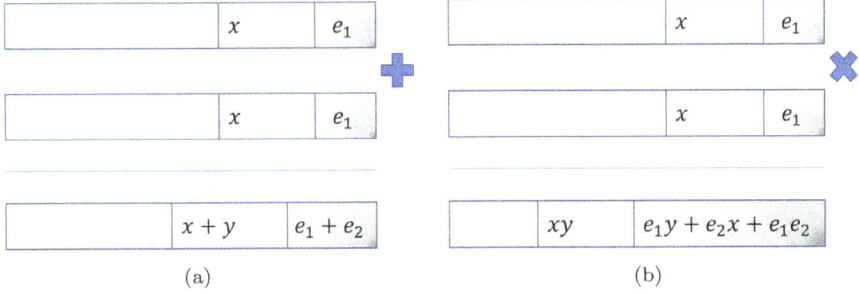

Fig. 2.4 CKKS addition and multiplication. (**a**) Addition. (**b**) Multiplication

holds also when decrypting the results of adding or multiplying two ciphertexts:

$$(x + te_1) + (y + te_2) = x + y + t(e_1 + e_2) \equiv x + y \pmod{t} \quad (2.1)$$

$$(x + te_1)(y + te_2) = xy + t(e_1 y + e_2 x + e_1 e_2) \equiv xy \pmod{t} \quad (2.2)$$

This encryption style supports exact integer operations as long as $x + y < t$ and $xy < t$ and as long as the error, which grows in the higher part of the ciphertext, does not wrap around, i.e., while the error is kept smaller than the ciphertext modulus q. Interestingly, when considering modular arithmetic modulo t, we can even ignore the requirements that $x + y > t$ or $xy > t$ because of the final reduction, which yields the right results even when the higher bits of the computations are blended with the error part of the ciphertext.

The B/FV-style placement is slightly more complex and we omit it for brevity. We refer the reader to the original papers to learn more about the exact process. For completeness, we illustrate a simplified version of it in Fig. 2.3 and mention that except for the different location of the data and the error, the correctness of the addition and multiplication are maintained as long as the error does not grow so much that it overflows into the data area, which would destroy the data.

The benefit of the B/FV and BGV error handling methods is that they allow to control the growth of the error and to prevent it from destroying the plaintext data. Once the error grows too much, a special method called bootstrapping can be invoked to reduce the error to its original size. This is why these schemes are called exact schemes. Significant advancements were made following the introduction of the bootstrap operation by Craig Gentry in 2009 [20], which was considered at first as a heavy operation. Indeed today, bootstrapping for various HE schemes may take less than a second on commodity hardware. Nevertheless, it is still much slower compared to other HE operations, and therefore a solution designer should carefully decide where and when to use it. Similar to noise handling, we expect compilers to handle the bootstrapping placement for the users. Nonetheless, while compilers can decide on the best location to call the bootstrap operations for a given function, an HE solution designer can manipulate the function in a way that eventually reduces the number of required bootstraps. We elaborate more on that in Chap. 4. More information about the bootstrapping process can be found in [13, 30] for TFHE and FHEW, [8, 24] for BGV, and [3, 5, 9, 26–29] for CKKS.

Figure 2.4 illustrates the error handling when using the CKKS scheme that supports approximated arithmetic. Here, the error is blended into the lower bits of the data. To avoid having the error destroy the data, the cleartext is multiplied by a predefined scale factor Δ during the encoding phase that precedes the encryption of the data. For example, if the data is $x = 1.344$ and the error is $e_1 = 12$, just adding the two yields the value 13.344, where extracting the error (without knowing it) is impossible. In contrast, if we set $\Delta = 1,000$, then $x\Delta = 1344$ and $x\Delta + e_1 = 1,344 + 12 = 1,356$. While we cannot extract the error, we can still divide by the scale during the decoding phase and receive the result 1.356 which is only 0.9% away from the original data. Using a larger scale will reduce the effect of the error on the data. We can now appreciate why this scheme is called an approximated FHE scheme, because even the early encoding operation already destroys some fraction of the data.

In approximated HE schemes such as CKKS that involve scaling the input during an encoding phase, the multiplication of two numbers becomes slightly more challenging. Consider having the two inputs $x = 1.2$ and $y = 3.4$ with

2.4 Noise

a product of $xy = 4.08$. If we first scale the inputs during encoding using $\Delta = 10^3$, then their product would be $\Delta^2 xy = 10^6 xy$, which after decryption and decoding (dividing by Δ) results in $10^3 xy = 4{,}080$, which is not what we expected. To mitigate this issue, such FHE schemes include a rescaling operation that, informally, divides the ciphertext by the scale Δ. Consequently, a rescaling operation should be invoked after multiplications to maintain the correctness of the results. In practice, performing a rescaling operation after every multiplication is not optimal, and applications may prefer to defer these operations to a later point of the computation or even to after the decryption. The downside is that this approach requires bookkeeping of the scale at every point of the computation and eventually dividing by the accumulated scale at the point of decoding. In the example above, dividing by $\Delta^2 = 10^6$ would be required during decoding rather than dividing by $\Delta = 10^3$. While this commonly used bookkeeping practice helps optimizing the application runtime, it should be handled with care to avoid overflowing beyond the ring moduli and to maintain the precision of the computation.

2.4.2 Typical Operations: Multiplications, Additions, and Rotations

So what operations are supported by FHE schemes? We already mentioned that additions and multiplications are supported between two ciphertexts or between a ciphertext and a plaintext. We can then split the Eval function into smaller building blocks such as Add and Mul. These operations can receive as input different types of encoded and maybe encrypted cleartext values depending on the scheme, such as Boolean values, finite field elements, vectors of s finite field elements, or vectors of s floating-point numbers. For brevity, we provide a single definition that captures all such types of inputs. For schemes that do not have single instruction multiple data (SIMD) support, i.e., that do not support operations on vectors, we set $s = 1$ and assume that Eq. 2.7 trivially holds. We denote by $+$ and $\cdot$ the addition and multiplication of tensors of elements, respectively, and denote by $\oplus$ and $\odot$ the addition and multiplication of ciphertexts with ciphertexts (or ciphertexts with plaintexts), respectively. Finally, we denote the i'th element of a tensor or a ciphertext using square brackets, e.g., $A[2]$ is the second element of A.

Definition 2.8 (Basic HE Operations) Let V_1 and V_2 be two cleartext vectors of s slots, and let C_1 and C_2 be two ciphertexts that encrypt them, respectively, and let α be a scalar. Then, for $0 \leq i, j < s$, the following holds:

$$V_1[i] + V_2[i] = \text{Dec}(C_1 \oplus c_2)[i] \tag{2.3}$$

$$V_1[i] \cdot V_2[i] = \text{Dec}(C_1 \odot c_2)[i] \tag{2.4}$$

$$V_1[i] \cdot V_2[i] = \text{Dec}(C_1 \odot V_2)[i] \tag{2.5}$$

$$\alpha V_1[i] = \text{Dec}(\alpha C_1)[i] \tag{2.6}$$

$$V_1[i] = \text{Dec}(\text{Rot}_j(C_1))[i - j \bmod s] \tag{2.7}$$

The above definitions hold for exact schemes. However, when using approximated schemes, the additional homomorphic noise should also be considered. This is done by changing the equality symbols (=) to ($\approx$) symbols or by adding a small ϵ to the equations above. For example, Eq. 2.3 will be modified to

$$V_1[i] + V_2[i] + \epsilon = \text{Dec}(C_1 \oplus c_2)[i] \tag{2.8}$$

and the rest of the equations are modified similarly.

While we listed all five equations under one definition, there is a huge performance and complexity difference between them. Additions are often the fastest operations, and then multiplying a ciphertext by a scalar, where in some cases integer scalar multiplication is faster than rational scalar multiplication. The list continues with multiplying a ciphertext by a plaintext, and finally performing ciphertext-ciphertext multiplications or ciphertext rotations that often require complex key switching operations and are the slowest operations. Their exact performance is hardware dependent. In some algorithms that are discussed in this book, we leverage the fact that some operations are faster than others and attempt to replace heavy operations with lighter ones.

In the rest of the book, we use the term scalar product or scalar multiplication to refer to the result of multiplying a ciphertext by a scalar (i.e., a single plain value, either integer, real, or complex). Note that both ciphertext-ciphertext and ciphertext-plaintext products can be considered non-scalar because they involve element-wise operations on vectors. However, in this book, we generally use the term non-scalar product to refer to the slower operation of multiplying two ciphertexts. When we refer to ciphertext-plaintext products, this is explicitly clarified in the text. Finally, we can use the term or (non-)scalar product to refer to the operation itself or to its results. For example, we can say the scalar product of 5 and c is $5c$, but we can also say that an algorithm uses m scalar products.

Finally, using the above definitions, we can construct additional primitives such as ShiftR and ShiftL.

Definition 2.9 (Shift Operations) Let V be a cleartext vector of s slots, and let $C = \text{Enc}(V)$ be its encryption. Then, for $1 < i < s$, a shift (left/right) operations is defined by

$$\text{ShiftL}_i(C) = \text{Rot}_{-i}(C) \odot ML_i \tag{2.9}$$

$$\text{ShiftR}_i(C) = \text{Rot}_i(C) \odot MR_i \tag{2.10}$$

where $ML_i = 0^{s-i}||1^i$ and $MR_i = 1^i||0^{s-i}$, i.e., cleartext masks with $s - i$ zeros and i ones in the leftmost and rightmost slots, respectively.

2.4 Noise

2.4.2.1 Limited Product Depth and Bootstrap

Schemes such as BGV and CKKS have a limit on the number of multiplications that can be performed on a ciphertext, known as the "multiplication chain index" (a.k.a. modulus chain index) or CIdx for short. This limit is set by the application developer to achieve the desired level of security and performance. Every ciphertext starts with a CIdx $= L$, which is the configured limit. After every multiplication of two ciphertexts with CIdx of x and y, the result has a CIdx of $\min(x, y) - 1$. Once a ciphertext's CIdx reaches 0, it can no longer be multiplied, unless a costly Bootstrap operation is performed to increase its CIdx, or even reset it back to L. The application designer should thus plan for computation paths that minimize the number of Bootstrap invocations. See Chap. 4 for more discussion on how depth and invoking Bootstrap.

2.4.3 Key Types

It is tempting to think that HE schemes similar to other public key encryption (PKE) schemes only involve one pair of keys, the secret key and the public key. In practice, the situation is quite different. More keys are needed to provide the functionalities presented above. For example, to perform FHE multiplications, dedicated keys called evaluation keys, multiplication keys, or sometimes even key switching keys are required. Some schemes require bootstrapping keys to perform the bootstrapping operation and some also require rotation keys (a.k.a. Galois keys), which are needed for rotating a ciphertext that encrypts a vector in a SIMD manner. Specifically, rotating by different numbers of slots requires different rotation keys.

Example 2.10 Rotating a given ciphertext by eight slots to the left is done by either using a rotation key that is associated with a rotation of eight slots to the left or by calling the rotation function twice, each time using a different rotation key, for example, by using the keys that are associated with one of the pairs (1,7), (2,6), or (3,5). Another alternative is to use the rotation key associated with rotating four slots to the left twice. In practice, the number of options does not end here; because rotations are cyclic, one can also use a "negative rotation" or a rotation to the right. For example, applying a rotation to the right by two slots and then applying a rotation to the left by ten slots may still result in a rotation of eight slots to the left.

One design choice that an application developer should make is how many rotation keys are needed for its application and which minimal rotation key set covers the various rotations used by the applications. Minimizing the number of rotation keys is critical as it reduces the setup time for an application, its bandwidth consumption and storage requirements, while also reducing cache misses due to reusing the same rotation keys. On the other hand, having fewer rotation keys may increase the number of serial rotations done on a ciphertext, which in turn increases the latency of the computation. The trade-off between fast setup time and fast online computations should therefore be taken into account.

To get a feeling of the key sizes required by an FHE application, consider Table 2.1, which reports the empirical CKKS key sizes, generated with Microsoft Seal [36] using different FHE configurations. As can be seen, the FHE keys often consume more space compared with other keys of maybe simpler cryptographic primitives, e.g., AES or RSA. For example, when an application involves ciphertext rotations, the required rotation keys can be several GBs in size. Here, Seal only generated the power of two rotation keys. In contrast, when using a scheme such as TFHE that does not involve rotations (while also using smaller N values compared to CKKS), the size of the keys is much reduced.

Remark 2.11 In practice, the size of the secret keys can be smaller than that reported in Table 2.1, as only a 256-bit seed is required for reproducing the keys when needed.

Example 2.12 Different algorithms exist for reducing the number of required rotation keys for different applications. Let c be an input ciphertext that encrypts a vector C of $n = \ell k$ slots, where $\ell \approx k \approx \sqrt{n}$, and consider the following task: sum up all the values in c using a naïve algorithm that accumulates one element at a time in slot 0 (and in fact also in all other slots). The computation is thus $s = \sum_{i=0}^{n-1} \text{Rot}_i(c)$. This requires $n-1$ rotations and therefore $n-1$ rotation keys. Instead, let us leverage the baby-step giant-step algorithm of [23] and replace the above equation with

$$s = \sum_{i=0}^{n-1} \text{Rot}_i(c) = \sum_{i=0}^{\ell} \sum_{j=0}^{k} \text{Rot}_{ki+j}(c) = \sum_{i=0}^{\ell} \text{Rot}_{ki}\left(\sum_{j=0}^{k} \text{Rot}_j(c)\right) \quad (2.11)$$

This requires only $\sqrt{n}$ keys to achieve the same goal. We see, thus, that a good rotation policy can reduce the number of required keys and therefore should be considered when constructing an FHE application. In Sect. 7.2, we will see another algorithm that achieves even better results.

Table 2.1 Size of CKKS keys, generated by Microsoft Seal [36], for different polynomial degrees N and multiplication levels L; see Sect. 2.5.1

Key	$N = 1,024$ $L = 1$	$N = 2,048$ $L = 1$	$N = 4,096$ $L = 3$	$N = 8,192$ $L = 5$	$N = 16,384$ $L = 10$	$N = 32,768$ $L = 20$
Secret key	8 KB	16 KB	98 KB	327 KB	1.3 MB	5.2 MB
Public key	16 KB	32 KB	196 KB	654 KB	2.6 MB	10.4 MB
Evaluation key	N/A	N/A	395 KB	2.6 MB	23 MB	200 MB
Rotation keys	N/A	N/A	8.7 MB	63.2 MB	615 MB	5.6 GB

2.5 Advanced: Intro to HE Math and Its Underlying Security Assumptions

This section includes some advanced HE topics that may help readers gain further understanding of the lower-level implementation details of some modern HE schemes. Readers who are satisfied with only gaining a high-level understanding of HE can safely skip this part and move directly to Chap. 3. In addition, the following books and papers are recommended for further reading about the technicalities of the FHE domain [2, 4, 12, 22, 25].

2.5.1 LWE and Ring-LWE

Modern FHE relies on the computational hardness of solving large lattice-based problems such as the learning with errors (LWE) or the ring LWE (R-LWE) problems that we list below. The science of lattices is fascinating, and it involves much mathematical and algorithmic research. However, only a shallow familiarity with the concept, if at all, is required for understanding higher-level concepts of the FHE domain. To this end, we only allocate a small section to describe it and refer the readers to [31, 33] for further reading about the topic. Definition 2.13 explains what a lattice is.

Definition 2.13 (Lattice) For a basis $\mathbf{B} = B_0, \ldots, B_{n-1} \in \mathbb{R}^n$, a lattice $\mathcal{L}(\mathbf{B}) \subset \mathbb{R}^n$ is the set of all integer linear combinations of vectors from $\mathbf{B}$, i.e.,

$$L = \left\{ \sum_{0 \leq i < n} a_i B_i : a_i \in \mathbb{Z} \right\} \tag{2.12}$$

Interestingly, more than one basis can be used to form the same lattice. Figure 2.5 illustrates a two-dimensional lattice that can be generated from two different bases: $\mathbf{B}_1$ or $\mathbf{B}_2$. The yellow dot in the figure indicates a near lattice point. It is a point that is close enough to a lattice point, but associating it with this lattice point is not always easy. Based on this figure, we provide the definition of two famous problems: the shortest vector problem (SVP) (Definition 2.14) and closest vector problem (CVP) (Definition 2.15).

Definition 2.14 (Shortest Vector Problem (SVP)) Given a lattice $\mathcal{L}(\mathbf{B})$ for a basis $\mathbf{B} \in \mathbb{R}^n$, and an ℓ−norm, find the shortest nonzero vector V in $\mathcal{L}$, using the ℓ-norm, i.e., $V = \arg\min_{V \in L \setminus \{0\}} \|V\|_\ell$

Definition 2.15 (Closest Vector Problem (CVP)) Given a lattice $\mathcal{L}(\mathbf{B})$ for a basis $\mathbf{B} \in \mathbb{R}^n$, an ℓ−norm, and a vector E, find the closest vector V in $\mathcal{L}$ to E, using the ℓ−norm, i.e., $V = \arg\min_{V \in L \setminus \{0\}} \|V - E\|_\ell$

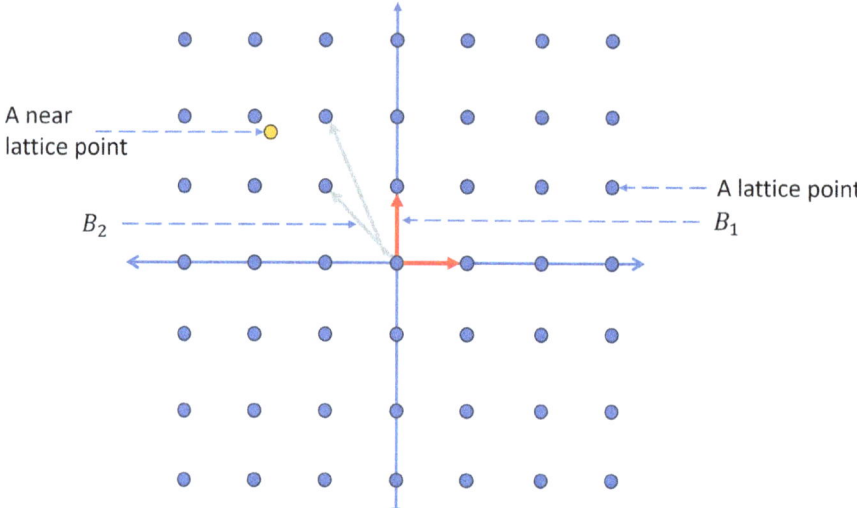

Fig. 2.5 A two-dimensional lattice illustration. Red arrows ($\mathbf{B}_1$) form one basis of the lattice and the gray arrows form an alternative basis ($\mathbf{B}_2$) of it

The relation between these problems as well as the relation to other lattice problems can be further studied in [31, 33].

The LWE problem shares many similarities with the SVP and CVP problems and in fact, there are some reductions between the three. Let us start with reviewing the LWE problem. Consider a matrix $A \in \mathbb{Z}_q^{m \times n}$, a vector $B \in \mathbb{Z}_q^m$, and a secret $S \in \mathbb{Z}_q^n$, where $\mathbb{Z}_q$ is the ring of integers modulo $q \geq 2$. If $B = AS$, it is easy to see that when A and B are known, recovering S can simply be done using simple Gaussian elimination on the linear system of equations $AS = B$.

Above, A can be considered as the lattice basis, S is the integer combination of the basis vectors, and B is the resulting lattice point. However, if instead of B, we provide an adversary with the knowledge of a near lattice point, i.e., a noisy vector $B' = B + E$ for some vector $E \in \mathbb{Z}_q^m$ with small coefficients. Then, finding S or B given only A and B' can be hard to compute.

Let χ_e and χ_s be the distributions over $\mathbb{Z}_q^m$ for selecting E and S with small coefficients, respectively. Denote by $x \xleftarrow{\$} S$ that x is selected uniformly at random from a set S and by $y \leftarrow \chi$ that y is selected according to the distribution χ. Then the LWE problem is.

Definition 2.16 (LWE) Let $A \xleftarrow{\$} \mathbb{Z}_q^{m \times n}$, $E \leftarrow \chi_e$, $S \leftarrow \chi_s$, and $B = AS^T + E$. Given the pair (A, B), recover S.

For defining the R-LWE problem, we recall two mathematical terms polynomial rings and cyclotomic polynomials. A polynomial ring is a mathematical structure (a ring) over the set of polynomials with coefficients in another ring (e.g., $\mathbb{Z}_q$). The

2.5 Advanced: Intro to HE Math and Its Underlying Security Assumptions

nth cyclotomic polynomial for $n \in \mathbb{Z}_+$ is the unique irreducible polynomial with integer coefficients that divides $x^n - 1$ but not divides $x^k - 1$ for any $k < n$.

The R-LWE problem is similar to the LWE problem except that instead of working with elements from $\mathbb{Z}_q$, we consider polynomials with coefficients in $\mathbb{Z}_q$ of the polynomial ring $\mathcal{R}_q = \mathbb{Z}_q[X]/\Phi_m(X)$, where $\Phi_m(X)$ is the m-th cyclotomic polynomial. That is, we first choose the polynomials $A, S, E \in \mathcal{R}_q$, where S and E have small coefficients according to some predefined distributions, and then provide the pair $(A, B) = (A, AS + E)$ to an adversary and ask them to recover S. Here, χ_e, χ_s act over the ring of polynomials $\mathcal{R}_q$ instead of over $\mathbb{Z}_q^m$ as before.

Definition 2.17 (Ring-LWE) For $a \xleftarrow{\$} \mathcal{R}_q$, $E \leftarrow \chi_e$, $S \leftarrow \chi_s$, and $B = AS + E$. Given the pair (A, B), recover S.

2.5.2 Encryption-Decryption Example

We can now use the terminology of the R-LWE problem from above, to describe the encryption and decryption methods of the BGV, B/FV, and CKKS schemes. The reason we chose these three schemes is that they share similar key generation methods. In particular, to generate the secret and public key pair, the user generates a secret key $S \leftarrow \chi_s$ and selects a polynomial $A \xleftarrow{\$} \mathcal{R}_q$ and a secret error $E \leftarrow \chi_e$. Subsequently, the following equation derives the public key, where the modulo reduction is performed coefficient-wise over the coefficients of the polynomials:

$$P = (P_0, P_1) =$$
$$= \begin{cases} (AS + E, -A) \pmod{q} & \text{B/FV} \\ (-(AS + tE), A) \pmod{q} & \text{BGV} \\ (AS + E, -A) \pmod{q} & \text{CKKS} \end{cases}$$

To encrypt a plaintext message μ using the public key P, the encryptor chooses the secret error terms $E_0, E_1 \leftarrow \chi_e$ and a random polynomial $U \leftarrow \mathcal{R}_q$ and computes

$$ct = \text{Enc}(\mu) = (CT_0, CT_1) =$$
$$= \begin{cases} (\lfloor \frac{q}{t} \rfloor \mu + UP_0 + E_0, UP_1 + E_1) \pmod{q} & \text{B/FV} \\ ((\mu \bmod t) + UP_0 + tE_0, UP_1 + tE_1) \pmod{q} & \text{BGV} \\ (\mu + UP_0 + E_0, UP_1 + E_1) \pmod{q} & \text{CKKS} \end{cases}$$

Here, the plaintext space for B/FV and BGV is $\mathcal{R}_t$ and the ciphertext space is $\mathcal{R}_q^2$, while for CKKS the plaintext space is $\mathcal{R}_q$ and the ciphertext space is $\mathcal{R}_q^2$.

Note how the encryption definition matches the description of the noise from Sect. 2.4. In B/FV the noise E_0, E_1 is added in the ciphertext's lower part, while the message is shifted by $\lfloor \frac{q}{t} \rfloor$ to its upper part. In contrast, in BGV the error is multiplied by t, while the message is left in the ciphertext's lower part. Finally in CKKS, the error is blended with the plaintext.

To decrypt a ciphertext CT using the secret key S, the decryptor performs

$$\mu = \text{Dec}(CT) = \begin{cases} \lfloor \frac{t}{q}(CT_0 + CT_1 S \pmod{q}) \rceil \pmod{t} & \text{B/FV} \\ (CT_0 + CT_1 S \pmod{q}) \pmod{t} & \text{BGV} \\ (CT_0 + CT_1 S) \pmod{q} & \text{CKKS} \end{cases}$$

For more details on the mathematics behind the scheme, please refer to the standard or the original papers that presented these schemes.

2.6 Lab Exercises: Using HE

The following lab exercise is designed to allow the readers to experiment with the FHE technology. For that, the readers should first download and install some FHE library of their choice such as HElayers [1]. Modern libraries can ease the work of selecting the FHE scheme parameters by offering default parameter choices that meet a 128-bit security level. In the exercises below, we recommend using these configurations unless indicated otherwise.

1. Select any FHE scheme and use the library default parameters. Write a computer program that fixes two integers a, b. Encrypt them to a and b, respectively. Evaluate the following functions on a, b:

$$f_1(x, y) = x + y \qquad f_2(x, y) = xy$$
$$f_3(x, y) = 5x + 6y \qquad f_4(x, y) = 3x^3 - x^2 + 5$$

   Decrypt and print the results.
2. Repeat Exercise 1 with different scheme parameters. For each parameter choice, print the size of the ciphertext and the time it took to perform the computation. What can you deduce?
3. Consider the BGV scheme with a plaintext modulus of $t - 1$. Fix a number a, encrypt it, and compute its inverse in the field $\mathbb{F}_t$.
4. Consider the TFHE scheme, and write a Boolean function that receives three Boolean inputs x, y, z, encrypts them, and computes the Boolean circuit $x\bar{y} + y\bar{z} + xyz$, where $\bar{x}$ indicates a logical-not gate of x and multiplication indicates the logical-and gate.

5. Repeat Exercise 4 using BGV, by selecting the relevant scheme parameters to allow Boolean computation.
6. Repeating Exercise 4 using CKKS, did you get the expected results?
7. Use B/FV or CKKS and encrypt the vectors

$$(0, 1, 2, 3, 4, 5, 6) \qquad (7, 8, 9, 10, 11, 12, 13).$$

Compute the functions f_1-f_4 from Exercise 1 on the encrypted data and decrypt the results. In which scheme do you observe errors? What were their magnitudes?
8. Repeat Exercise 6 but this time apply $f_1()$ n times on the inputs, where $n = 10, 100, 1000$ and $1{,}000{,}000$ times. What was the error magnitude this time?
9. Start from a ciphertext that encrypts the value 0 using a scheme of your choice. Add the ciphertext to itself and repeat this doubling process 30 times. Decrypting the result, did you get the expected results?

References

1. Aharoni, E., Adir, A., Baruch, M., Drucker, N., Ezov, G., Farkash, A., Greenberg, L., Masalha, R., Moshkowich, G., Murik, D., Shaul, H., Soceanu, O.: HeLayers: A tile tensors framework for large neural networks on encrypted data. In: Privacy Enhancing Technology Symposium (PETs) 2023 (2023). https://petsymposium.org/popets/2023/popets-2023-0020.php
2. Albrecht, M., Chase, M., Chen, H., Ding, J., Goldwasser, S., Gorbunov, S., Halevi, S., Hoffstein, J., Laine, K., Lauter, K., Lokam, S., Micciancio, D., Moody, D., Morrison, T., Sahai, A., Vaikuntanathan, V.: Homomorphic Encryption Standard, pp. 31–62. Springer International Publishing, Cham (2021). https://doi.org/10.1007/978-3-030-77287-1_2
3. Bae, Y., Cheon, J.H., Cho, W., Kim, J., Kim, T.: Meta-bts: Bootstrapping precision beyond the limit. In: Proceedings of the 2022 ACM SIGSAC Conference on Computer and Communications Security, CCS '22, p. 223–234. Association for Computing Machinery, New York (2022). https://doi.org/10.1145/3548606.3560696
4. Bergerat, L., Boudi, A., Bourgerie, Q., Chillotti, I., Ligier, D., Orfila, J.B., Tap, S.: Parameter optimization and larger precision for (T)FHE. J. Cryptolo. **36**(3), 28 (2023). https://doi.org/10.1007/s00145-023-09463-5
5. Bossuat, J.P., Mouchet, C., Troncoso-Pastoriza, J., Hubaux, J.P.: Efficient bootstrapping for approximate homomorphic encryption with non-sparse keys. In: Canteaut, A., Standaert, F.X. (eds.) Advances in Cryptology – EUROCRYPT 2021, pp. 587–617. Springer International Publishing, Cham (2021). https://doi.org/10.1007/978-3-030-77870-5_21
6. Brakerski, Z.: Fully homomorphic encryption without modulus switching from classical GapSVP. In: Safavi-Naini, R., Canetti, R. (eds.) Advances in Cryptology – CRYPTO 2012, vol. 7417 LNCS, pp. 868–886. Springer, Berlin (2012). https://doi.org/10.1007/978-3-642-32009-5_50
7. Brakerski, Z., Gentry, C., Vaikuntanathan, V.: (Leveled) fully homomorphic encryption without bootstrapping. ACM Trans. Comput. Theory **6**(3), 1–36 (2014). https://doi.org/10.1145/2633600
8. Chen, H., Han, K.: Homomorphic lower digits removal and improved fhe bootstrapping. In: Nielsen, J.B., Rijmen, V. (eds.) Advances in Cryptology – EUROCRYPT 2018, pp. 315–337. Springer International Publishing, Cham (2018). https://doi.org/10.1007/978-3-319-78381-9_12

9. Chen, H., Chillotti, I., Song, Y.: Improved bootstrapping for approximate homomorphic encryption. In: Ishai, Y., Rijmen, V. (eds.) Advances in Cryptology – EUROCRYPT 2019, pp. 34–54. Springer International Publishing, Cham (2019). https://doi.org/10.1007/978-3-030-17656-3_2
10. Cheon, J., Kim, A., Kim, M., Song, Y.: Homomorphic encryption for arithmetic of approximate numbers. In: Proceedings of Advances in Cryptology - ASIACRYPT 2017, pp. 409–437. Springer, Cham (2017). https://doi.org/10.1007/978-3-319-70694-8_15
11. Cheon, J.H., Han, K., Kim, A., Kim, M., Song, Y.: A full RNS variant of approximate homomorphic encryption. In: Cid, C., Jacobson Jr., M.J. (eds.) Selected Areas in Cryptography – SAC 2018, pp. 347–368. Springer International Publishing, Cham (2019). https://doi.org/10.1007/978-3-030-10970-7_16
12. Cheon, J.H., Costache, A., Moreno, R.C., Dai, W., Gama, N., Georgieva, M., Halevi, S., Kim, M., Kim, S., Laine, K., Polyakov, Y., Song, Y.: Introduction to Homomorphic Encryption and Schemes, pp. 3–28. Springer International Publishing, Cham (2021). https://doi.org/10.1007/978-3-030-77287-1_1
13. Chillotti, I., Gama, N., Georgieva, M., Izabachène, M.: Faster fully homomorphic encryption: Bootstrapping in less than 0.1 seconds. In: Cheon, J.H., Takagi, T. (eds.) Advances in Cryptology – ASIACRYPT 2016, pp. 3–33. Springer, Berlin (2016). https://doi.org/10.1007/978-3-662-53887-6_1
14. Chillotti, I., Gama, N., Georgieva, M., Izabachène, M.: TFHE: fast fully homomorphic encryption over the torus. J. Cryptol. **33**(1), 34–91 (2020). https://doi.org/10.1007/s00145-019-09319-x
15. Daniel, J., Benaloh, C., Benaloh, J.D.C.: Verifiable secret-ballot elections. Ph.D. Thesis, Yale University, New Haven, CT (1987)
16. Drucker, N., Moshkowich, G., Pelleg, T., Shaul, H.: BLEACH: cleaning errors in discrete computations over CKKS. J. Cryptol. **37**(1), 3 (2023). https://doi.org/10.1007/s00145-023-09483-1
17. Ducas, L., Micciancio, D.: FHEW: Bootstrapping homomorphic encryption in less than a second. In: Oswald, E., Fischlin, M. (eds.) Advances in Cryptology – EUROCRYPT 2015, pp. 617–640. Springer, Berlin (2015). https://doi.org/10.1007/978-3-662-46800-5_24
18. Elgamal, T.: A public key cryptosystem and a signature scheme based on discrete logarithms. IEEE Trans. Inf. Theory **31**(4), 469–472 (1985). https://doi.org/10.1109/TIT.1985.1057074
19. Fan, J., Vercauteren, F.: Somewhat practical fully homomorphic encryption. In: Proceedings of the 15th international conference on Practice and Theory in Public Key Cryptography, pp. 1–16 (2012). https://eprint.iacr.org/2012/144
20. Gentry, C.: A fully homomorphic encryption scheme. Ph.D. Thesis, Stanford University, Palo Alto, CA (2009). https://crypto.stanford.edu/craig/craig-thesis.pdf
21. Goldwasser, S., Micali, S.: Probabilistic Encryption & How to Play Mental Poker Keeping Secret All Partial Information, pp. 173–201. Association for Computing Machinery, New York (2019). https://doi.org/10.1145/3335741.3335749
22. Halevi, S.: Homomorphic Encryption, pp. 219–276. Springer International Publishing, Cham (2017). https://doi.org/10.1007/978-3-319-57048-8_5
23. Halevi, S., Shoup, V.: Faster homomorphic linear transformations in HElib. In: Shacham, H., Boldyreva, A. (eds.) Advances in Cryptology – CRYPTO 2018, pp. 93–120. Springer International Publishing, Cham (2018). https://doi.org/10.1007/978-3-319-96884-1_4
24. Halevi, S., Shoup, V.: Bootstrapping for HElib. J. Cryptol. **34**(1), 7 (2021). https://doi.org/10.1007/s00145-020-09368-7
25. Halevi, S., Polyakov, Y., Shoup, V.: An improved rns variant of the bfv homomorphic encryption scheme. In: Matsui, M. (ed.) Topics in Cryptology – CT-RSA 2019, pp. 83–105. Springer International Publishing, Cham (2019). https://doi.org/10.1007/978-3-030-12612-4_5
26. Han, K., Hhan, M., Cheon, J.H.: Improved homomorphic discrete fourier transforms and fhe bootstrapping. IEEE Access **7**, 57361–57370 (2019). https://doi.org/10.1109/ACCESS.2019.2913850

References

27. Jung, W., Kim, S., Ahn, J.H., Cheon, J.H., Lee, Y.: Over 100x faster bootstrapping in fully homomorphic encryption through memory-centric optimization with gpus. IACR Trans. Cryptogr. Hardw Embed. Syst. **2021**(4), 114–148 (2021). https://doi.org/10.46586/tches.v2021.i4.114-148
28. Jutla, C.S., Manohar, N.: Sine series approximation of the mod function for bootstrapping of approximate HE. In: Dunkelman, O., Dziembowski, S. (eds.) Advances in Cryptology – EUROCRYPT 2022, pp. 491–520. Springer International Publishing, Cham (2022). https://doi.org/10.1007/978-3-031-06944-4_17
29. Lee, Y., Lee, J.W., Kim, Y.S., Kim, Y., No, J.S., Kang, H.: High-precision bootstrapping for approximate homomorphic encryption by error variance minimization. In: Dunkelman, O., Dziembowski, S. (eds.) Advances in Cryptology – EUROCRYPT 2022, pp. 551–580. Springer International Publishing, Cham (2022). https://doi.org/10.1007/978-3-031-06944-4_19
30. Micciancio, D., Polyakov, Y.: Bootstrapping in FHEW-like cryptosystems. In: Proceedings of the 9th on Workshop on Encrypted Computing & Applied Homomorphic Cryptography, WAHC '21, pp. 17–28. Association for Computing Machinery, New York (2021). https://doi.org/10.1145/3474366.3486924
31. Micciancio, D., Regev, O.: Lattice-Based Cryptography, pp. 147–191. Springer, Berlin (2009). https://doi.org/10.1007/978-3-540-88702-7_5
32. Paillier, P.: Public-key cryptosystems based on composite degree residuosity classes. In: Stern, J. (ed.) Advances in Cryptology — EUROCRYPT '99, pp. 223–238. Springer, Berlin (1999). https://doi.org/10.1007/3-540-48910-X_16
33. Peikert, C.: A decade of lattice cryptography. Found. Trends Theor. Comput. Sci. **10**(4), 283–424 (2016). https://doi.org/10.1561/0400000074
34. Rivest, R.L., Adleman, L., Dertouzos, M.L.: On data banks and privacy homomorphisms. Found. Secure Comput. **4**, 169–180 (1978)
35. Rivest, R.L., Shamir, A., Adleman, L.: A method for obtaining digital signatures and public-key cryptosystems. Commun. ACM **21**(2), 120–126 (1978). https://doi.org/10.1145/359340.359342
36. Microsoft SEAL (Release 3.5). Microsoft Research, Redmond, WA (2020). https://github.com/Microsoft/SEAL

Chapter 3
Modern HE: Security Models

Abstract Using HE as a security primitive requires a careful understanding of the security guarantees it provides for applications, where different applications use it in various ways for which they may require different security models. This chapter introduces the security aspects of using HE. It starts by describing some basic cryptographic concepts and then provides examples of using these in some HE-based applications.

3.1 Security Models and Assumptions When Using FHE

Using FHE in an application should serve a purpose, i.e., it should provide some security guarantees, specifically confidentiality for the data used within the application. However, FHE does not come for free; even with the major progress in this domain, it still adds some latency and memory overheads to the system. An application developer should carefully understand its needs and weigh the security requirements against the above overheads while choosing wisely where and how to apply FHE. For that, a designer should start by classifying the data that goes through the application by asking the following questions: Who owns the data? In which computation should it be used? How can it be leaked? Subsequently, the designer should generate a threat model from which it can derive a security model that explains what protection should be provided by HE to the application.

The first step in generating the threat model is to identify the entities involved in the FHE process. For example, common FHE implementations involve the following entities:

- **Data owner**—The entity that owns the data to be processed such as the training set when considering an FHE-based training of a ML model application or the query data for an FHE-based secure inference application.
- **Function owner**—The entity that owns the function that will be evaluated under FHE on the data of the data owner. For example, the model architecture and weights of a machine learning model.

- **Server**—The entity that owns the compute resources, such as CPU, graphical processing unit (GPU), memory, and software stack, which are utilized to perform the respective function on the data. An example of a server is a cloud provider.
- **Client**—The entity that owns the compute resources that are used for encrypting the data and for decrypting the results of the processing of the data.

Once all the entities in the system are identified, the next step is to classify the data and functionality using guidelines as in [14], which allow us to distinguish classified information from public data. Moreover, they allow us to learn how much time every piece of data should remain secure. For example, assume that an algo-trading company predicts the prices of stocks next week. It decides to classify this data as sensitive; otherwise, clients will not pay money to purchase these predictions. Here, the company should ensure that the data is kept encrypted for at least 1 week, but it has no need to keep it a secret further than that because after a week the stock price becomes public knowledge. Another example is credit card numbers; hiding them after their validity date expires is often unneeded. Clearly, maintaining data confidentiality for a longer time requires better security guarantees from the underlying encryption scheme. In addition, classifying the data as sensitive or not assists us in minimizing the encryption costs. Encrypting only sensitive data pieces reduces metrics such as bandwidth, storage, and latency when it comes to FHE-based applications.

3.1.1 Basic Cryptographic Definition

Calling a cryptographic scheme secure or saying that it provides security or confidentiality is a bit vague or even ambiguous. What is the meaning of saying that a scheme is unbreakable? Does it mean that no one will ever be able to recover a secret key from a set of ciphertexts? Does it mean that by observing a ciphertext no one can learn something about the encrypted plaintext? Clearly, using brute-force techniques, one can search for all possible keys and at some point even succeed in finding the right one. But, as we saw before, time plays an important factor when defining security. For example, having a large key space of size 2^{128} can cause a brute-force attack to be unpractical. Even if an attacker uses 2^{64} parallel machines, it still needs to perform 2^{64} operations on every one of them. In contrast, a key space of 2^{64} can easily be explored using $2^{32} \approx 1M$ machines.

To avoid these definitions' ambiguity, when dealing with FHE, we consider several cryptographic definitions (that may also apply to other PKE schemes):

- Indistinguishability under chosen plaintext attack (IND-CPA)
- Indistinguishability under non-adaptive chosen ciphertext attack (IND-CCA1)
- Indistinguishability under adaptive chosen ciphertext attack (IND-CCA2)

3.1 Security Models and Assumptions When Using FHE

Table 3.1 Cryptographic indistinguishability games

IND-CPA	IND-CPA^D
$b \xleftarrow{\$} \{0, 1\}$	$b \xleftarrow{\$} \{0, 1\}$
$(sk, pk, ek) \leftarrow \text{KeyGen}(k)$	$(sk, pk, ek) \leftarrow \text{KeyGen}(k)$
$st, m_0, m_1 = \mathcal{A}^{\text{Enc}_{pk}(\cdot), \text{Eval}_{ek}(\cdot)}$	$st, m_0, m_1 = \mathcal{A}^{\text{Enc}_{pk}(\cdot), \text{Dec}^1_{sk}(\cdot), \text{Eval}_{ek}(\cdot)}$
$c = \text{Enc}_{pk}(m_b)$	$c = \text{Enc}_{pk}(m_b)$
$b' = \mathcal{A}^{\text{Enc}_{pk}(\cdot), \text{Eval}_{ek}(\cdot)}(st, c)$	$b' = \mathcal{A}^{\text{Enc}_{pk}(\cdot), \text{Eval}_{ek}(\cdot)}(st, c)$
Return $b == b'$	Return $b == b'$
IND-CCA1	**IND-CCA2**
$b \xleftarrow{\$} \{0, 1\}$	$b \xleftarrow{\$} \{0, 1\}$
$(sk, pk, ek) \leftarrow \text{KeyGen}(k)$	$(sk, pk, ek) \leftarrow \text{KeyGen}(k)$
$st, m_0, m_1 = \mathcal{A}^{\text{Enc}_{pk}(\cdot), \text{Dec}_{sk}(\cdot), \text{Eval}_{ek}(\cdot)}$	$st, m_0, m_1 = \mathcal{A}^{\text{Enc}_{pk}(\cdot), \text{Dec}_{sk}(\cdot), \text{Eval}_{ek}(\cdot)}$
$c = \text{Enc}_{pk}(m_b)$	$c = \text{Enc}_{pk}(m_b)$
$b' = \mathcal{A}^{\text{Enc}_{pk}(\cdot), \text{Eval}_{ek}(\cdot)}(st, c)$	$b' = \mathcal{A}^{\text{Enc}_{pk}(\cdot), \text{Dec}_{sk}(\cdot), \text{Eval}_{ek}(\cdot)}(st, c)$
Return $b == b'$	Return $b == b'$

These definitions are described via cryptographic "games" that are listed in Table 3.1. Every game is executed by a challenger who challenges an adversary $\mathcal{A}$ to break the indistinguishability property of ciphertexts. The goal of $\mathcal{A}$ is to learn some information from the challenged ciphertext that encrypts one of two messages, at least enough to deduce some information about the underlying plaintext message. If $\mathcal{A}$ can do so with probability $\frac{1}{2} + negl(k)$, where $negl(k)$ is a negligible function in the security parameter k, it means that $\mathcal{A}$ indeed gained some knowledge from the ciphertext (and not just randomly guessed the solution) and thus wins the game. Here, the function $negl(k)$ is defined as follows: for every (nonzero) polynomial function $poly()$, there exists some k_0 such that $|negl(k)| < |\frac{1}{poly(k)}|$ for all $k > k_0$.

A critical restriction on $\mathcal{A}$ is that it must run in probabilistic polynomial time (PPT). In other words, $\mathcal{A}$ must complete the game and output a guess within a number of time steps that are polynomial in some security parameter. This security parameter helps us tune the time that the scheme maintains the confidentiality of the encrypted data. In practice, we choose this parameter to be very large, e.g., 2^{128}, to ensure the confidentiality of the data for, informally, an indefinite time.

The IND-CPA, IND-CCA1, IND-CCA2, and IND-CPA^D games have almost the same steps, where the difference lies in the capabilities of the adversary.

1. At the beginning of every game, the challenger makes a random selection of a bit $b \in \{0, 1\}$ and stores it for a later use.
2. The challenger generates a public and secret key pair, pk and sk respectively, according to some security parameter k as well as the evaluation key ek and publishes pk and ek to the adversary $\mathcal{A}$ while retaining sk.
3. $\mathcal{A}$ is allowed to perform a polynomially bounded number of queries to some given functionality known as "oracle" as well as other operations over arbitrary data. Here lies the main difference between the games.

- In the IND-CPA game, $\mathcal{A}$ only gets access to encryption and evaluation oracles that allow it to encrypt messages and evaluate some functions over ciphertexts.
- In the IND-CPA^D game, $\mathcal{A}$ also gets access to a decryption oracle $\text{Dec}^1_{sk}(\cdot)$ that upon receiving a *valid* ciphertext c that was either the output of the encryption oracle or the output of the evaluation oracle returns its decryption. This can be viewed as if the challenger decrypts (valid) ciphertexts for $\mathcal{A}$.
- In the IND-CCA1 and IND-CCA2 games, $\mathcal{A}$ has extended capabilities, and it is now allowed to query the decryption oracle also for messages that are not necessarily valid ciphertexts, e.g., it can provide random data to this oracle and observe the "decrypted" results.

4. Once ready, $\mathcal{A}$ is expected to submit two chosen plaintexts denoted as m_0 and m_1 to the challenger while also storing all the data learned during its work in some *state* parameter st.
5. The challenger encrypts only one of the messages and returns $\text{Enc}_{pk}(m_b)$ to $\mathcal{A}$.
6. $\mathcal{A}$ may perform any number of additional computations using the supported oracles over the ciphertext, the state, and also using additional data.
7. Now, following the above preparation, the adversary needs to guess the value of b, i.e., which of the two messages m_0, m_1 was encrypted by the challenger.
8. $\mathcal{A}$ wins the game if and only if, statistically, it manages to guess the right value of b more than 50% of the times.

Even though an adversary knows the content of the messages m_0, m_1 and pk, the probabilistic characteristic (semantic security) of $\text{Enc}_{pk}(\cdot)$ implies that the encryption of m_b may have many valid ciphertexts. Thus, using the encryption oracle for encrypting m_0 and m_1 multiple times, and comparing the resulting ciphertexts with the challenge ciphertext, does not provide any non-negligible advantage to the adversary.

Modern FHE schemes are IND-CPA secure, whereas no IND-CCA1 secure FHE schemes are known at the time of writing this book. Moreover, note that the main difference between IND-CCA1 and IND-CCA2 is in the last guessing step, where $\mathcal{A}$ is allowed to use the decryption oracle after receiving the challenge ciphertext c for every ciphertext except c. However, $\mathcal{A}$ is not allowed to decrypt c because then it would trivially know the value of b and the original message. It can use the encryption oracle to encrypt m_0 into $c' = \text{Enc}_{pk}(m_0)$ and ask the decryption oracle to decrypt the ciphertext $c'' = c - c'$. If the decrypted value is 0, then $\mathcal{A}$ guesses that $b = 0$; otherwise, it guesses $b = 1$. This leads to the commonly known result that no FHE scheme can be IND-CCA2 secure.

In 2020, Li and Micciancio [23] introduced a new security notion called IND-CPA^D and showed that this notion does not hold for approximate HE schemes such as CKKS. Specifically, in CKKS when used incorrectly, a key recovery attack is possible. The idea is that if an adversary observes both a valid ciphertext and its decryption, it can learn some information from the error that is now blended with the decrypted message. That is, the error leaks some data about the secret key. To avoid this, the scheme needs to be changed to include an error management subsystem, and the decryption method should include a truncation of the error before releasing

3.1 Security Models and Assumptions When Using FHE

the decrypted value. In this way, the scheme is transformed to be an exact scheme, which is claimed to be IND-CPA^D secure. In 2024 [6, 8] showed that also exact FHE schemes such as BGV, B/FV, and TFHE are not IND-CPA^D unless their implementations include an error management utility that controls the accumulated noise.

It is interesting to see how the cryptographic notions are translated to real-world applications.

Example 3.1 (IND-CPA) A data owner generates a set of FHE keys. It sends the public and evaluation keys to the server, similar to the challenger who submitted the keys to the adversary. Now, if an attacker is able to control the server, it can use it to perform encryption and evaluation operations, i.e., it has encryption and evaluation oracles. If the scheme is IND-CPA when observing ciphertexts that were encrypted by the data owner, the attacker cannot learn anything about the encrypted data. While the IND-CPA property guarantees the confidentiality of the data, it does not guarantee its integrity. For example, an attacker can still replace one ciphertext with another to alter the overall computation.

Example 3.2 (IND-CPA^D) Continuing from the above example, just this time, the server performs some computation and sends the resulting ciphertexts back to the data owner who decrypts the data and immediately uploads the decrypted data to its website. Here, an application developer may innocently believe that the IND-CPA property of the FHE scheme is enough to maintain the confidentiality of the uploaded encrypted data from the server. However, as shown by [23], this is not the case. An attacker who eavesdrops the user communication can observe the traffic of valid ciphertexts. Additionally, from the website, it learns the decryption values of these ciphertexts. In that sense, the attacker has a decryption oracle of valid ciphertexts provided (unplanned) by the data owner. This becomes even worse if the attacker gains the ability to inject ciphertexts into the aforementioned decryption oracle. As shown in [23], this allows it to mount a key recovery attack on FHE schemes that are not IND-CPA^D secure and to extract the secret keys. To mitigate such an attack when using an IND-CPA but not IND-CPA^D secure FHE scheme, the application designer should ensure that the decryption output is not uploaded to a public location before applying some manipulations on it or to limit the number of ciphertexts that are allowed to be decrypted under a specific secret key.

3.1.2 Privacy: Who Hides What from Whom? Data Owner Versus Function Owner and Client Versus Server

The following are some typical use cases for FHE in which some or all the above entities need to be considered (Fig. 3.1).

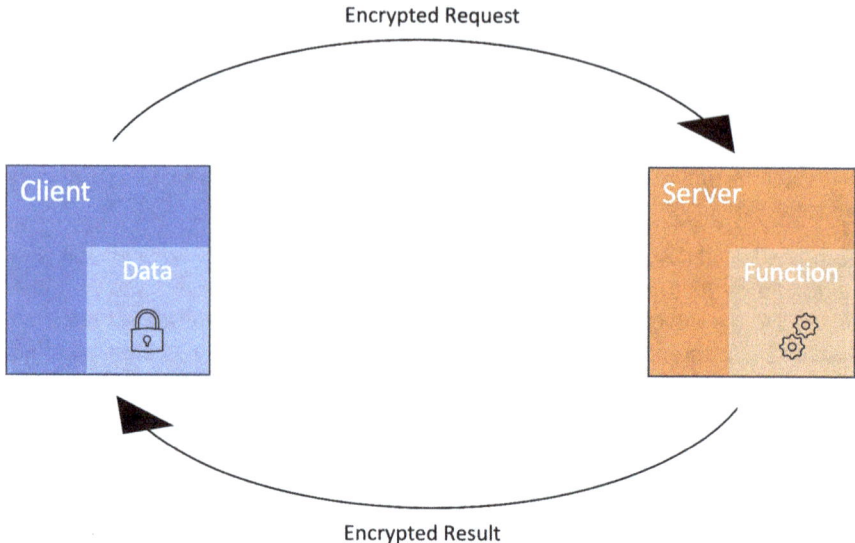

Fig. 3.1 Common FHE use case entities

Outsourcing of Compute
There are several reasons why an organization may wish to outsource compute, such as cost considerations and the lack of sufficient compute resources within the organization. Typically, an organization (the data owner and function owner) leverages computation resources of a cloud provider (the server) to perform some function over its data. However, if the data includes some private features, then the organization will not be allowed or capable to leverage the server in this way due to imposed regulations and security concerns. Thus, the data owner would want to hide the private data from the server and, moreover, from anyone outside of the data owner's trusted perimeter. In this case, a simple solution would be to encrypt the data within the organization's trusted perimeter using FHE, transmit the ciphertexts to the server for processing under encryption, retrieve the encrypted results of the processing, and finally decrypt the results within the organization's trusted perimeter.

Protection of IP Deploying software that embodies IP in an untrusted and uncontrolled environment creates an inherent risk for reverse engineering. In essence, the function owner wants to hide the function from the data owner, the client, or anyone outside of the function owner's trusted perimeter. In some cases, FHE can be applied to significantly reduce the risk depicted above.

A simple example is a deep neural network (DNN) in which the weights are encrypted using FHE and yet the DNN retains its functionality and ability to process incoming data albeit preventing the explicit replication of the model by a potential attacker. The aforementioned method will not mitigate adversarial attacks such as model extraction attacks (see Sect. 3.4). To mitigate such attacks, one needs to add additional layers of defense and essentially create a controlled environment in which the number and type of requests to the model are controlled and monitored.

Using Sensitive Data in an Untrusted Service
Some of the main inhibitors for monetizing data are regulations, security, and privacy restrictions on sensitive data such as PII, sensitive personal information (SPI), and personal health information (PHI). In many cases, organizations cannot use external services due to the risk of sensitive data leakage. Consider, for example, a service provider that seeks to expand its reach and provide services to regulated industries, specifically to consume their sensitive data while preserving their privacy. In this case, FHE can be used to encrypt the incoming organization's data (data owner) before it is sent to the service provider (function owner). The processing will be performed on encrypted data, and the results will be transmitted back to the data owner to be decrypted in their trusted environment. In this case, the data owner is hiding the data from the function owner, as well as from everyone else outside of the data owner's trusted perimeter.

Learning Using Sensitive Data
Sometimes sensitive data is essential for increasing the efficacy of a given function. A simple example would be taking a classifier trained on open data and increasing its accuracy by fine-tuning it with sensitive data. When the classifier and the sensitive data are owned by different entities, one needs to consider the security and privacy of both the classifier and the sensitive data. In this case, FHE can be used to protect the privacy of both, and thus the data owner is hiding the data from the function owner and vice versa.

Generating Insight from Multiple Parties' Data
In many areas, there is a growing need for sharing data and insights between organizations, for common business objectives or for other reasons. Here the major concern is leakage of sensitive data from one organization to the other, or the exposure of sensitive information to any other unauthorized party. There are a variety of technologies that aim to mitigate these concerns, such as anonymization, masking, and differential privacy. However, these types of solutions usually come at the expense of significantly reducing the utility and fidelity of the shared data and thus reducing the efficacy of generated insights. FHE, and more precisely multi-party FHE (MP-FHE), can provide the necessary privacy protection of the shared data and at the same time retain its utility and fidelity. In this case, multiple data owners are hiding their data from each other and from anyone else outside their trusted perimeter. See a detailed discussion of multi-party HE in the following section.

3.1.3 Threat Model and Adversary Capabilities

Having the different scenarios in mind, we continue to define the application threat model that may use FHE for different purposes, for different entities, and with different expectations of its security guarantees. For every such use case, it needs to define the communication protocol between the parties and ensure that it is secure. Different protocols have different leakage profiles and they protect against different adversaries. In what follows, the *input* of a party is the data they have before the protocol starts. The *output* of the party is the data they are supposed to get from the protocol. The set of messages a party "sees" during the protocol is called their *view*.

In Sect. 3.1.1, we already discussed the difference between computationally bounded adversaries with bounded computational resources as opposed to *computationally unbounded adversaries*. We mentioned that an FHE scheme can be secure only against adversaries that run in PPT; otherwise, a simple brute-force attack is always possible.

Another commonly used way to classify adversaries is as follows:

- **Honest but curious adversaries.** (a.k.a. *semi-honest*) These adversaries must follow the protocol but can try to learn as much as they can from the intermediate values and the output, i.e., their view of the protocol.
- **Malicious adversaries.** These adversaries may deviate from the protocol and execute code different than what was specified. For example, where the protocol instructs to train one epoch of the model, the adversary may choose to skip this step or even multiply the model by 0. Protecting against malicious adversaries can be done using, e.g., zero-knowledge proofs, which force the adversaries to prove at each step that they have followed the required protocol. But today, verifiable FHE is considered impractical and thus we do not discuss it in this book.
- **Covert adversaries.** These adversaries are malicious adversaries that can deviate from the protocol, but they have an incentive (e.g., monetary incentive) not to do so.

It turns out that for using FHE, most applications will need to assume that adversaries are semi-honest or covert, where the latter requires justification for the incentives of the adversaries. In this book, we assume semi-honest adversaries.

Remark 3.3 It might be argued that in the commonly used scenario, the data owner only cares about confidentiality, which FHE schemes provide. Thus, the threat model can assume malicious adversaries. However, we emphasize that confidentiality is not the sole requirement of a security protocol; correctness is also essential. For instance, consider a server that deletes all messages from the client upon arrival and eventually returns the encryption of 0. While this protocol provides confidentiality for the data owner's information, it is clearly useless. Since FHE is malleable (see Sect. 3.2), a malicious adversary can manipulate the returned value, causing the protocol to yield incorrect results. Consequently, a threat model

3.1 Security Models and Assumptions When Using FHE

cannot assume a malicious adversary without employing additional cryptographic machinery.

Finally, the threat model should clearly state whether the security model captures a collusion of several entities, e.g., several data owners or a data owner and the server may collude against a victim data owner.

Remark 3.4 In some cases, adversaries may feed the protocol with false input, even if they are honest-but-curious. Authenticating the input of the parties is outside of the scope of this document.

3.1.4 Multi-party FHE

It has been shown that FHE primitives can be used to instantiate a MPC schemes. Several methods have been demonstrated for extending existing FHE schemes such as B/FV and CKKS, so as to support computations that secure the privacy of the data of multiple parties. These methods are now supported by publicly available FHE tools. This section discusses the main features of MPC based on FHE, how it can be implemented, and some of the differences between this method and the other main methods for implementing MPC.

Figure 3.2a shows one of the common privacy-preserving analytics use cases discussed in this book. Here, a single party encrypts its private data with its own FHE encryption key before sending it to the cloud server, which proceeds to train a corresponding model under FHE. Later the party encrypts and sends new private inference query parameters to the server, which performs the inference under FHE and sends the encrypted result back to the party. The party can now decrypt the result using its secret key.

Figure 3.2b shows a useful extension of this use case, in a setup where multiple parties collaborate to train a single model based on their private data and then perform inference with this model on new private inference data. The model is a "birds-eye" model that encapsulates and generalizes information from the multiple parties. For example, multiple banks may want to collaborate to train a shared model for detecting fraudulent transactions using more varied types of historic cases originating from the different banks. In such a case, the multiple parties would still require that their private training and inference data be kept secret, both from the server and from any other party.

Basic FHE schemes, such as B/FV or CKKS, cannot directly support the above multi-party use case. If, for example, all the parties encrypt their private data using a shared public key PK_shared, then the shared model is also encrypted under PK_shared, as well as the inference results, which are sent back to the different parties. The question arises: Who would be allowed to keep the secret key SK_shared corresponding to PK_shared? *Someone* must keep SK_shared because this key is required to decrypt the inference results. But anyone holding SK_shared would also

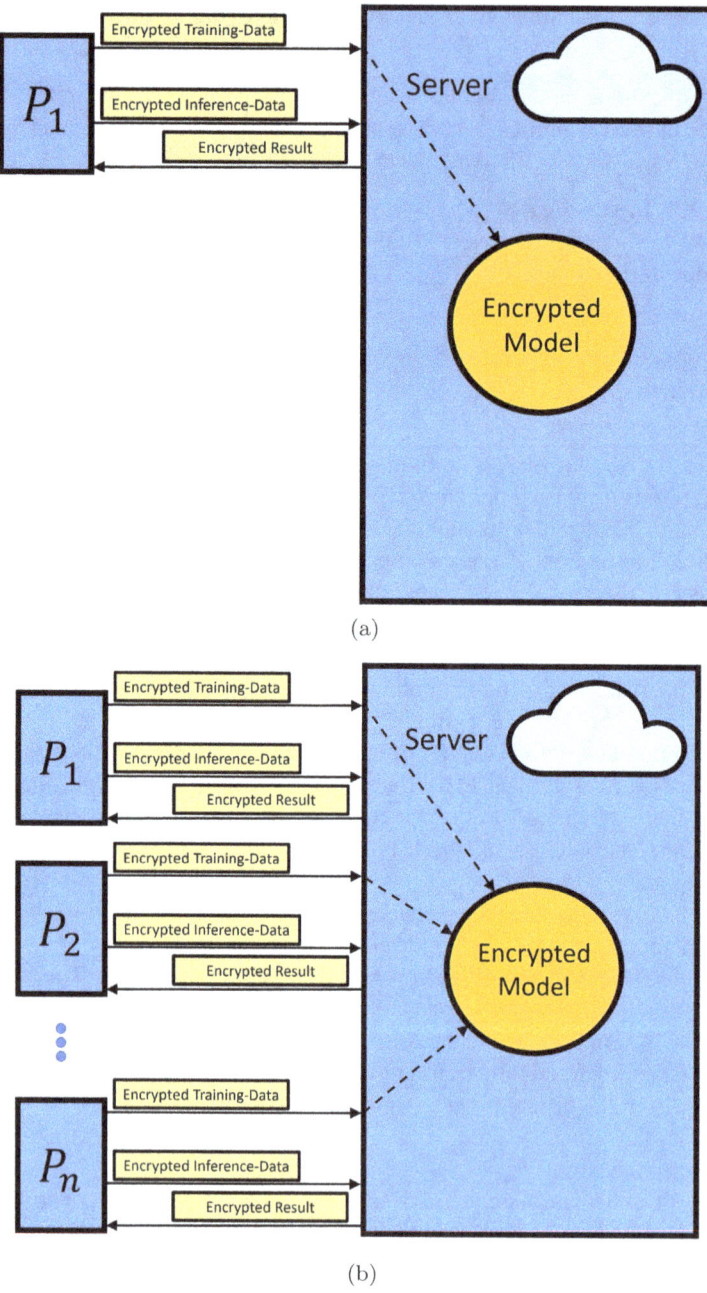

Fig. 3.2 Single-party and multi-party privacy-preserving analytics. (**a**) Single-party privacy-preserving analytics. (**b**) Multi-party privacy-preserving analytics

3.1 Security Models and Assumptions When Using FHE

be able to decrypt the private data of the parties either by itself or through collusion with the server.

The work of [4] from 2022 includes a summary of the various families of multi-party FHE schemes. For example, several multi-key FHE (MK-FHE) schemes allow each party to add its own layer to a ciphertext while still enabling FHE computations over the "onion-like" ciphertexts. In order to decrypt such a ciphertext or the result of the subsequent computation, all the parties that added layers would need to participate in the decryption to "peel-off" their respective layers. This approach has the benefit that it requires no special setup phase among the parties, and the set of parties involved in the multi-party computations is very flexible and can change dynamically during and between different computations. However, the space requirements of the process increase at least linearly with the number of parties, and evaluation complexity increases at least quadratically.

The [4] review also describes the approach of some *threshold* MP-FHE schemes that require interactive setup and decryption steps among the parties. At the same time, the computation step itself can be done independently by a cloud server, with space and time performance that is independent of the number of parties. For example, in the schemes proposed in [5] and [27], multiple parties first participate in a protocol in order to generate shared public encryption and evaluation keys based on their multiple private secret key shares. In the context of ML, the shared public key is then used to encrypt the training and inference data of the parties and is thus also the key in which the computed model and inference results are encrypted. The only way to decrypt ciphertexts encrypted under the shared public key is to let all the original parties authorize the decryption by participating in a decryption protocol.

The following description provides more detail on these multi-party FHE schemes and shows how they can be used to implement the multi-party use case of Fig. 3.2b.

Generation of a Shared Public Encryption Key

First, each party generates its own secret key and uses it to generate its share of a joint public key. The parties then share their public key shares via a dedicated protocol, and the shares are aggregated to create the single shared public key PK_{shared}.

Example 3.5 (Multi Party BFV) In the single-party B/FV scheme, the secret key SK is a polynomial with very small coefficients ($-1, 0,$ or 1), and the public key PK is a pair of polynomials (p_0, p_1) such that $p_0 = SKp_1 + e$, where e is a small noise in the form of a polynomial with small coefficients (see Sect. 2.5.1). A ciphertext is also a pair of polynomials (c_0, c_1) such that SK can be used to decrypt the message as $M = c_0 + c_1 SK$.

In the MP-FHE scheme of [27], each party P_i generates its own secret key SK_i. The parties agree on some common public random polynomial P, and then each party generates its own share of the public key $PK_i = -SK_i P + e_i$. The PK_i shares are then sent to and aggregated by some designated aggregator (which does not necessarily need to be one of the parties). The aggregation is basically the sum of the shares, so

$$PK_{\text{shared}} = (p_0 = \sum PK_i + \sum e_i, p_1 = P). \tag{3.1}$$

Let $SK_{\text{shared}} = \sum SK_i$; then indeed $p_0 = SK_{\text{shared}} \, p_1 + e$ as required by B/FV public keys. The noise in PK_{shared} is linear in the number of parties N, but this is still usable since N is usually much smaller than the modulus q. Finally, the aggregated PK_{shared} is made public or sent directly to any party that is allowed to encrypt data.

Note, however, that SK_{shared} is never directly computed and is therefore never accessible to any of the parties. Thus, no one can by itself decrypt a message encrypted with PK_{shared}, and an alternative decryption protocol is used, with the participation of all the parties, as will be shown below.

Remark 3.6 In the following paragraphs, the term *party* will be limited to the abovementioned parties, i.e., to the parties that contributed their secret shares to generate PK_{shared}. This is in order to distinguish them from the server that may be employed to perform the FHE computations or from occasional users of the model that may not be members of the original set of parties.

Generation of Shared Public Evaluation Keys

The parties also participate in similar protocols to generate the other public keys necessary for the required FHE computations. This includes the relinearization key needed for FHE multiplications and the switching keys needed for the required rotations. The public encryption key and various public evaluation keys are also shared with the server that is assigned with performing the training and inference computations.

Training

The shared public encryption key PK_{shared} can now be used by all the parties to encrypt their training data. The server receives training data from multiple sources but all of the data is encrypted under the same public key. The server can now use the data to train a shared model under FHE, which will also be encrypted under the same public key.

Inference

Any party that wishes to make predictions using the shared model on private inference inputs can encrypt the inference inputs using the shared public encryption key and send this encrypted input to the server. In fact, anyone with access to the public encryption key can make such an encrypted inference request. This includes any occasional users that are not among the original parties above. In any case, the server can now perform the inference computations based on the (encrypted) model and the inputs that are both encrypted with the same shared public encryption key. The inference result, also encrypted under the same shared public key, is sent back to the querying user.

3.1 Security Models and Assumptions When Using FHE

Decryption

At this point, the querying user is holding the encrypted inference result, but it cannot decrypt it on its own, even if the user is one of the original parties that generated the keys. Instead, the user must ask all the parties to authorize the decryption of the result by participating in a dedicated decryption protocol. In this protocol, each party computes its public share of the decryption using its share of the secret key. All the shares of the decryption are now communicated as part of the decryption protocol and aggregated to create the complete decrypted message.

Example 3.7 (Multi Party B/FV—Cont.) Suppose every party P_i owns its own share of the secret key SK_i, such that

$$SK_{\text{shared}} = \sum SK_i$$
$$PK_{\text{shared}} = (p_0 = \sum PK_i + \sum e_i, p_1 = P). \tag{3.2}$$

Here, every party can compute its own share of the required decryption as $h_i = c_1 SK_i + e_i$. The h_i shares can now be sent to and aggregated by some designated aggregator, typically the original owner of the ciphertext C. The aggregation is basically the sum of the shares, so $M = c_0 + \sum h_i$.

Remark 3.8 The steps of generating the shared public keys and then the decryption step involve communication of public key shares or decryption shares among the parties. The security of the data in these shares relies on the hardness of R-LWE. This means that the shares can be communicated via a public authenticated channel, and a cloud server can be employed to perform the necessary aggregations if needed.

Communication Considerations

The best topology in terms of optimizing the total communication is a *star* topology, where all the parties send their shares asynchronously to a single "leader" who aggregates the shares to create the keys or the decrypted plaintext and then sends the aggregated result to the target party or parties. With such a topology (as opposed, e.g., to a clique topology where all the parties send their shares to be aggregated by all the other parties), the communication volume of every party is fixed and independent of the number of parties or, in other words, does not increase with an increase in the number of parties. Only the leader, who is not necessarily one of the original parties, has a communication volume that increases linearly with the number of parties.

As described above, a single "round" of back-and-forth communication between the parties and the leader is necessary in the protocols used for generating the public encryption key and for decryption. Generating the shared public evaluation keys may take a few more rounds of communication, though the number of rounds is small and fixed, and there is an ongoing effort in the FHE community to reduce the number of such rounds and their communication overhead.

Note that all the communication overhead is limited to the key generation steps that need to take place just once and then for every decryption step. The training and inference steps do not involve any communication and can be carried out locally on a cloud server. Furthermore, the FHE operations used in the training and inference steps are identical to those used in single-party FHE training and inference. This means that all the single-party training and inference legacy code can be used without modification also in a multi-party FHE setup.

We can now summarize the main properties of the above MP-FHE approach (as advanced, e.g., by Asharov et al. [5], Mouchet et al. [27], and Kwak et al. [22]):

There is a one-time synchronous key setup phase with a small number of rounds. The volume of communication per party in this phase is independent of the number of parties and on the circuits to be computed (except for the "leader" who communicates with all the parties). Every decryption requires another synchronous phase where the volume of communication per party is again independent of the number of parties (except for the party that holds the decrypted ciphertext, which needs to send it to all the other parties for decryption). This can be compared to some non-FHE MPC methods such as GC [21] that involves a one-time communication of potentially very large data related to the evaluated function (circuit), which is then followed by the sending of the encrypted inputs for each computation of the circuit.

There is also an asynchronous ("offline") FHE evaluation phase that does not involve any communication. A cloud server that has access to the encrypted data of the multiple parties (either because these parties use the cloud server for storage of their respective encrypted private data or after they send their data for the purpose of performing some specific computation) can perform the computations independently, with time and space performance that is independent of the number of parties.

This can again be compared to some non-FHE MPC methods such as Linear secret sharing scheme (LSSS) which involve many rounds of interaction for computing the gates of the computed circuit. The communication in each round is small, but the implication is that every party needs to stay online during the computation, and the offloading of computation tasks by a client to a server in the manner supported by MP-FHE is thus not possible with LSSS.

The multi-party computation process is generic in the sense that the same protocols are carried out independent of the circuit being computed. The parties only need to send their encrypted input data to the server and then participate in the decryption of the result, and there is no circuit-dependent data or keys that need to be sent among the parties.

As with every FHE-based design, the security model of these MP-FHE schemes only protects against semi-honest adversaries, but not against malicious ones.

It has been demonstrated in practice that LSSS can be extended using zero-knowledge proofs so as to be secure against malicious adversaries. Similar extensions using zero-knowledge proofs may be possible in the future for MP-FHE schemes, but such adaptations are not yet practical with the current state of the art.

3.1.5 Federated Learning with HE

Another approach for having a collaborative training process between multiple parties who do not want to share sensitive information with each other is called FL (see an extensive review in [26]). The main difference between FL and MP-FHE protocols is that in FL the parties prefer to avoid uploading some sensitive data to the server even when it is encrypted. The goal however stays the same: collaboratively train a model. In FL, one designated party acts as an *aggregator* and it orchestrates the learning process. Each of the parties trains a local model based on its own data generating what is known as a *model update*. The model updates are shared with the aggregator who aggregates them to produce a global model.

The advantages of FL make it useful when data cannot be shared for the training process either because of privacy, regulatory (e.g., GDPR [13] restrictions), or communication constraints. While not sharing the data mitigates the risk of data breaches, it does not guarantee the privacy of the data (see more on this below). An FL algorithm with a proof that some information on the input of the data owner does not leak is called privacy-preserving federated learning (PPFL).

An FL solution involves the following:

- **(At least two) data owners**—participants in the learning process that hold portions of the data.
- **(At least one) model owner**—the party that eventually receives the trained model. The model owner may be some (or all) of the data owners, the aggregator (in a centralized learning setup), or a third party.
- **An aggregator**—the entity that orchestrates the training process (getting model updates from the data owners and aggregating them to generate one global model).

An FL is defined according to one of the following data partitions (see also Fig. 3.3):

- **Horizontal partition.** Data is horizontally partitioned when all the data owners hold records with the same features in their datasets with the same respective labels.
- **Vertical partition.** Data is vertically partitioned if each data owner has a different subset of the features for the same record. In this partition, before the training starts, the parties must align their datasets so the records will appear in the same order in all datasets. See Sect. 3.1.6.1.
- **Hybrid partition.** Data may be partitioned in a hybrid way (vertically and horizontally), where each data owner has some features of some records.

In the case of vertical data partition, we have an additional role:

- **Active party.** A designated party that carries a special role in the FL protocol, for example, determining the records on which the training will take place. At different times of the protocol, the role of the active party may be played by different parties.

Data owner 1					Data owner 2			
Name	Feature 1	Feature 2	Feature 3		Name	Feature 1	Feature 2	Feature 3
Alice	1.5	4.6	6.2		David	7.3	3.5	7.4
Bob	5.3	3.5	6.3		Eve	3.3	9.4	4.6

(a)

Data owner 1					Data owner 2		
Name	Feature 1	Feature 2	Feature 3		Name	Feature 4	Feature 5
Alice	1.5	4.6	6.2		Alice	7.3	3.5
Bob	5.3	3.5	6.3		Bob	3.3	9.4

(b)

Fig. 3.3 Two partition types. (**a**) A horizontal data partition, where each data owner has the same set of features for different records. (**b**) A vertical data partition, where each data owner has a different set of features for the same records

3.1.6 Vanilla Federated Learning

Generally speaking, a FL training process involves two stages: pre-processing and training.

3.1.6.1 Pre-processing: Dataset Alignment

Before training a FL model, data owners run pre-processing steps on their data, in addition to the typical pre-processing steps required to train any ML model. For example, in horizontal FL, for models trained on tabular data, it is necessary to ensure that the features are in the same order for all parties.

An additional step is to align the datasets of different data owners. For that we need to identify matching records according to some unique ID, or using entity resolution (ER) methods; see Sect. 3.1.6.5. In horizontal partitions, this step reduces biases caused by repetitive records, but it is not compulsory. When the data is vertically partitioned, this step is compulsory to align features of the same record to appear in the same index of the datasets.

3.1.6.2 Training

A vanilla FL training process is an iterative process, illustrated in Fig. 3.4. The figure presents a training process between two data owners and an aggregator. Every

3.1 Security Models and Assumptions When Using FHE 53

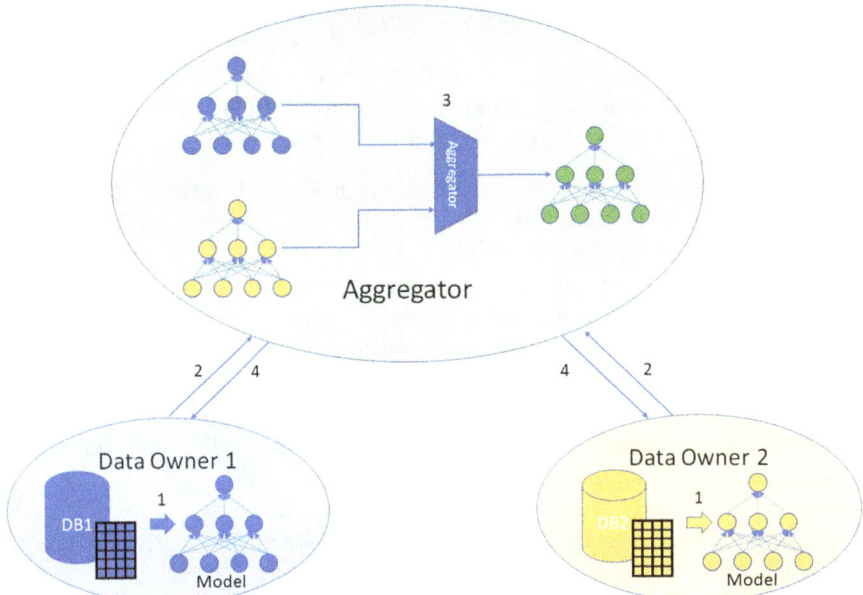

Fig. 3.4 An illustration of a vanilla FL process

iteration includes the following steps:

1. Every data owner trains (or updates) a local model based on its own dataset.
2. Every data owner sends its local model to the aggregator.
3. The aggregator aggregates the local models by taking their weighted average (see also Sect. 3.1.6.3).
4. A model update is sent to the data owners who update their local copies of the model.

3.1.6.3 Biases and How to Avoid Them

The model aggregation operation that is performed in Step 3 above may lead to a bias in the results. For example, if the datasets of the data owners have different magnitudes, then a simple averaging of the models by the aggregator may create a bias toward the data owner with the smaller dataset. To address this bias, the aggregator can consider a weighted average, where the weight of a data owner is derived from the number of records in her dataset.

3.1.6.4 Security Model and Attacks

Except for the adversaries and their capabilities, the threat model should also define the allowed or disallowed leakage from the FL protocol. Below are some examples for such leakage.

- **Size of input.** The size of the input a party has may be private information. This information is being used (or sent over the wire), for example, when the aggregator computes the weighted average or during the pre-processing alignment phase between the data owners. In that case, padding or cryptographic primitives, such as HE, can be used to hide the input size.
- **Size and members of the intersection.** When the parties engage in an aligning pre-processing step, the information whether a record belongs to another party's input may be private. For example, if the parties are expert clinics, then even the information that a person appears in several datasets reveals something on the patient medical condition.
  One option to protect against other parties learning the records in the intersection is to encrypt the output of the alignment protocol where the (secret) key needed to decrypt this output is not available to the data owner.
- **Intermediate models.** An intermediate model may leak information on the input it was trained on. To avoid leaking intermediate models, they need to be encrypted or masked when sent to the aggregator (to be aggregated) and also when sent to the data owner (for training on the next epoch).
- **The final model.** Recently, [25] showed that given a ciphertext and its decryption with the CKKS HE scheme, the secret key can be recovered. In the context of FL, this means that parties that have in their views both the encrypted final model and the decrypted final model (as output) may recover the secret key and decrypt other ciphertexts in their view. To protect against this attack, when using CKKS, random noise needs to be added to the final model before revealing it, as was suggested, e.g., in [24].
- **Decrypting the final model.** We note that eventually the final model needs to be decrypted. If the secret key for the decryption is held by one party, it is imperative that this party does not see any unmasked encrypted messages. As the last sentence suggests, one way to do it is by masking all the messages that are sent to the party with the secret key. Another way is to use threshold FHE or multi-key FHE to split the decrypting process between at least two parties (see Sect. 3.1.4).

3.1.6.5 Related Tools

- **Private set intersection (PSI).** private set intersection (PSI) is a protocol involving at least two parties that hold two lists of IDs. Every party wishes to find the IDs shared by all parties without leaking its own (unshared) IDs. There are several PSI protocols addressing different threat models and different settings. See, for example, [15, 16, 20]. The output of these protocols is a plaintext indicator vector for every party, indicating its shared IDs. Depending on the threat model, if this is something that we do not want to leak, we can run these protocols under FHE so that the output is an encrypted indicator vector that cannot be decrypted without the secret key, which is available only to a designated party. See, for example, the work of [7] describing how PSI can be implemented with FHE.

- **Entity resolution (ER).** The field of ER deals with identifying and handling different occurrences of the same entity in two datasets of two different parties. record linkage (RL) methods are used to identify and link records of different databases when they refer to the same entity. The problem of matching records in two or more datasets without revealing additional information to the parties beyond the set of matches is called privacy-preserving record linkage (PPRL) [10] or blind data linkage (BDL) [9]. A survey of PPRL methods is available in [17], and [2] presents an efficient and practical method for performing PPRL combining PSI and local sensitive hash (LSH) functions and that runs in linear time with respect to the number of records in the compared data sets.
- **Differential privacy (DP).** DP has been termed the gold standard in privacy protection standards, the idea having first been published in 2006 [11]. In short, DP uses random noise to preserve the privacy of individuals and their sensitive attributes in a dataset while still allowing population trends to be accurately observable. The distribution of noise used is specially calibrated to guarantee that the results of an analysis made on two databases different in a single record are statistically indistinguishable, therefore hiding the presence/absence of a single individual in the dataset.

3.2 Security: Against Modification (Integrity, Malleability)

The benefit of using HE is that it protects the confidentiality of the data while in use. However, HE, by itself, cannot guarantee the integrity of the data. Inherently, HE has an issue called malleability. This means that an untrusted entity can manipulate ciphertexts in a way that decrypting them results in valid plaintexts (or cleartexts). Specifically, operations such as multiplying by a scalar, adding two ciphertexts, and multiplying two ciphertexts are possible even without the knowledge of the secret key. The next example demonstrates why this is problematic.

Example 3.9 Consider a company that processes its employees' paychecks encrypted using HE in the cloud. A malicious employee who gets access to the cloud storage server may observe the encryption of its own salary and decide to change it. Even though the data is encrypted and the employee cannot see the underlying value, the employee would still be able to homomorphically multiply it by some scalar, e.g., 10. As a result, the employee would succeed in increasing its expected salary by a factor of 10, without anyone being able to detect this.

FHE schemes are inherently malleable. Thus, the security model should outline strategies to limit adversaries' access to ciphertexts. For instance, storing ciphertexts in secure locations such as TEEs ensures integrity. Another consideration is the exposure of the public and evaluation keys. It is tempting to assume that because a key is called public, it can be revealed online without consequences. However, this is not the case. Leveraging the public key, adversaries can generate valid ciphertexts and replace them with genuine ciphertexts unnoticed due to the semantic security of FHE. Alternatively, it can use these ciphertexts to manipulate computations by adding or multiplying them with other valid ciphertexts. Furthermore, while addition requires no special key, other operations such as multiplications, rotations, and bootstrapping do. Adversaries with access to evaluation or bootstrapping keys can execute these operations.

3.3 Client-Aided Designs

Until now we only considered the confidentiality provided by FHE, but when designing a solution, other metrics such as latency should also be evaluated. Interestingly, in some cases, writing an efficient client-server FHE application is still considered a hard task, where one approach for simplifying it is to get some assistance from the client. The server may ask the client to aid it by performing small computations that are much faster to perform on plaintext than under encryption. For example, performing a division, taking the square root, or even just estimating the size of an integer on the client side can significantly improve the performance. To this end, the server performs many computations under encryption until it reaches a function that the client can do better. It then hands the current computation state to the client who decrypts the state, performs the computation, and re-encrypts the data for further computation on the server side. Figure 3.5 illustrates this process. Let's explore a concrete example.

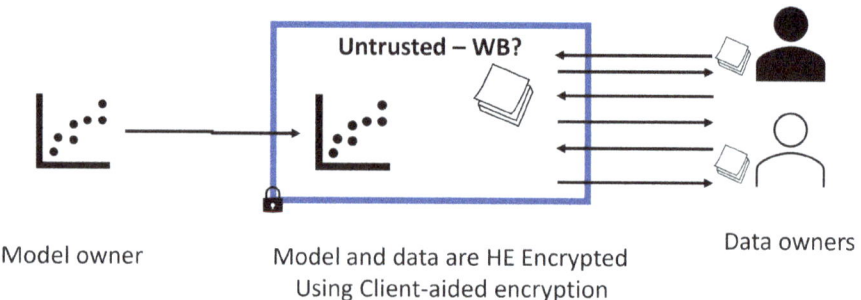

Fig. 3.5 An illustration of a client-aided FHE solution

3.4 Other Privacy Risks

Example 3.10 Consider an FL training application where all users train local NN models and send their encrypted gradients to the server. The server performs the desired aggregation under encryption and returns the updated encrypted model to all users. The users can now decrypt the updated model and run another training iteration on it, and the whole process continues until convergence.

Note that even though the server cannot see the model or the gradients because they are encrypted with FHE, the users see the intermediate models and can use this to learn information about the data of their collaborators. Another disadvantage of such client-aided designs is that the client is bothered with tasks that were meant to be assigned to the server. This forces the client to remain online for the necessary interactions.

Another example of a client-aided solution is described in [19]; there the need for a ReLU polynomial approximation is bypassed by asking the client to perform the ReLU each time the activation function is reached by the server. This has the added benefit of cleaning up noise, and reducing the effective multiplication depth, thus removing the need for costly bootstrapping. As the final stage of the NN usually involves some non-polynomial Softmax function or thresholding, many solutions omit these functions from the server model and perform these calculations with the help of the client.

3.4 Other Privacy Risks

Interestingly, security and trustworthiness mean different things to different audiences. Specifically, when talking with someone from the HE domain and mentioning security, most often we consider the confidentiality achieved for the data or model while processed at the untrusted third-party server. However, the situation is different when talking with data scientists. Here, trustworthiness and hence security refer to privacy attacks that a malicious user can trigger to extract data about a trained model, the training information, or even other users. It is thus important to understand the boundaries of what a HE-based solution can provide.

Consider again the simple case of two (or more) entities: a model owner who trains a model over private training sets and data owners who would like to use the services of the model owner to perform prediction, classification, or other tasks over their data. When analyzing the trustworthiness of a model, we often consider two types of attacks: white-box attacks, where the adversary is able to see intermediate steps of the inference computation, and black-box attacks, where adversaries only get to provide input and observe the returned output.

In most cases, HE allows a solution designer to consider only black-box adversaries. See Fig. 3.6 for an example, in which both the model owner and the data owners encrypt their data and use HE to evaluate the model on the encrypted samples. The evaluation process can happen everywhere (dependent on who holds the HE private keys), e.g., locally at the model owner side or at a not-so-trusted cloud

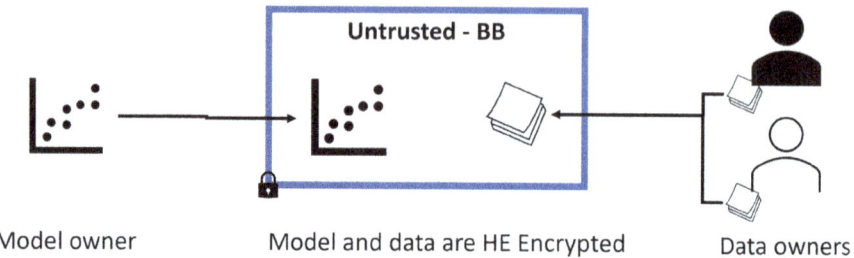

Fig. 3.6 An illustration of the FHE black box

server. Due to HE, whoever does the computation cannot see the input models and samples or the resulting output.

Surveying the state-of-the-art (SOTA) privacy attacks, mitigation, and threat models is beyond the scope of the book. The interested reader should consider the following surveys [18, 28] to learn more about the subject. For completeness, we briefly describe below four main attack classes: reconstruction, membership inference, property inference, and model extraction.

- **Reconstruction attacks**, a.k.a. *attribute inference* or *model inversion attacks*, aim to reconstruct a training sample or its label from the original and sometimes confidential training set. Here, the adversary, who has access to some output labels and also some knowledge of the target sample features, attempts to recover other sensitive features or the entire original sample.
- **Membership inference attacks.** This attack was introduced by Shokri et al. [29]. It assumes that the adversary holds knowledge of the model's output prediction while attempting to determine whether an input sample was part of the training set.
- **Property inference attacks.** Here, the adversary's goal is to extract information (properties) that were not explicitly revealed by the model. Consider, for example, the task of identifying whether a patient has cancer. An adversary can attempt to deduce if the training subjects were mostly men or women even if this feature was not explicitly given as an input to the training procedure. Leaking such information may put at risk the privacy of the patients.
- **Model extraction attacks.** In these attacks, the adversary gets black-box access to the model, and by querying the model multiple times, it attempts to generate a local duplication of the model, i.e., to steal the model. While this attack does not directly cause privacy issues for the training subjects, it may turn it inefficient and less desirable for a model owner to set up such a service, to begin with.

The above privacy attacks can be practical even if the model, samples, and computation are encrypted, i.e., even when considering the model as a black box. Consequently, we remind the readers that achieving a secure and trustworthy solution is possible only when considering a holistic approach, which uses HE to consider only black-box attacks while also considering other mitigation strategies against privacy attacks.

3.5 Business Use Cases with Privacy Requirements

In Sect. 3.1.2, we briefly presented ideas and reasons for using HE-based solutions. Here, we discuss in more detail some more concrete use cases that businesses and industries may consider.

3.5.1 Fraud Detection

Fraud is a prevalent problem in the financial system. It is estimated that transactional fraud alone in 2023 will reach $32B and will grow to an estimated $43B by 2028. A key driver of the above trend is the continuous increase in digitization, globalization, and the availability of sophisticated attacking capabilities alongside the slow adoption of advanced and robust security and privacy controls and technologies. Some of the most common types of transactional fraud are credit card and online banking fraud where criminals steal credit card information and gain access to bank accounts, respectively. FHE can be a powerful tool to mitigate such transactional fraud use cases as will be shown below.

A traditional transactional fraud detection solution would normally rely on validating various data elements in a transaction such as the transfer amount, target, and location of the transaction initiator against a known baseline. Nevertheless, criminals have been known to overcome this approach, and thus the ability to also use highly sensitive information in the analysis of a transaction is important for achieving a more robust and effective fraud detection capability.

As an example, we take an existing NN model used for detecting online banking fraud and retrain it under encryption with highly sensitive and personal data, generating a more accurate and robust model. The incoming transactions are thus analyzed using the enriched model and transactions that are suspected as fraudulent are rejected.

The following represents the basic flow shown in Fig. 3.7:

1. When creating a bank account, the account holder generates a secret and public key pair using the bank's mobile application, encrypts highly sensitive personal data using the public key, and sends it via the mobile app to the bank.
2. The bank as part of its fraud detection training process (Step I) augments historical transaction data with the highly sensitive personal data creating an enriched FHE encrypted model.
3. Whenever a new transaction comes in (Step II), the bank runs a preliminary analysis using the encrypted enriched model and creates an encrypted indicator vector where the value 0 indicates a suspected fraudulent transaction.
4. The result is sent back to the user where the bank's mobile app performs the decryption and marks the transaction invalid in case of a fraud.

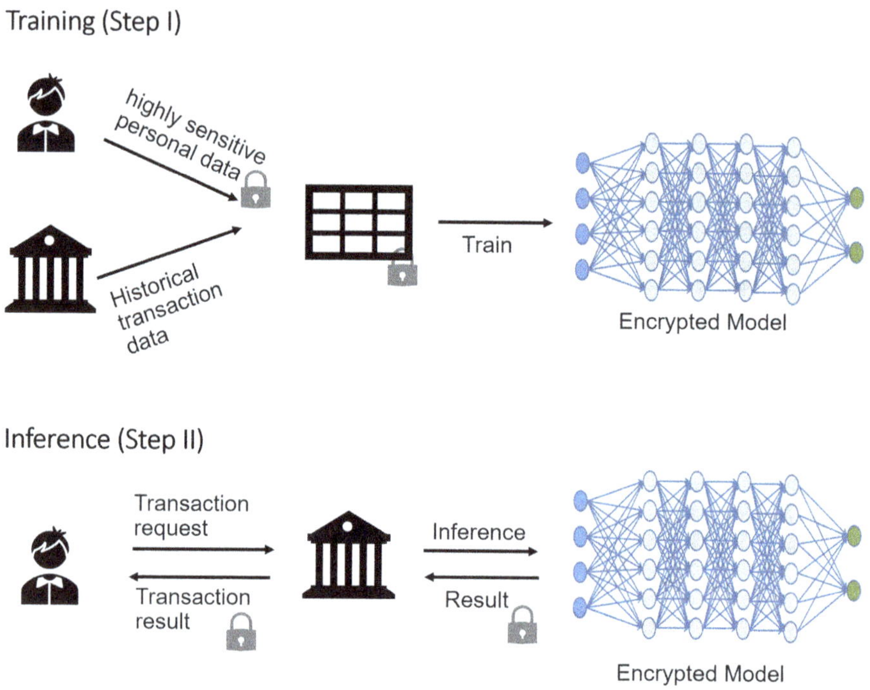

Training (Step I)

Inference (Step II)

Fig. 3.7 Fraud detection using FHE

3.5.2 Loan Approval

Loan approval heavily relies on the ability to access sensitive personal information. The heart of the loan approval process lies in the ability of the lender, typically a bank, to accurately assess the ability of the borrower to repay the loan. This in turn directly affects the lender's willingness to provide the loan and/or the respective associated interest rate. At the same time, borrowers are hesitant to provide and expose highly sensitive information as part of their initial search for adequate loan providers and before a loan has been approved.

FHE provides a solution in which a client looking for a potential lender can safely provide highly sensitive information to the lender in a privacy-preserving manner.

As an example, a borrower can provide a potential lender with information such as income, debt, education, employment details, health details, and assets all under encryption. The lender can use, e.g., an XGBoost model to analyze the request outputting an encrypted result which is returned to the potential borrower who in turn decrypts it and finds out whether a loan request would be approved. If so, the potential borrower can continue the process with the lender, which may require a more thorough verification of the borrower's details by the lender.

3.5 Business Use Cases with Privacy Requirements

Classification (Step I)

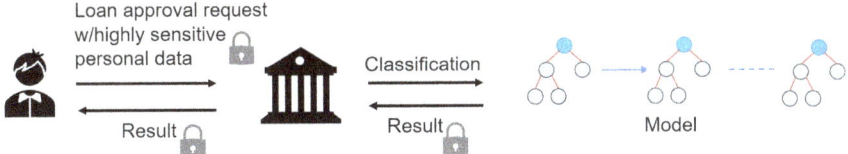

Validation (Step II)

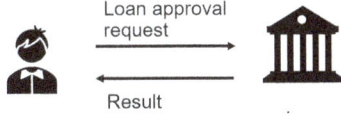

Fig. 3.8 Privacy-preserving loan approval

The following represents the basic flow shown in Fig. 3.8:

1. When applying for a loan, the potential borrower generates a secret and public key pair (SK, PK) using a trusted application, encrypts highly sensitive personal data using the public key, and sends it (Step I) via the trusted app to the bank.
2. The bank classifies the encrypted request using a pre-trained XGBoost model and returns an encrypted loan approval score. For example, a positive score may indicate the expected interest for an approved loan, and the value 0 would indicate that the loan will probably not be approved given the information provided to the bank.
3. The borrower decrypts the result using their secret key and follows up with sending all the necessary information to the bank in the clear for validation, if they choose so.

3.5.3 Medical: Image Classification and Anomaly Detection

Image classification of healthcare information is usually highly sensitive and regulated. The US federal law, health insurance portability and accountability act (HIPAA), regulates the privacy rules and US national standards. The purpose of these regulations is to protect individuals' medical records or any other individually identifiable health information, as such private information plays a key role in the processing of medical data.

In some cases, it is beneficial to utilize third-party processing and IP. An example for this is the usage of a cloud-based solution for detecting lung disease in individuals using X-rays. The strict HIPAA regulation makes it difficult to share sensitive health information outside a controlled and trusted environment. FHE can provide a solution in which the sensitive data can be securely shared with third parties outside a trusted controlled environment while still allowing for the generation of valuable insights.

In the case of the lung disease analysis, individual X-ray images are encrypted using a public key and are then sent for classification to a cloud service that hosts a NN model that attempts to determine if a disease is present. The encrypted results of the analysis are sent back to the trusted and controlled client environment where they are decrypted, and further evaluation and processing can be performed.

3.5.4 Biometric: Authentication

Biometric authentication is today a powerful and useful method to authenticate an individual. It is mostly used in mobile devices where an individual is authenticated against the previously enrolled feature vector of their fingerprint, face, or voice. Thus, the process of authentication is performed on the device, which has great advantages in terms of usability as the individual is not required to use or remember a password but rather use their own biometric data for quick and easy access to the device or application. This type of biometric authentication can be termed as client-side authentication because the authentication is performed on a client device controlled by the user issuing the authentication request.

One of the main caveats with client-side authentication is the fact that the code and data required for running the authentication process are performed on the client device and therefore susceptible to comprise by bad actors. Moreover, enterprises and governments use biometric information for multiple purposes such as the granting of access to restricted facilities or access to highly sensitive applications. This means that biometric information needs to be stored and processed on a remote server in a centralized location. This type of biometric authentication, coined, server-side biometric authentication, has an inherent vulnerability, which is, that in the event of server compromise, a large amount of biometric information relating to many individuals that may be leaked. Such was the case in the alleged breach of India's biometric information database, potentially putting more than 1B individuals' private data at risk [1].

We can leverage the power of FHE to mitigate the potential breach of a centralized server storing many individuals' biometric information. Namely, provide a secure server-side authentication solution in which the biometric information stored in the database remains encrypted, including during processing, so that in the event of a breach in the server, the leaked data will be useless as it is encrypted.

The following informal example describes an FHE-based biometric authentication protocol, also shown in Figs. 3.9 and 3.10.

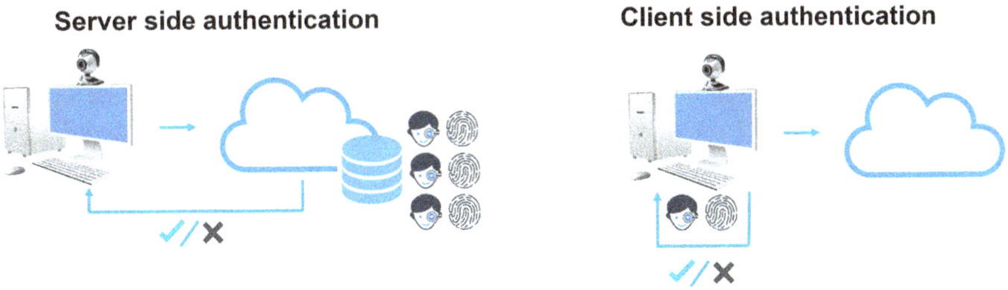

3.5 Business Use Cases with Privacy Requirements

Server side authentication   **Client side authentication**

Fig. 3.9 Biometric authentication overview. (**a**) Server-side authentication. (**b**) Client-side authentication

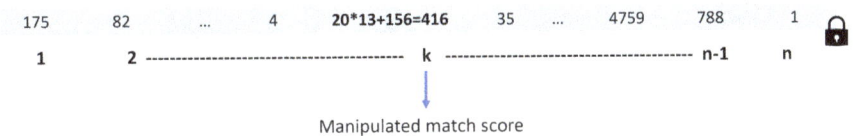

Manipulated match score

Fig. 3.10 Biometric authentication server-side data manipulation

Enrollment

1. A user enrolls onto the server-side biometric authentication solution by generating a secret and public key (SK, PK) pair of an exact HE scheme, where the plaintext space is $\mathbb{Z}_t$. They encrypt their facial feature vector and upload the ciphertext alongside the public key to the server.
2. The server stores (c_0, PK) in its database and associates them with the user.

Authentication

1. The client encrypts its facial feature vector using PK to generate c_1 and sends it to the server for authentication.
2. The server homomorphically compares the feature vector in the database c_0 with the newly sent feature vector c_1, using an Euclidean distance algorithm, producing a match score m.
3. The server generates uniformly at random two random numbers $r_1, r_2 \in \mathbb{Z}_t$ and multiplies the match score by r_1 and adds r_2.
4. The server generates $n - 1$ additional random numbers in $\mathbb{Z}_t$ and encrypts them using the requester's public key. Together with the manipulated match score, the server creates an array of n numbers, where the location of the manipulated match score in the array is randomly selected. The server then sends the array of n numbers back to the client.
5. The client decrypts the numbers using the respective secret key and sends the decrypted numbers back to the server.
6. The server verifies that the $n - 1$ random numbers are in their expected places and then verifies the manipulated match score by subtracting r_2 and dividing by r_1 and comparing it against some acceptance threshold th.
7. Only if all the numbers have been verified the client is authenticated.

Remark 3.11 (Informal Security) The server is semi-honest and thus follows the protocol. The client can be malicious and may attempt to fool the server by bypassing the authentication protocol. The facial-recognition system described is not bulletproof, and an attacker that carefully crafts the feature vectors has a probability p_0 to fool the system either because it selected the correct features or because it hit a false-positive event, where p_0 depends on the threshold th and the number of compared features. In addition, the server response after decryption is indistinguishable from random because it involves $n-1$ random values and another masked value. The probability p_1 of using a random manipulation on this data without being detected by the server is low and depends on the values of t, n and th. Consequently, the probability of forging the system is $p = \max\{p_0, p_1\}$.

3.5.5 Client-Facing Service: ELECTRON

ELECTRON is an EU H2020-funded research project involving multiple companies, including IBM. The aim of the project is to deliver a next-generation electrical power and energy system (EPES) platform, capable of empowering the resilience of energy systems against cyber, privacy, and data attacks [12].

Moreover, as critical infrastructure systems are increasingly susceptible to cyberattacks, it is imperative to leverage next-generation technologies for collecting and analyzing the vast amounts of data generated by these systems. There are several goals for this. The first goal is to mitigate and reduce the response time for any ongoing attacks. The second goal is to statistically reduce the mean time to detect (MTTD) a security incident to a minimum.

To this end, the ability to use advanced AI techniques and to securely collect and analyze vast amounts of data is crucial. Naturally, collecting large amounts of potentially sensitive data from distributed critical systems raises obvious concerns over the privacy of the collected data. In fact the actual action of collecting sensitive data from a large distributed system for the purpose of analyzing it in a centralized location inherently increases the overall system's attack surface. This is because it allows adversaries to have the opportunity to compromise the data, either in transit or where it resides while waiting to be analyzed.

In the ELECTRON project, IBM Research has implemented (as part of the HElayers FHE library [3]) an advanced privacy-preserving anomaly detection algorithm that homomorphically computes over encrypted event logs provided by the edge device, in this case a charging station for electrical vehicles. The system builds an encrypted behavioral model of the normal physical behavior of the charging stations by analyzing encryptions of benign and normal time-series data coming from the charging stations. Incoming encrypted logs are then evaluated by the behavioral model with the purpose of detecting any anomalies or deviations from the norm and the expected next signal in the time-series data. The training of the time-series model, the predictions, and the anomaly detection was done as described in Sect. 10.2.3.

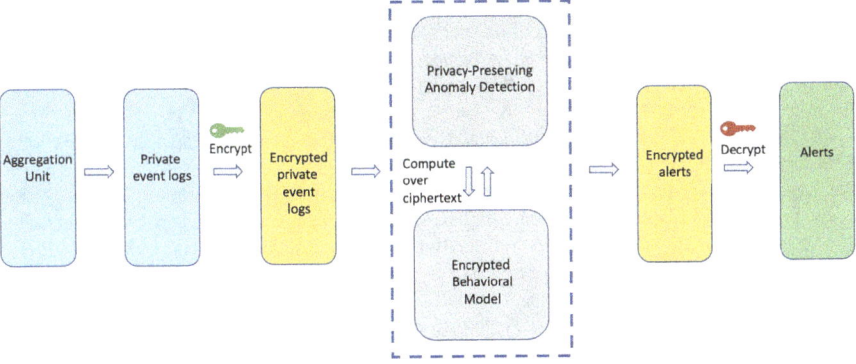

Fig. 3.11 An illustration of a privacy-preserving anomaly detection system used by ELECTRON

The output of the anomaly detection unit is an encrypted anomaly alert report that is sent to a trusted location (typically the owner of the private event logs) for decryption and visualization on a centralized dashboard. In addition, with the help of other partners in the project and to ensure that the encrypted data being sent and processed has not been tampered with, the system signs and then authenticates all the FHE encrypted communication. This is imperative since the FHE encryption key for performing the encryption of the logs and the resulting alert report is a public key and potentially accessible by a malicious actor. Without the authentication and signing protocol, a malicious actor could send malicious or fake encrypted logs to the anomaly detection unit or send malicious or fake alert reports to the centralized dashboard.

Figure 3.11 illustrates the basic components and flow of the privacy-preserving anomaly detection implementation in the ELECTRON project.

References

1. Massive Aadhaar Data Breach of 815 Million Indians: Here's How to Keep Your Details Safe (2023). https://business.outlookindia.com/news/massive-aadhaar-data-breach-of-815-million-indians-heres-how-to-keep-your-details-safe. Last Accessed on 03 Jan 2024
2. Adir, A., Aharoni, E., Drucker, N., Kushnir, E., Masalha, R., Mirkin, M., Soceanu, O.: Privacy-Preserving Record Linkage Using Local Sensitive Hash and Private Set Intersection, pp. 398–424. Springer International Publishing, Cham (2022). https://doi.org/10.1007/978-3-031-16815-4_22
3. Aharoni, E., Adir, A., Baruch, M., Drucker, N., Ezov, G., Farkash, A., Greenberg, L., Masalha, R., Moshkowich, G., Murik, D., Shaul, H., Soceanu, O.: HeLayers: A tile tensors framework for large neural networks on encrypted data. In: Privacy Enhancing Technology Symposium (PETs) 2023 (2023). https://petsymposium.org/popets/2023/popets-2023-0020.php
4. Aloufi, A., Hu, P., Song, Y., Lauter, K.: Computing blindfolded on data homomorphically encrypted under multiple keys: An extended survey. arXiv preprint arXiv:2007.09270 (2020). https://arxiv.org/abs/2007.09270

5. Asharov, G., Jain, A., López-Alt, A., Tromer, E., Vaikuntanathan, V., Wichs, D.: Multiparty computation with low communication, computation and interaction via threshold FHE. In: Pointcheval, D., Johansson, T. (eds.) Advances in Cryptology – EUROCRYPT 2012, pp. 483–501. Springer, Berlin (2012). https://doi.org/10.1007/978-3-642-29011-4_29
6. Checri, M., Sirdey, R., Boudguiga, A., Bultel, J.P.: On the practical CPAD security of "exact" and threshold FHE schemes and libraries. Cryptology ePrint Archive, Paper 2024/116 (2024). https://eprint.iacr.org/2024/116
7. Chen, H., Laine, K., Rindal, P.: Fast private set intersection from homomorphic encryption. In: Proceedings of the 2017 ACM SIGSAC Conference on Computer and Communications Security, CCS '17, p. 1243–1255. Association for Computing Machinery, New York (2017). https://doi.org/10.1145/3133956.3134061.
8. Cheon, J.H., Choe, H., Passelègue, A., Stehlé, D., Suvanto, E.: Attacks against the INDCPA-D security of exact FHE schemes. Cryptology ePrint Archive, Paper 2024/127 (2024). https://eprint.iacr.org/2024/127
9. Churches, T., Christen, P.: Blind data linkage using n-gram similarity comparisons. In: Dai, H., Srikant, R., Zhang, C. (eds.) Advances in Knowledge Discovery and Data Mining, pp. 121–126. Springer, Berlin (2004). https://doi.org/10.1007/978-3-540-24775-3_15
10. Clifton, C., Kantarcioundefinedlu, M., Doan, A., Schadow, G., Vaidya, J., Elmagarmid, A., Suciu, D.: Privacy-preserving data integration and sharing. In: Proceedings of the 9th ACM SIGMOD Workshop on Research Issues in Data Mining and Knowledge Discovery, DMKD '04, pp. 19–26. Association for Computing Machinery, New York (2004). https://doi.org/10.1145/1008694.1008698
11. Dwork, C., McSherry, F., Nissim, K., Smith, A.: Calibrating noise to sensitivity in private data analysis. In: Halevi, S., Rabin, T. (eds.) Theory of Cryptography, pp. 265–284. Springer, Berlin (2006). https://doi.org/10.1007/11681878_14
12. ELECTRON: rEsilient and seLf-healed EleCTRical pOwer Nanogrid (2024). https://electron-project.eu/
13. EU General Data Protection Regulation: Regulation (EU) 2016/679 of the European parliament and of the council of 27 April 2016 on the protection of natural persons with regard to the processing of personal data and on the free movement of such data, and repealing directive 95/46/EC (general data protection regulation). Official J. Europ. Union **119**, 1–88 (2016). http://data.europa.eu/eli/reg/2016/679/oj
14. Force, J.T.: Guide for mapping types of information and systems to security categories. Technical Report, National Institute of Standards and Technology (2024)
15. Freedman, M.J., Hazay, C., Nissim, K., Pinkas, B.: Efficient set intersection with simulation-based security. J. Cryptol. **29**(1), 115–155 (2016). https://doi.org/10.1007/s00145-014-9190-0
16. Freedman, M.J., Nissim, K., Pinkas, B.: Efficient private matching and set intersection. In: Cachin, C., Camenisch, J.L. (eds.) Advances in Cryptology - EUROCRYPT 2004, pp. 1–19. Springer, Berlin (2004). https://doi.org/10.1007/978-3-540-24676-3_1
17. Gkoulalas-Divanis, A., Vatsalan, D., Karapiperis, D., Kantarcioglu, M.: Modern privacy-preserving record linkage techniques: an overview. IEEE Trans. Inf. Forens. Secur. **16**, 4966–4987 (2021). https://doi.org/10.1109/TIFS.2021.3114026
18. Hu, H., Salcic, Z., Sun, L., Dobbie, G., Yu, P.S., Zhang, X.: Membership inference attacks on machine learning: a survey. ACM Comput. Surv. **54**(11s), 1–37 (2022). https://doi.org/10.1145/3523273
19. Juvekar, C., Vaikuntanathan, V., Chandrakasan, A.: GAZELLE: A low latency framework for secure neural network inference. In: 27th USENIX Security Symposium (USENIX Security 18), pp. 1651–1669. USENIX Association, Baltimore (2018). https://www.usenix.org/conference/usenixsecurity18/presentation/juvekar
20. Kolesnikov, V., Matania, N., Pinkas, B., Rosulek, M., Trieu, N.: Practical multi-party private set intersection from symmetric-key techniques. In: Proceedings of the 2017 ACM SIGSAC Conference on Computer and Communications Security, CCS '17, pp. 1257–1272. Association for Computing Machinery, New York (2017). https://doi.org/10.1145/3133956.3134065

21. Kreuter, B., Shelat, A., Shen, C.H.: Billion-Gate secure computation with malicious adversaries. In: 21st USENIX Security Symposium (USENIX Security 12), pp. 285–300 (2012). https://www.usenix.org/system/files/conference/usenixsecurity12/sec12-final202.pdf
22. Kwak, H., Lee, D., Song, Y., Wagh, S.: A unified framework of homomorphic encryption for multiple parties with non-interactive setup. Cryptology ePrint Archive (2021). https://eprint.iacr.org/2021/1412
23. Li, B., Micciancio, D.: On the security of homomorphic encryption on approximate numbers. In: Canteaut, A., Standaert, F.X. (eds.) Advances in Cryptology – EUROCRYPT 2021, pp. 648–677. Springer International Publishing, Cham (2021). https://doi.org/10.1007/978-3-030-77870-5_23
24. Li, B., Micciancio, D., Schultz, M., Sorrell, J.: Securing approximate homomorphic encryption using differential privacy. In: Dodis, Y., Shrimpton, T. (eds.) Advances in Cryptology – CRYPTO 2022, pp. 560–589. Springer Nature Switzerland, Cham (2022). https://doi.org/dp-ckks1
25. Li, Y., Purcell, M., Rakotoarivelo, T., Smith, D., Ranbaduge, T., Ng, K.S.: Private graph data release: a survey. ACM Comput. Surv. **55**(11), 1–39 (2023). https://doi.org/10.1145/3569085
26. Ludwig, H., Baracaldo, N.: Federated Learning: A Comprehensive Overview of Methods and Applications. Springer International Publishing, Cham (2022). https://books.google.co.il/books?id=wAJ6EAAAQBAJ
27. Mouchet, C., Troncoso-Pastoriza, J., Bossuat, J.P., Hubaux, J.P.: Multiparty homomorphic encryption from ring-learning-with-errors. Proc. Priv. Enh. Technol. **2021**(4), 291–311 (2021). https://petsymposium.org/popets/2021/popets-2021-0071.pdf
28. Rigaki, M., Garcia, S.: A survey of privacy attacks in machine learning. ACM Comput. Surv. **56**(4), 1–34 (2023). https://doi.org/10.1145/3624010
29. Shokri, R., Stronati, M., Song, C., Shmatikov, V.: Membership inference attacks against machine learning models. In: 2017 IEEE Symposium on Security and Privacy (SP), pp. 3–18. IEEE, Piscataway (2017). https://doi.org/10.1109/SP.2017.41

Chapter 4
Approaches for Writing HE Applications

Abstract Designing algorithms for applications over HE presents challenges stemming from the special characteristics of HE. In this chapter, we review these special characteristics and the challenges they bring.

- The set of operations that an algorithm under HE can use is different from the set of operations available for a plaintext algorithm. For example, given two encrypted numeric messages, a HE algorithm cannot determine which message has a bigger value. The inability to compare values has a dramatic effect on how algorithms should be designed for HE.
- As discussed in Chap. 2, some noise is added during the encryption process. This noise grows with every operation performed on the ciphertext, which requires some attention to keep it from growing too much.
- In some HE schemes, operations are performed on vectors in a SIMD manner. Leveraging this requires some adaptation of the algorithms.

In this chapter, we discuss techniques to address and mitigate these challenges.

4.1 Comparison Model Versus Circuit Model

A computation model defines what operation algorithms can perform and how the algorithm can change its behavior depending on the input or other parameters. There are several computation models considered in computer science. One is the comparison model in which the algorithm can change its behavior based on comparisons it makes on inputs or intermediate values. For example, the behavior of a sorting algorithm (e.g., the pattern in which the memory is accessed) depends on the order in which the values appear in the input.

Another model is the circuit model in which the algorithm has the same behavior for all inputs. To see the classic motivation behind the circuit model, we can imagine a hardware circuit printed on silicone. In this case, the circuit is composed of gates, where each gate has at least one input wire and a single output wire. The way the wires are connected determines the functionality the circuit performs. Clearly, once the circuit is printed on silicone, it cannot be modified and it has the same behavior for all inputs. Since in HE we also cannot change the behavior of the code based on the inputs (as we explain below), there is some similarity between hardware design and HE algorithmic design. Indeed, as we see below, ideas from hardware design can be "imported" to designing algorithms under HE.

4.1.1 FHE Is a Circuit Model

As noted above, FHE code cannot change its behavior based on the inputs. Intuitively, the change in behavior means that some information on the input has leaked (specifically, the information that tells the algorithm how the behavior should change). For example, consider the following code:

```
x := 1
y := 3
if (x < y)
    minSqr := x*x
else
    minSqr := y*y
```

Specifically, consider the line `if (x < y)`. This line considers the values in the variables x and y and executes a different piece of code depending on whether x is smaller than y. In more details, when the program is executed, the program counter (PC) either continues to the block executing `minSqr := x*x` or the one executing `minSqr := y*y`. Unfortunately, this is not possible under FHE. To see why, assume it was possible. That is, given ciphertexts x and y, there is some way to execute the line `minSqr := x*x` or the line `minSqr := y*y`. Then, by observing the PC, we can tell whether $x < y$ or not. This contradicts the semantic security of FHE. Moreover, by encrypting different values of y and running this code iteratively, we can extract the value of x. We conclude that any conditional branching that depends on values of ciphertexts is impossible to implement under FHE as long as we wish to preserve its inherent semantic security. In the next section, we will discuss a general recipe for how to modify a code that includes conditional branching to a code that uses multiplexing to compute the same output.

4.1.2 A General Recipe

Since an FHE program executes the same operations oblivious to the input, we can draw the operations as gates in a circuit. This is very similar to hardware where gates are physically printed on a piece of silicone. Although the values in a hardware circuit are not encrypted, it is still impossible to implement the equivalent of branching, i.e., modify the next gate that should execute according to an input-dependent condition. Instead, a common practice in hardware is to use multiplexing

4.1 Comparison Model Versus Circuit Model

(Mux) to perform something equivalent to branching. A Mux gate has two inputs x and y and one control bit c. The output of the Mux gate is either x or y, depending on the control bit:

$$\text{Mux}(a, b, c) = \begin{cases} a, & \text{if } c = 0 \\ b, & \text{if } c = 1 \end{cases} \tag{4.1}$$

In our example above, the code would transform to

```
minSqr := Mux(y*y, x*x, isGreater(x,t)),
```

where isGreater(x, y) is a bit that equals 1 (true) if $x < y$ and 0 (false) otherwise. In other words, a common practice in hardware design is to compute the output of both branches (y*y and x*x, in our example) and use $c = $ isGreater(x, y) as a bit that controls whether that code takes effect.

This leads to a general recipe to implement branching in FHE. Even though it is impossible to have a control bit in the plain, we can still compute a ciphertext holding the value $c = (x < y)$, that is,

$$c = \begin{cases} 1, & \text{if } x < y \\ 0, & \text{otherwise.} \end{cases}$$

See Sect. 6.1 for a discussion on how $isGreater(x, y)$ can be computed and implemented in CKKS. There is also some work discussing the implementation of this operator in BGV/BFV, for example, [33, 43]. Multiplexing can then be implemented as $\text{Mux}(a, b, c) = (1 - c) \cdot a + c \cdot b$.

We next consider two examples that show how algorithms behave differently in the circuit and comparison models. First, we consider a sorting algorithm and show that following the general recipe yields an algorithm whose size is of $O(n!)$. Then we consider information retrieval and show that it has a lower bound of $\Omega(n)$ in the circuit model.

Example 4.1 (Sorting) Consider an algorithm, $\mathcal{A}lg$, for sorting an array, A, of n values. There are $n!$ different permutations of A, and the objective of $\mathcal{A}lg$ is to find the permutation that sorts A. In any comparison-based sort, we start with all possible $n!$ permutation, and (in a best-case scenario) with every comparison we make, we rule out half of them. After $\log n! = O(n \log n)$ iterations, we find the permutation that sorts A. This is, in a nutshell, the proof that comparison-based sorting takes $\Omega(n \log n)$ time. If we apply the general recipe on $\mathcal{A}lg$, i.e., at each decision point, we compute both branches and use Mux, we end up considering all possible permutations. In other words, in the circuit model and using the general recipe described above, $\mathcal{A}lg$ will require $O(n!)$ gates to compute.

There is, of course, a better way to sort an array in circuit model borrowing ideas from the hardware domain. Sorting networks can sort arrays more efficiently in circuit model because the code they execute does not change its behavior depending on the input and therefore does not need to use Mux to simulate conditional branching.

See, for example, Bitonic [8], Batcher [9], and Ajtai, Komlós, and Szemerédi (AKS) sort [3] sorts. Both Bitonic and Batcher sorting networks have depth of $O(\log^2 n)$ and size of $O(n \log^2 n)$. AKS sorting network has depth of $O(\log n)$ and size of $O(n \log n)$. Although asymptotically superior, the multiplicative constant of AKS makes it less efficient in practice. In a recent work [32], the authors considered a sorting network that uses k-way sorting, that is, unlike Batcher and AKS that use building blocks that sorts two numbers, they used building blocks that sorts k number. This changed the depth from $O(\log_2^2 n)$ to $O(k \log_k^2 n)$ which is better, for example, when $k = 5$.

We conclude with the realization that although any algorithm can be implemented under FHE, algorithms that work well in comparison model and converted to circuit model do not necessarily work well.

Example 4.2 (Direct Addressing) When performing direct addressing, we are given an array A with n entries and an input i. The goal is to return the value in the i-th entry, $A[i]$. In plaintext, given i the value $A[i]$ can be read in sublinear time (using only comparisons, it is $O(\log n)$, and if direct addressing is allowed, it can be done in constant time). When i is encrypted, the problem is also referred to as private information retrieval (PIR). Suppose that an algorithm $\mathcal{A}lg$ solves PIR under FHE, that is, it outputs $A[i]$ given an encrypted i. First, it is obvious that $A[i]$ must also be a ciphertext; otherwise, given $A[i]$, some information leaks on i. For example, if values in A are unique, then i can be determined given $A[i]$. Even if the values in A are not unique, knowing $A[i]$ still induces some nonuniform distribution on the possible values of i. This would contradict the semantic security of FHE.

BMX [11] showed that $\mathcal{A}lg$ has a lower bound $\Omega(n)$ on its running time. Intuitively, if $\mathcal{A}lg$ does not read the value of $A[c]$ before outputting $A[i]$, an adversary that tracks the memory access pattern can conclude that $i \neq c$, which again contradicts the semantic security of FHE. This means $\mathcal{A}lg$ needs to (at least) read every element in A.

We consider here two ways to solve PIR with FHE. In this section, we consider a generalization of the Mux-ing idea and in the next section, we consider a solution based on indicator vectors. We start here with a generalization of the Mux-ing idea:

$$PIR_n(i, A) = \sum_j Eq(i, j) \cdot A[j], \qquad (4.2)$$

where Eq is a function whose value is 1 if $i = j$ and 0 otherwise:

$$\mathrm{Eq}(i, j) = \begin{cases} 1, & \text{if } i = j \\ 0, & \text{otherwise.} \end{cases} \qquad (4.3)$$

See Chap. 6 for a discussion on how to implement Eq in CKKS.

We say the way Eq. 4.2 computes $PIR(i, A)$ is an extended n-way Mux. Specifically, in the classic two-way Mux, we have A an array of two elements and

4.2 The Computation Depth Challenge

$i \in \{0, 1\}$. Then these identities hold: $Eq(i, 0) = 1 - i$ and $Eq(i, 1) = i$ and Eq. 4.2 simplifies to $PIR_2(i, A) = (1 - i) \cdot A[0] + i \cdot A[1]$.

4.2 The Computation Depth Challenge

Another challenge when designing an algorithm for HE is the noise that grows with the computation. As mentioned in Sect. 2.4.2.1, leveled homomorphic encryption can perform only a limited number of multiplications before exhausting the multiplication chain index. In fully homomorphic encryption, this number is made unlimited by using bootstrapping. Since bootstrapping is an expensive operation (see the cost of bootstrapping below), it is important when designing algorithms to consider not only the size (number of operations) but also the chain indexes of values. To consider chain indexes, we first need to properly define the depth of a circuit (Fig. 4.1):

Definition 4.3 (Depth) The depth of a circuit is the maximum number of gates on a single path in the circuit.

An FHE circuit can be represented using a directed a-cyclic (hyper) graph (DAG) with a finite depth, which is the longest path in the graph, i.e., the path that has the maximum number of gates from the circuit inputs to a circuit output.

The concept of circuit depth is closely related to the "critical path," which sets a lower bound to the execution time of the circuit. Asymptotically, the critical path and the path defining circuit depth are the same. In concrete instantiations, they may be different because different operations may take different times to execute.

In FHE, the depth has additional implications coming from the noise added to the message by the encryption scheme. We remind that the noise that is added to messages during encryption is essential to make the scheme secure (see Sect. 2.5.1). In most modern schemes, the multiplication operation increases the noise by a significantly larger amount than the addition operation. Thus, in the context of FHE, we are usually interested in the **multiplicative depth** of a circuit.

Definition 4.4 (Multiplicative Depth) The multiplicative depth of a circuit is the maximal number of multiplication gates on a path in the circuit.

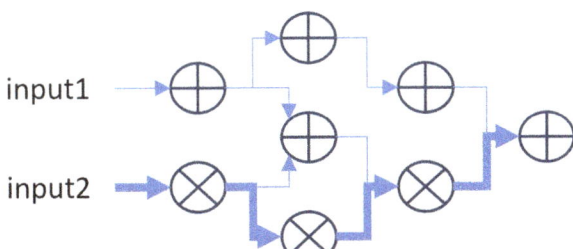

Fig. 4.1 An arithmetic circuit of size eight gates (five addition gates and three multiplication gates) has depth 4 (highlighted path). The multiplication portion of the depth is only 3

As an example, we consider the CKKS [16] scheme, where the noise may overlap the least-significant bits (LSBs) of the message. As discussed in detail in Sect. 2.4.1.1, here as well, the number of multiplication operations that can be performed is limited, unless bootstrapping is applied on the ciphertexts. Even when bootstrapping is available and the number of multiplication is unlimited, the noise still roughly doubles ([16]) with multiplication. Since the noise in CKKS may overlap the LSB of the message, its accuracy decreases when the noise grows. In some cases, the noise can be reduced. For example, if the message is known to be an integer, the technique of [26] can be used to reduce the noise. The technique in [26], which they call BLEACH, applies an iterative method that converges to whole numbers. The noise cannot be eliminated entirely because the BLEACH process itself adds noise but an invariant of having a small noise can be kept.

Another case where the noise is reduced was noted in [18]. Algorithms that have a negative feedback (e.g., gradient descent) are less affected by the added noise.

In LHE schemes or FHE schemes instantiated as LHE schemes, we can control how much the noise can grow before overlapping (partially or completely) with the encrypted message. This can be done by, e.g., choosing the relevant parameters for the FHE key, which translates to the number of sequential multiplication gates the scheme can execute.

4.2.1 Cost of Bootstrap

As mentioned in Sect. 2.4, a bootstrap operation can "reset" a ciphertext to allow more multiplications. This operation is significantly more expensive than others in the circuit. For example, at the time of writing this book, running a CKKS multiplication gate on a single Intel(R) Xeon(R) 2.20 GHz core takes an order of milliseconds, while a bootstrap operation takes an order of seconds with the same key. To improve the performance of an algorithm, the number of bootstraps needs to be reduced. Obviously, the greedy algorithm is to defer the bootstrap operation as much as possible, but this is not necessarily the optimal solution. In fact, finding the optimal solution is an NP-hard problem [12].

4.2.2 TFHE

An exception is the TFHE scheme [19] where the bootstrap operation is cheaper and is applied after each multiplication operation. Indeed with TFHE, depth is not an issue to consider when designing an algorithm. However, TFHE is still not the dominant scheme because it does not support SIMD and therefore is slower (amortized) compared to when SIMD is well utilized.

4.3 Techniques for Addressing the Challenges of HE

After listing the challenges, we turn to discuss ways to address them.

4.3.1 Indicator Vectors

Largely speaking, an indicator vector of size n is a binary vector $\chi \in \{0, 1\}^n$, of which the values serve as indications. For example, in Eq. 4.2, we have $\chi[j] = Eq(i, j)$ indicating whether the j-th element is the element that needs to be returned. We can therefore rewrite Eq. 4.2 to use indicator vectors:

$$PIR_n(i, A) = \sum_j \chi[j] \cdot A[j]. \tag{4.4}$$

Next we show how indicator vectors can be used to utilize the SIMD feature of HE.

4.3.2 Comparing Bitwise Representations

The previous section considered an indicator vector whose j-th position indicates whether $i = j$. We now consider more complicated conditions. We first continue the PIR example from before and use a conjunction of multiple (simple) conditions to improve the implementation of Eq. 4.2. Then we consider the more complicated example of decision trees.

Continuing the PIR example of the previous section, we break the condition $\chi[j] = Eq(i, j)$ to $\chi[j] = \chi_1[j] \cdot \ldots \cdot \chi_{\log n}[j]$ (here we consider i as fixed within a single query and also assume that n is a power of 2), where

$$\chi_b[j] = \begin{cases} 1, & \text{if } Bit_b(i) = Bit_b(j) \\ 0, & \text{otherwise,} \end{cases} = 1 + Bit_b(i)Bit_b(j) - Bit_b(i) - Bit_b(j)$$

where $Bit_b(a)$ is the b-th bit of a in its binary representation. and then we can write

$$PIR_n(i, A) = \sum_j \chi_1[j] \cdot \chi_2[j] \cdots \chi_{\log n}[j] \cdot A[j]. \tag{4.5}$$

Recall that j is known in plaintext since it is the iterator index of the sum. When comparing Eqs. 4.4 to 4.5, we see there are n invocation $Eq(i, 1), \ldots, Eq(i, n)$ compared to $\log n$ invocation of $Bit_1(i), \ldots, Bit_{\log n}(i)$. When i is given in binary, this is expected to be more efficient than using the Eq function.

4.3.3 Traversing a Tree

We now turn to a more complicated case where we traverse a tree based on conditions made at the tree nodes. Decision trees and search trees are two examples for such trees. For simplicity, let T be a full binary decision tree with n inner nodes, $v_1, \ldots, v_n$ and $n+1$ leaves, $\ell_1, \ldots, \ell_{n+1}$, where each inner node v_i is associated with a condition C_i and each leaf is associated with a label. When traversing the tree (starting at the root), we check the condition at the current node and continue right if it holds and left if it does not until we reach a leaf. At a leaf, we output the label associated with it.

As argued above, using Mux to implement the tree traversal leads to computing all conditions in all inner nodes. Another way to find the leaf that should be reached is by squashing the tree, i.e., considering all n paths to all n leaves and computing for each leaf the conditions along the path that leads to it. This can be rearranged to test all conditions $C_1, \ldots, C_n$ and then for each leaf, ℓ_l compute the product $\prod \hat{C}_{i_j}$, where $C_{i_1}, C_{i_2}, \ldots$ are the conditions on the path from the root to ℓ and $\hat{C}_{i_j}$ is either C_{i_j} or $\neg C_{i_j}$ depending whether the path to ℓ_l turns left or right at v_i.

This can also be rewritten using indicator vectors. We set $2n$ indicator vectors, $\chi_1, \ldots, \chi_n, \hat{\chi}_1, \ldots, \hat{\chi}_n \in \{0, 1\}^{n+1}$, where χ_i and $\hat{\chi}_i$ correspond to the condition C_i. Each is an indicator vector with a slot for each leaf. Specifically, the jth slot of χ_i indicates whether ℓ_j can be reached if C_i is true, that is, C_i is not on the path to ℓ_j or C_i to be true to reached ℓ_j. Similarly the jth slot of $\hat{\chi}_i$ indicates whether ℓ_j can be reached if C_i is false. See Fig. 4.2. The leaf reached when traversing the tree is given by

$$\ell = \prod_i (C_i \cdot \chi_i + \hat{C}_i \cdot \hat{\chi}_i).$$

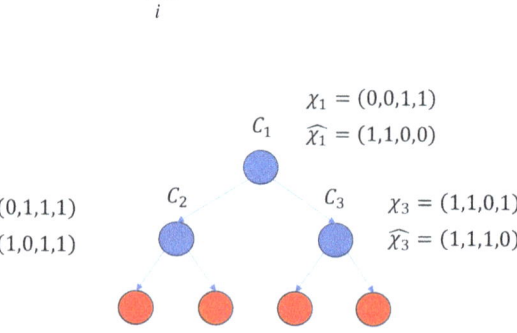

Fig. 4.2 An example of a tree with four leaves $\ell_1, \ldots, \ell_4$ and three inner nodes associated with three conditions C_1, C_2, C_3. Each condition is associated with two indicator vectors. For example, C_2 is associated with $\chi_2 = (0, 1, 1, 1)$ and $\hat{\chi}_2 = (1, 0, 1, 1)$. The vectors indicate that if C_1 is true, then ℓ_1 may be reached and ℓ_2 cannot, while if C_1 is false, ℓ_1 cannot be reached and ell_2 may. Both vectors indicate ℓ_3 and ℓ_4 can be reached because C_2 is not on the path to these nodes

Here C_i and $\hat{C}_i$ are vectors with one value broadcast in them, that is, $C_i = (1, \ldots, 1)$ if the condition holds and $C_i = (0, \ldots, 0)$ if it doesn't.

If the conditions C_i are given in plaintext (e.g., in random forests) and the input is encoded in binary, there are heuristics that reduce the amount of computation, e.g., [21].

4.3.4 Copy and Recurse

Copy and recurse is a method to implement a plaintext algorithm $\mathcal{A}lg$ that traverses an r-ary tree T (i.e., every inner node has r children) where $\mathcal{A}lg$ continues into at most $\xi < r$ children of each inner node and computes some function $f(v)$ on each node v it visits. A naïve implementation of $\mathcal{A}lg$ under FHE performs $f(v)$ on all nodes of T (and uses a mask on the output), but using copy and recurse, the number of times $f(\cdot)$ is applied is the same (in terms of O notation) as that of the plaintext version. A motivation to copy and recurse is range searching and range counting (see example below).

As a simple explanation, consider the case of a binary search tree. Every inner node has two children (i.e., $r = 2$) and the search recurses into one child (i.e., $\xi = 1$). See Fig. 4.3 (top) for a depiction. In this case, the plaintext algorithm recurses into one of the children depending on the condition at the root. Under FHE we use a Mux gate to create a **copy** of the subtree the plaintext recurses into as we now detail. The inputs to the Mux gate are the two subtrees, and the selector of the Mux gate that selects between the inputs is the condition at the root. We say this is a **copy** because a Mux gate is defined as $\text{Mux}(in_1, in_2, select) = (1 - select) \cdot in_1 + select \cdot in_2$, which creates a new copy of either in_1 or in_2. See Fig. 4.3 (bottom).

When the general case (i.e., $r > 2$ and $\xi \geq 1$) is implemented, the selector is replaced with an indicator vector that indicates which $\xi < r$ children need to be recursed into. The Mux gate is replaced with FHE code that given the indicator vector copies the said subtrees.

Since the size of subtrees diminishes exponentially, the total overhead of copying is linear in the size of the tree. The advantage of this method is that the expensive code executed at the nodes is called only a sublinear number of times. More specifically, if t is the time spent at nodes and $\xi = \xi(r) = r^c < r$, then the running time of traversing a tree is $O(t \cdot n^{c+\epsilon} + n)$.

Example 4.5 (Range Counting) In the *range searching* problem, we are given a finite set of points $P \subset \mathbb{R}^d$ and a volume (called range) $\gamma \subset \mathbb{R}^d$ and wish to quickly compute some function $f(P \cap \gamma)$. This problem is related to database queries because some database queries can be reduced to range searching. For example, a database with the columns sugar level, age, and LDL level can be represented by $P \subset \mathbb{R}^2$, where a record with sugar level s, age a, and low-density lipoprotein (LDL) level l is represented by a point $(a, l) \in P$ (s is an additional

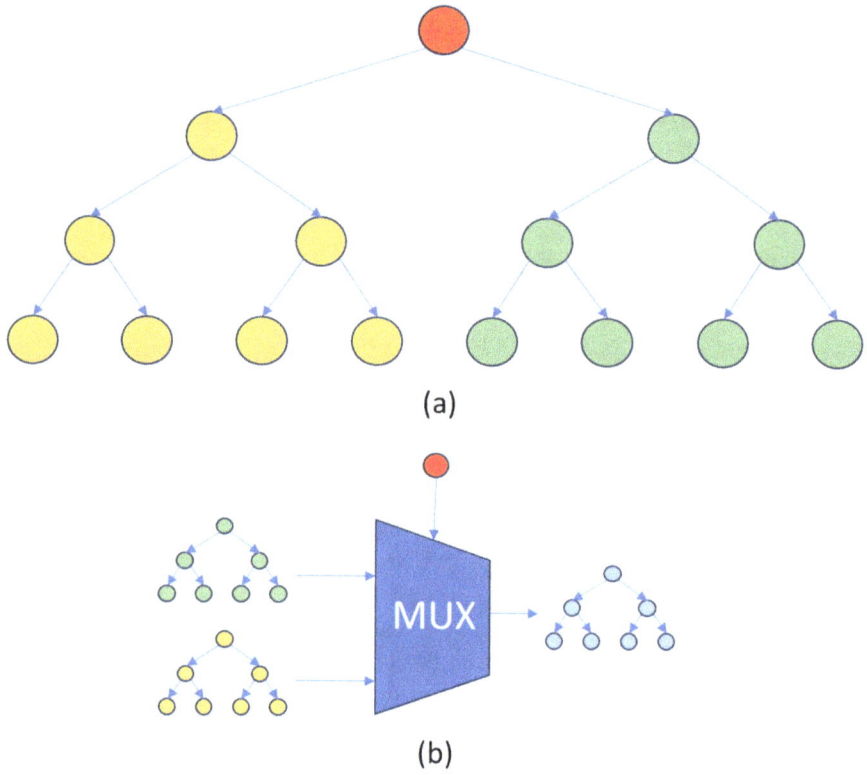

Fig. 4.3 An example of the copy-and-recurse method applied on a (binary) search tree. The search tree is shown at the top. Upon checking the condition in the root (shown in red), the plaintext algorithm would continue in the right subtree (shown in green) or the left subtree (shown in yellow). The copy phase is shown at the bottom. The two subtrees are given as inputs to a Mux gate with the condition from the root used as the selector of the Mux. The output is a copy of a subtree (depending on the selector) and the FHE algorithm continues in recursion in a single subtree

datum associated with the point). Consider, for example, the query:

```
SELECT AVERAGE(sugar level) FROM patients
  WHERE (age > 30) and (age < 40)
    and (LDL level > 130) and (LDL level < 160);
```

This query can be answered by averaging the sugar levels associated with points in P that fall inside the axis-parallel rectangle $\gamma = \{(x, y) \mid 30 < x < 40 \text{ and } 5 < y < 10\}$. Common examples for the function f are the following:

- Count, i.e., $f(P \cap \gamma) = |P \cap \gamma|$.
- Averaging, i.e., $f(P \cap \gamma) = \text{Sum}_{p \in P \cap \gamma} P / |P \cap \gamma|$.
- Standard deviation.

4.3 Techniques for Addressing the Challenges of HE

The naïve implementation checks for every point $p \in P$ whether $p \in \gamma$ for a total of $O(t \cdot n)$ operations, where $n = |P|$ and t is the number of operations to check whether $p \in \gamma$. When working with plaintexts, efficient solutions group points together and efficiently check how γ relates to the group (e.g., by comparing γ to its boundary). If the group is contained in γ, it is accounted in its whole and the algorithm continues in recursion into groups that are only partially contained in γ. These solutions rely heavily on branching to skip entire groups of points. As mentioned above, under FHE it is impossible to branch based on a condition that depends on the input.

In a recent work [37], the authors addressed the problem of range searching. First, they noticed that when running a range searching algorithm, a significant amount of time is spent on determining (under FHE) how a range interacts with a point or a set of points. Then they used a data structure called partition trees (a search tree with certain properties as we explain below) together with a method they called *copy and recurse*. With this method, they were able to reduce the number of times a range is checked against a set of points to the number of nodes visited when using partition trees in plaintext. This comes with an overhead cost of copying portions of the partition. Their experiments, made with CKKS, show that this leads to a significant improvement in running time.

4.3.5 Communicating Indicator Vectors

In the sections above, we discussed how indicator vectors are used to simplify computation. We now discuss how indicator vectors can be communicated efficiently. Specifically, in Sect. 4.3.3 we discuss a vector that indicates which (single) leaf was reached in a search tree. In this case, the indicator has a single 1. In other cases, for example, when the vector indicates which records in a database match a filter, there can be multiple 1s. Naively, the entire vector can be transmitted; however, this is wasteful when the number of 1s is known to be significantly smaller than the number of elements in the vector. In what follows, we discuss how indicator vectors can be transmitted efficiently.

Given an indicator vector χ with n elements, it can be naively transmitted by transmitting all n elements. We say χ is s-sparse if it contains at most s 1s. For the case of s-sparse vectors (where s is public), it was shown in [5] how to efficiently transmit χ. Their method used coresets based on group testing algorithms. A more recent work is [27], which uses bloom filters. Another approach due to [4] considers an encoding that returns one location of (a value of) 1. Unlike linear codes (see [23, 39]), to generate this encoding, nonlinear operations are needed. Their approach can be modified and used repeatedly, so in the first round, the encoding returns the location of the first 1 and in subsequent rounds the locations of subsequent locations of 1 are returned.

4.4 Non-polynomial Computation

In general, arithmetic over FHE schemes includes only additions and multiplications. This means that any polynomial function can be computed directly and that non-polynomial functions can be approximated using polynomial approximations as we discuss in Part II. Here, we mention another approach to evaluate these functions using lookup tables (LUTs). A lookup table (LUT) of size n is a data structure that stores values $f(x_1), \ldots, f(x_n)$ and can quickly return $f(x_i)$ given x_i. LUTs are common in both software and hardware design.

Example 4.6 The advanced encryption standard (AES) encryption algorithm requires performing an inverse operation in a mathematical field of characteristic 256. While computing the inverse using modular arithmetic in hardware is considered a slow operation, having an LUT of only 256 values speeds up the computation.

LUTs have received a lot of attention in FHE since they extend the available tools that can be used in algorithmic design. In FHEW and TFHE, LUTs were incorporated into the bootstrapping process to allow what is known as programmable bootstrapping (PBS), where after bootstrapping a message i, we end up with a message whose value is $f(i)$ for an arbitrary function f [19, 20, 34].

Since the table used in PBS is limited, approaches such as tree-based LUT have been used [14, 31]. Specifically, [14] with many input ciphertexts can be viewed as traversing a search tree that performs logarithmic number of comparisons between the input and the table keys.

In BGV and B/FV, similar approaches were used, e.g., [25, 40]. Here, the bits of index i were used with a series of Mux gates to retrieve $f(i)$.

In the case of CKKS, a LUT implementation should also consider the noise that accumulates in the ciphertext. In [26] the authors suggested a cleanup method (BLEACH) for CKKS, for the case where the messages are known to be integers, a condition that is naturally met by the indexes to the LUT. Later in [17], the authors considered several additional approaches to the LUT problem, where one approach is similar to the copy and recurse technique [37] described in Sect. 4.3.4.

4.5 Masking and Two-Server Model

In this section, we show how FHE can be combined with masking to compute certain functions more efficiently. This uses a different security model that involves two non-colluding servers, one of which has the secret key and can thus decrypt ciphertexts and perform more complicated operations on them. Such a model was considered, for example, in [7, 29].

This model is depicted in Fig. 4.4. The user uploads their encrypted input to server S_1 and their secret key to S_2. The two servers then collaborate in a secure

4.5 Masking and Two-Server Model

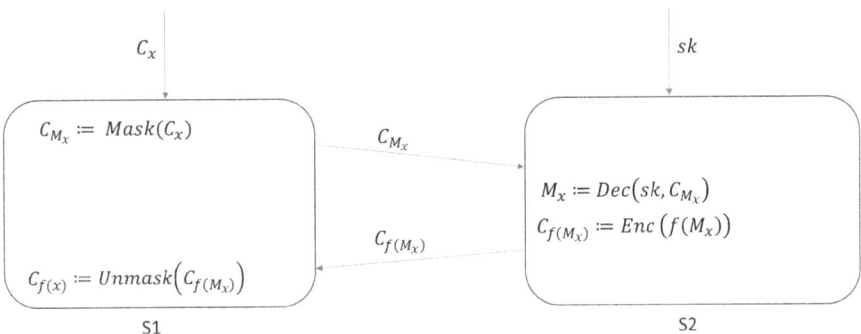

Fig. 4.4 A protocol involving FHE and two non-colluding servers, S1 and S2. The protocol uses FHE and masking, where a masked value is indistinguishable from random. The view of S1 includes only encrypted values and the view of S2 includes only masked values

protocol. For example, in [7, 29] the servers solve a set of linear equations $Ax = y$ as part of training a linear regression model. To find x, the servers compute A^{-1} and then set $x = A^{-1}y$. See this example in more detail in Sect. 10.2.1.1.

To keep the security of the input (and intermediate values), S_1 and S_2, use masking during the protocol as we explain below. Intuitively, for the protocol to be secure, S_1 views only encrypted messages and S_2 views only masked (decrypted) messages. *Masking* a value x is the action of adding or multiplying a random value r so that $x + r$ or $x \cdot r$ do not leak any information about x. For example, this can be done if x and r are taken from a finite ring and $x + r$ is uniformly distributed. For example, consider the ring $\mathbb{Z}_n$ (i.e., integers modulo n), for some $n \in \mathbb{Z}$. When $r \leftarrow \mathbb{Z}_n$ is chosen uniformly, then $x + r \mod n$ is uniformly distributed and therefore leaks no information about x. Specifically, in [30] the authors use Paillier and in [7] the authors use BGV. In both cases, the two servers are used to invert a $d \times d$ matrix A, an operation that is otherwise hard to compute under FHE.

To conclude, this security model can be used to compute operations that are inefficient to compute using only FHE. To use this model, the developer needs to show the following:

1. How the input can be masked to be indistinguishable from random
2. How to manipulate the masked value such that later the mask can still be removed under FHE and result in the desired output

To show the security of the protocol, we need to prove that the mask (typically, $x + r$ or $x \cdot r$) is indistinguishable from a random value. There are examples proving that when the messages space is a finite ring, that is, (1) it is finite (of size $|M|$) (2) $Pr[x + r = m] = \frac{1}{|M|}$, for any possible message m in the message space. These examples were shown for BGV, B/FV, and TFHE. The question of how to use masking with CKKS has been addressed in some special cases (see [46]), but in the general case, it is still open.

4.6 Loops

As mentioned above, when running under FHE, the code cannot make decisions that are based on input values. Therefore, loops whose end conditions depend on such values are impossible to implement under FHE. Below we survey some practices to implement loops.

4.6.1 Loop Unrolling

We start with the simplest type of loops:

$$\text{for } i = 1, \ldots, c$$

where $c \in \mathbb{N}$ is a parameter given in the plain. The condition in this type of loop does not depend on the inputs, and the loop is executed in the same way with the same number of iterations regardless of the inputs. To implement these loops in FHE, we perform what is known as *loop unrolling*, i.e., we rewrite the code to explicitly have all iterations of the loop. Obviously, this practice is impossible if the ending condition depends on the inputs.

4.6.2 Loops with Input-Dependent Conditions

We continue our discussion with loops whose stopping condition depends on the input. First, we execute c iterations of the loop, where c is an upper bound on the number of iterations that need to be executed. Next, we draw some inspiration from hardware design and add a ciphertext which we call *enable*. We set $enable = 1$ before starting to execute the loop, and at each iteration, we set $enable = enable \cdot (1 - cond)$, where $cond$ is the ending condition of the loop, so that $cond = 1$ if the ending condition is met and $cond = 0$ otherwise. Looking at the value of messages in *enable* through the different iterations of the loop, we see that in the first iterations, we have $enable = 1$ and in the latter iterations, we have $enable = 0$, where the changing point happens when the ending condition is met. We then use *enable* to enable or disable the actions taken inside the loop.

Example 4.7 (Loop in a Circuit Model) The following loop sets $x = 2^a$:

```
x := 1
while (a > 0) {
    x := x * 2
    a := a - 1
}
```

4.7 Exploiting SIMD

can be rewritten as (assuming $a \leq 64$)

```
x := 1
enable := 1
for (i := 0; i < 64; ++i) {
    enable := enable * isGreater(a, 0)
    x := enable*(x * 2) + (1 - enable)*x
    a := enable*(a - 1) + (1 - enable)*a
}
```

where $isGreater(a, b) = 1$, if $a > b$ and 0 otherwise. We do note that specifically, in this simple example, this can be optimized to

```
x := 1;
for (i := 0; i < 64; ++i) {
    x := x + x * isGreater(a, i)
}
```

To conclude, we see there are two challenges with implementing a loop in FHE: (1) the ending condition may require a deep computation, especially if the ending condition involves values that change inside the loop (as in the first example); and (2) the number of iterations must be set to its maximum. We note that using an interactive protocol, an additional party that has the secret key can act as an oracle that reveals when the ending condition is met. However, revealing such information leaks information and should be considered carefully (see [6] for details).

4.7 Exploiting SIMD

In this section, we discuss the SIMD feature in the context of FHE. We refer the user to Chaps. 7–9 for a broader discussion on the matter.

SIMD is a parallel computing model where a single instruction is applied to multiple data. In SIMD computations, data items are vectors or arrays of constant size where operations are performed slot-wise. For example, if $x = (x_1, \ldots, x_s)$ and $y = (y_1, \ldots, y_s)$ are two vectors, then

$$x + y = (x_1 + y_1, \ldots, x_s + y_s) \tag{4.6}$$

and

$$x \cdot y = (x_1 \cdot y_1, \ldots, x_s \cdot y_s). \tag{4.7}$$

Usually, SIMD systems have an additional operation to rotate the elements of a vector:

$$\text{Rot}_i(x) = (x_{1+i}, \ldots, x_s, x_1, \ldots, x_i). \tag{4.8}$$

The SIMD concept was initially developed and employed in the context of vector processors; this approach enables efficient handling of repetitive and parallelizable tasks by exploiting the inherent parallelism in datasets. While SIMD has a history dating back to the era of vector computers, its relevance and application have endured and evolved, finding implementation in modern processors through instruction set extensions such as Intel's streaming SIMD extensions (SSE) and advanced vector extensions (AVX), as well as ARM's NEON.

The concept of SIMD in lattice-based cryptography first appeared in [38] where the authors observed that fast Fourier transform (FFT)/number theoretic transform (NTT) computation applied on a vector (of polynomial coefficients) are independent of each other and can be parallelized in a SIMD manner. In the context of FHE, a similar observation allows schemes such as BGV [15] and CKKS [16] to have messages that are vectors. The addition and multiplication operations are done slot-wise.

Although algorithms and implementations over FHE can be inspired by similar designs for vector processors, there are several differences that one should keep in mind when drawing such inspiration.

Polynomial Computation Computations that can be evaluated over FHE are limited to evaluating polynomials. Vector processors (as well as regular processors), on the other hand, can compute non-polynomial operations. An example of a non-polynomial operation is computing EQ A B which computes an indicator vector where each slot holds 1 if the corresponding slots in A and B are equal and 0 otherwise.

Branching Similarly to FHE, a code running on a vector processor cannot include branching based on a condition on individual slots. For example, the condition may hold for some of the slots and not hold for others in which case the (vector) processor needs to take both branches (for different subsets of slots). However, unlike FHE, a vector processor can branch based on a condition made on an entire vector. For example, a vector processor can run loop iterations until all slots in a vector are zero.

4.7.1 How SIMD Improves Performance

Using the SIMD feature significantly improves the performance of FHE applications. As a naïve example, consider a program whose input is $x_1, \ldots, x_d$ (for simplicity, we assume d is a power of 2) and outputs $\max x_i$. One possible implementation computes this maximum in a tournament manner, i.e., we iteratively compute the maximum of each pair: $x_i^{(1)} = \max(x_{2i-1}, x_{2i})$, for $i = 1, \ldots, d/2$ and then $x_i^{(2)} = \max(x_{2i-1}^{(1)}, x_{2i}^{(1)})$, for $i = 1, \ldots, d/4$ until we are left with a single value. Tournaments are favored because they have a small computation depth (see discussion above about the advantages of a small depth). We now detail three ways

4.7 Exploiting SIMD

to pack data into slots to utilize SIMD. In these examples, we denote by sc the number of slots.

Example 4.8 (Full Batching) Consider a batch of sc inputs, i.e., sc different inputs:

$$x_1^{(1)}, \ldots, x_d^{(1)},$$
$$\vdots \qquad (4.9)$$
$$x_1^{(sc)}, \ldots, x_d^{(sc)},$$

where we wish to compute $\max(x_1^{(i)}, \ldots, x_d^{(i)})$, for $i = 1, \ldots, sc$. In this case, we can have d ciphertexts $c_1, \ldots, c_d$ and map the i-th input set into the i-th slot of the ciphertext. That is, map $x_j^{(i)}$ into the i-th slot of c_j.

We then perform the tournament as in the non-SIMD case on $c_1, \ldots, c_d$. This effectively computes over d input sets in parallel.

Example 4.9 (No Batching) Consider a batch of a single input where $d = sc$. In this case, we keep a single ciphertext and map x_i to its i-th slot. We then compute $\max x_i$ in a tournament manner as we now elaborate:

$$\begin{aligned} c^{(1)} &= \max(c, \text{Rot}_1(c)) \\ c^{(2)} &= \max(c^{(1)}, \text{Rot}_2(c^{(1)})) \\ c^{(4)} &= \max(c^{(2)}, \text{Rot}_4(c^{(2)})) \\ c^{(8)} &= \max(c^{(4)}, \text{Rot}_8(c^{(4)})) \\ &\vdots \\ c^{(n/2)} &= \max(c^{(n/4)}, \text{Rot}_{n/2}(c^{(n/4)})) \end{aligned} \qquad (4.10)$$

where $\text{Rot}_i(c)$ is a ciphertext that has the slots of c rotated by i positions and max is a function that takes two ciphertexts and computes their slot-wise maximum.

The code above first computes the maximum of every consecutive pair of slots. Then it computes the maximum of every consecutive set of four slots and it continues until the last step, where it computes the maximum of all the slots.

Example 4.10 (Partial Batching) The two examples above describe two extreme cases. The next example that naturally comes to mind is when $1 < d, b < sc$, where b is the batch size and d is as above. In that case, we use a hybrid mapping that enjoys the best of both mappings mentioned above. Chapters 8–9 discuss the concept of tile tensors and how they can be used to achieve such hybrid mapping.

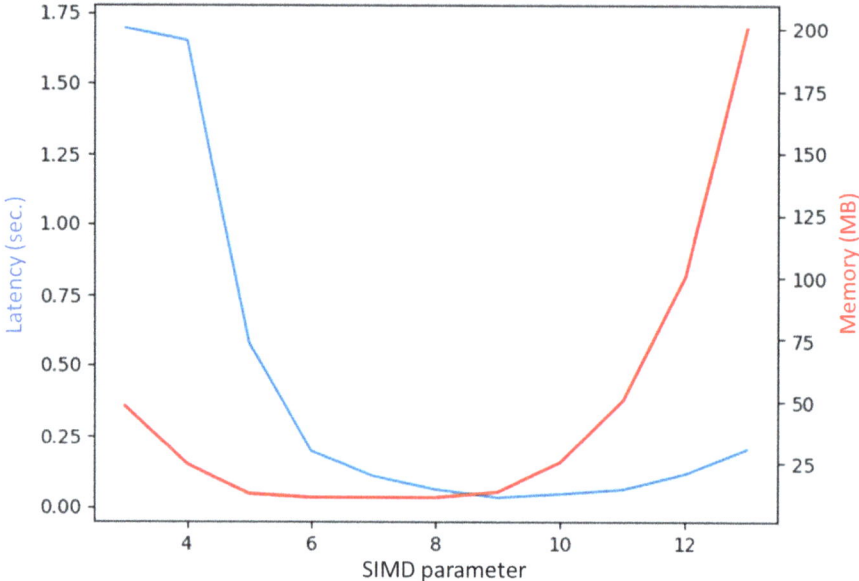

Fig. 4.5 A graph that shows how different packing schemes affect the running time and memory requirement. The x-axis represents different packing schemes that use SIMD differently. A value of $x = 0$ means full batching and a value of $x = 15$ means no batching. See Chap. 7 for more details. The blue line shows the latency, i.e., the time to compute a single batch. The red line shows the amount of memory needed

4.7.2 The Effects of Packing

Before delving into more technical discussion, we want to show how different packing affects the performance of a system. Consider the graph in Fig. 4.5. It shows how the *latency*, i.e., the time to compute a single batch and the memory changes with different packing schemes. The x-axis represents different packing schemes in a way that is detailed in Chap. 8. In a nutshell, the graph depicts various options for what the previous section referred to as hybrid packing. That is a packing where in each ciphertext, we pack a portion of multiple batches. The x-axis in the figure refers to $\log b$—the log of the number of elements from each batch in a ciphertext. For example, when $x = 0$, there is a single ($1 = 2^0$) element from each batch as in Example 4.8. When $x = \log sc$, there are sc elements of a batch in a ciphertext as in Example 4.9.

As the figure shows, the amount of memory and time needed to compute a computation varies dramatically for different packing schemes. For example, the latency ranges from 0.1 second to 1.7 seconds and the memory needs vary from 15 MB to 200 MB. Interestingly, in this experiment, the optimum for both cases is not achieved at the extreme points.

4.7.3 Comparing SIMD to Non-SIMD

We saw that the way we use the SIMD feature has a dramatic effect on the performance of the system. But does schemes that support SIMD are always faster than schemes that do not support SIMD? The answer to this question, of course, depends on the specific details and benchmarks of the schemes we are comparing as well as the type of computation we are performing. When circuits are deep but narrow, i.e., they do not involve many parallel computation, non-SIMD supporting schemes such as TFHE are usually faster. One reason is that these types of schemes need to handle less data; thus, a single operation of them is faster than a single operation in SIMD-supporting schemes such as CKKS or BGV. Furthermore, in the case of such circuits, these schemes are simpler to use because application designers can ignore questions related to packing. In contrast, for wide enough applications, it turns out that when the SIMD feature is used efficiently, it can outperform non-SIMD schemes.

4.7.3.1 Decrypting AES: CKKS Versus TFHE

As a first example, we consider the case of decrypting AES blocks under FHE, which is also referred to as transciphering or hybrid encryption [24] and is a benchmark that is commonly used for FHE. To understand the interest in this use case, we know that FHE ciphertexts have a very large overhead which makes communicating or storing them inefficient. A way to improve communication or storage efficiency is to encrypt messages using AES, which has no overhead, and providing the AES key encrypted with FHE. Then the messages can be AES-decrypted under FHE which effectively leaves them encrypted only with FHE.

A recent study [45] showed how to decrypt a single AES block under TFHE in 28 seconds on 16 cores. Another recent study [1] (see also [2]) showed that under CKKS 32,768 AES blocks can be decrypted on a server with 44 cores in an **amortized** time of 57 milliseconds. That is (after normalizing the different number of CPU cores), ×178 faster execution, per AES block. The improved running time per AES block comes from running multiple blocks in parallel using SIMD. The results are summarized in Table 4.1.

Table 4.1 Comparing the running time of AES decryption under CKKS and TFHE at the time of writing the book. The 3rd column shows the number of cores in the experiment. The fourth column shows the amortized time to decrypt an AES block and the fifth block shows the latency, i.e., the time it took the computation to complete

	Scheme	Cores	Amortized latency (sec)	Latency (sec)
[1]	CKKS	44	0.057	1860
[45]	TFHE	16	28	28

This example demonstrates that choosing whether to use a SIMD supporting scheme or not depends on the application. For the sake of comparison, let's normalize the latency of [1] to $1,860 * 44/16 = 5,115$ seconds according to the number of CPU cores used in the two experiments. The fact that the amortized latency is 178× faster does not kick in until an application needs to decrypt $5,115/28 = 182.7$ blocks (2.85 KB) in parallel. Consider, for example, an application that requires decrypting only 1 KB of data (64 AES blocks). Here, the latency of the TFHE implementation is $28 * 64 = 1,792$ seconds, while the latency of the CKKS implementation is still 5,115 seconds (2.85× slower). In contrast, when an application needs to decrypt 10 KB of data (640 AES blocks), the latency is 17,920 and 5,115 seconds, respectively (3.5× faster).

4.7.3.2 Game of Life

The game of life is a simple automaton described by Conway [28]. It describes a binary board, where each cell either has a value of 1 (also denoted *alive*) or 0 (also denoted *dead*). The automaton's simple rules define when a live cell dies (when it has too little or too many living neighbors) and when a cell is born (when it has a certain number of neighbors). Although simple it was proven that the automaton is Turing complete [42]. The work of [26] used game of life as an example to demonstrate a de-noising technique for CKKS when the messages are known to be integers. They compared their game-of-life running on different board sizes using CKKS and TFHE. We bring their comparison in Fig. 4.6. The figure shows that for small boards $n \leq 15$, when SIMD is not efficiently utilized, the TFHE implementation is faster, while for larger boards, SIMD is more efficiently utilized and CKKS becomes faster.

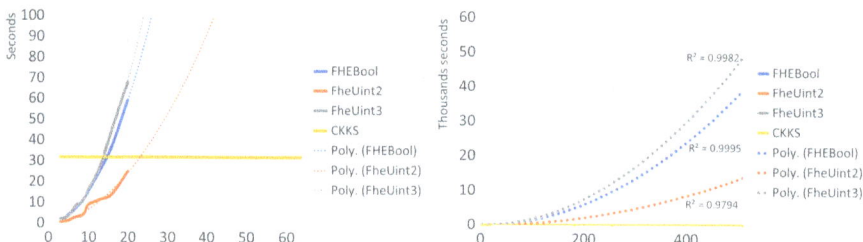

Fig. 4.6 Graphs taken from [26]. A comparison of game-of-life implementations with various board sizes. A value n on the x-axis represents a board of size $n \times n$. The curve labeled CKKS is the CKKS implementation. The other curves are various implementations of TFHE. The TFHE implementations were run with board sizes $n \leq 20$ and the extrapolated using quadratic extrapolation for larger boards. The graph shows that for small boards TFHE is faster but for large boards CKKS is faster

4.8 Polynomial Evaluation Methods: Horner, Paterson-Stockmeyer, and Others

Polynomials are the bread and butter of modern FHE, whose security is based on the hardness of the R-LWE problem, which deals with polynomial rings over finite fields (see Sect. 2.5.1). FHE arithmetic over ciphertexts is basically limited to addition, subtraction, and multiplication (along with complex conjugation in CKKS and some other nonarithmetic operations such as rotations). Thus, most FHE computations can in fact be considered as evaluations of some finite polynomial over the input ciphertexts. If the required computation involves functions that cannot be expressed with a finite polynomial (e.g., division, square roots, or trigonometric functions), then these functions must be approximated by a finite polynomial, as described in Chaps. 5 and 6. But whether the polynomial being computed is itself the required function or merely estimates it, we should in any case attempt to evaluate it in the most optimal manner. The following metrics should be considered when selecting the method to be used for evaluating a polynomial under FHE:

Performance. The performance of the various FHE operations differs among the different FHE schemes and also depends on the specific cryptographic setup and the hardware that runs them. However, addition and subtraction are typically relatively fast to compute under FHE, while products that often involve subsequent operations such as relinearization, rescaling (as in CKKS), or bootstrapping (as in TFHE) are much slower to compute. Rotations are also typically time-consuming.

In most FHE schemes, homomorphic multiplication of two ciphertexts (which in this book we call a non-scalar product—see Sect. 2.4.2) is slower to compute than a multiplication of a ciphertext with a plain scalar (termed a scalar product). This suggests that when selecting a method for evaluating polynomials, it is more important to optimize the number of products with the ciphertext encrypting the polynomial's variable than products with scalar coefficients. Thus, for example, in the polynomial $P(X) = 5X^2 + 2X$, the non-scalar product X^2, which needs to be computed over the encryption of X, is slower to compute than the scalar products with the coefficients 2 and 5.

Depth of products. This is a specific performance consideration relating to FHE multiplications. Section 2.4.2.1 introduced the concept of the maximal "product depth" or "multiplication chain index" supported by an FHE scheme, as the limit on the number of multiplications that can be performed on the same ciphertext before a costly and noisy bootstrap operation needs to be performed. Section 4.2 further defined the product depth of a specific algorithm or computation as the maximum number of multiplication gates on a path in the circuit that corresponds to the computation.

As stated above, since products are usually more time-consuming to compute under FHE than, say, additions, the *number* of products could also be an important performance metric. However, here we are referring to the *depth* of the products in the computation for evaluating the polynomial, which is an even more important consideration due to the high cost of bootstraps.

Numeric noise. FHE schemes that support floating-point arithmetic such as CKKS naturally incur numerical errors typical of approximate arithmetic due to the limited precision of the data representation and also due to the numerical stability of the evaluation method. In addition, homomorphic noise is deliberately added for security purposes during encryption and as part of the various arithmetic operations, for example, rotation operations may add new noise due to the noise present in the FHE switching keys. The amount of added noise may depend on the type of operation and on the magnitude of the encrypted data. Any noise that is introduced somewhere along the computation is later propagated and possibly enlarged in the subsequent dependent computations. Given some expression that needs to be computed, different alternative computation paths in the circuit may produce different levels of noise, so the numerical stability of the polynomial evaluation method should also be a consideration. We will see below that in some cases, the polynomial can be expressed in different forms that result in different noise levels. However, some polynomials may be inherently noisy to evaluate even in their best possible form and are better replaced with alternative polynomial approximations to the required function.

Referring to the above optimization targets, we describe below a number of polynomial evaluation methods and discuss their merits when used to evaluate polynomials under FHE. Table 4.2 summarizes the value of the above optimization targets reached by each of the presented methods.

4.8.1 The Naïve Method

The most naïve way to evaluate a polynomial $P(X) = \sum_{i=0}^{n} a_i X^i$ is to compute it term by term where every term $a_i X^i$ is computed with $i - 1$ non-scalar products by X and one scalar product by the coefficient a_i. The first term equals a_0 and involves no product, and thus, evaluating all the polynomial terms would take $\sum_{i=1}^{n} i = n(n+1)/2$ products with a maximal product depth of n and n additions to add up all the terms. This is one of the worst ways to evaluate a polynomial both in terms of the number of products and the product depth. It is given here just to emphasize that it should be avoided and as something to compare the following methods with.

4.8.2 Horner's Method

Horner's method (a.k.a. *Horner's rule*) for polynomial evaluation is based on the following equality:

$$P(X) = \sum_{i=0}^{n} a_i X^i = a_0 + X(a_1 + X(a_2 + \ldots + X(a_{n-1} + Xa_n)\ldots)) \quad (4.11)$$

4.8 Polynomial Evaluation Methods: Horner, Paterson-Stockmeyer, and Others

The above equality can also be described as in Algorithm 1.

Algorithm 1: Polynomial evaluation using Horner's rule

Input: A polynomial $P(X) = \sum_{i=0}^{n} a_i X^i$ and a value x.
Output: $p_x = P(x)$.
1 $p_x \leftarrow a_n$
2 **for** $i \leftarrow n - 1$ **to** 0 **do**
3 $\quad | \quad p_x \leftarrow p_x \cdot x + a_i$
4 **end**
5 **return** p_x

Algorithm 1 evaluates the polynomial using $n - 1$ non-scalar products and one scalar product, which is much better than the naïve method, but the product depth is still n.

4.8.3 Exponentiation by Squaring

When a polynomial term $a_i X^i$ has a degree that is a power of 2, i.e., $i = 2^m$ for some positive integer m, then it is possible to compute this term by m repeated squaring using just m products, with a product depth of m, or of $m + 1$ when also considering the final scalar product. For example, let $i = 8$, and then the term $a_8 X^8 = a_8 \left(\left(X^2 \right)^2 \right)^2$ uses only $\log_2 8 = 3$ square operations and one final scalar product by a_8. This is a great improvement over the naïve method that required i products and depth i to compute such a term. However, in practice we need to consider polynomials with many terms, not all of which are of a degree that is a power of 2. When the degree of the term is not a power of 2, then a reasonable approach is to use combinations of square operations and direct multiplications. Various such methods are described in [22, 35]. For example, the method of the Power() function defined in Algorithm 2 assumes that many different powers are used by the polynomial and caches pre-computed powers in order to reuse them later when computing new powers while using square operations where possible. A polynomial $P(X)$ of degree n can then be evaluated by calling the Power() function for all the nonzero terms of P. Subsequently, every term X^i is multiplied by the respective coefficient a_i and the terms are summed up. The cache used by the power function ensures that every needed power X^i for $i \in [2, n]$ is computed at most once. Every power is computed using a single product beyond the products used for the lower powers in the cache, so the total number of non-scalar products needed to compute all the powers of X for an entire n degree polynomial is $n - 1$. To this must be added n more scalar products with the nonzero coefficients of these terms.

The recursion of Power($cache, d$) is at most $\lceil \log_2 d \rceil$ deep, with an extra product depth for the product with the coefficient. Therefore, the product depth of any term is at most $\lceil \log_2 n \rceil + 1$.

Algorithm 2: Polynomial evaluation using squaring

Input: A polynomial $P(X) = \sum_{i=1}^{n-1} a_i X^i$ and a value x.
Output: $p_x = P(x)$.

1 $cache[0] \leftarrow 1$
2 $cache[1] \leftarrow x$
3 $p_x \leftarrow a_0 + a_1 x$
4 **for** $i \leftarrow 2$ **to** n **do**
5 $\quad$ **if** $a_i \neq 0$ **then**
6 $\quad\quad$ $x^i = \text{Power}(cache, i)$
7 $\quad\quad$ $p_x = p_x + a_i x^i$
8 $\quad$ **end**
9 **end**
10 **return** p_x

Input: $cache$ with previously computed powers of x and a required degree d.
Output: x^d. $cache$ is passed by reference so its modified content is also returned to the caller.

11 **Function** Power($cache, d$):
12 $\quad$ **if** $cache[d]$ is defined **then**
13 $\quad\quad$ **return** $cache[d]$
14 $\quad$ **end**
15 $\quad$ **if** d is a power of 2 **then**
16 $\quad\quad$ $cache[d] \leftarrow \text{Power}(cache, d/2)^2$
17 $\quad$ **else**
18 $\quad\quad$ $m = \lfloor \log_2 d \rfloor$
19 $\quad\quad$ $cache[d] \leftarrow \text{Power}(cache, 2^m) \cdot \text{Power}(cache, d - 2^m)$
20 $\quad$ **end**
21 $\quad$ **return** $cache[d]$
22 **end**

Example 4.11 Let $P(X) = X^8 - 4X^5 - 6X^2 + 3X + 1$ and suppose we wish to compute $P(X = 4)$ using Algorithm 2. We start by setting $cache[0] = 1$, $cache[1] = 4$, and $p_x = 1 + 3 * 4 = 13$. The next term is $-6X^2$, which requires one square operation to compute $x * x = 16$ and $p_x = 13 - 6 * 16 = -83$. The coefficients $a_3 = a_4 = 0$ and thus we do not need to compute the fourth and fifth terms. We compute the term X^5 by multiplying $cache[1] = x = 4$ that was computed before with $x^4 = 256$, which we recursively compute as $x^4 = x^2 \cdot x^2$ using $cache[2]$ that was also computed before. The results are added to the cache ($cache[5] = X^5 = 1024$) and we set $p_x = -83 - 4 * 1024 = -941$. Finally, we add the term $x^8 = (x^4)^2 = 65,536$ to p_x and get $P(X = x) = p_x = 65,536 - 941 = 64,595$.

4.8.4 Paterson and Stockmeyer Polynomial Evaluation Method

As mentioned above, when evaluating polynomials over a ciphertext variable x under FHE, the non-scalar products (e.g., when computing the powers of x) are slower to compute than scalar products (e.g., when multiplying the ciphertext power of x with the corresponding scalar coefficient). A similar situation occurs when computing polynomials over a matrix variable, because multiplying pairs of matrices is slower than multiplying a matrix with a scalar coefficient. For example, suppose we wish to evaluate the polynomial $P(X) = 5X^2 + 2X$ where X is an n by n matrix. The non-scalar product needed to compute X^2 involves n^3 multiplications, whereas the scalar product of the result with 5 involves only n^2 multiplications. The Paterson and Stockmeyer polynomial evaluation algorithm [41], originally designed for evaluating polynomials over matrices, makes this distinction between the two types of products and requires $\sim \sqrt{2n} + \log_2 n$ non-scalar products along with $n - \sqrt{\frac{n}{2}}$ scalar products. The maximal product depth is again $\lceil \log_2 n \rceil$.

Algorithm 3 shows the Paterson and Stockmeyer polynomial evaluation algorithm, and Algorithm 4 is an annotated version of the same algorithm. The non-scalar products are used in Steps 3, 4, and 7 of the algorithm (when computing bs and gs and the product of the gs values) and also in Step 20 of the ps function, adding up to $\sim k + m + n/2k + m$ non-scalar products. Taking $k \approx \sqrt{n/2}$ and $m \sim \lceil \log_2(n/k) \rceil$ (from Steps 1 and 2 of the algorithm), we get a total of $\sim \sqrt{2n} + \log_2(n)$ non-scalar products.

The deepest product chain comes from the last Step 20 of the ps function at the top level of the recursion, where G, Q, and S are all of depth $\sim log_2(k) + m \approx log_2(k) + log_2(n/k) = log_2(n)$. Note that C does not affect the depth at line 20 because it is more shallow than the other multiplied terms.

4.8.5 Evaluation by the Polynomial Roots

By the fundamental theorem of algebra, a polynomial $P(X)$ of degree n has n complex roots $r_1, r_2, \ldots, r_n$, counting multiplicity (i.e., some roots may be repeated). Thus, any $P(X)$ can be expressed as

$$P(X) = c(X - r_1)(X - r_2) \cdots (X - r_n) \qquad (4.12)$$

where $c, r_1, r_2, \cdots r_n$ are complex numbers.

If we are using an FHE scheme that supports complex arithmetic (such as CKKS), then after some pre-processing, P can be expressed as in Eq. 4.12. We can now compute the n products of this representation in a possibly unbalanced binary computation tree, which would involve n products, including $n - 1$ non-scalar products between the terms and one scalar product with c. The product depth is $\lceil \log_2(n+1) \rceil$.

However, the number of non-scalar products can be reduced, by computing the root terms in pairs, where, for example, $(X - r_1)(X - r_2)$ would be computed as $(X^2 - (r_1 + r_2)X + r_1 r_2)$, and similarly for all the other pairs of root terms. Thus, if n is even, then the entire polynomial can be computed as follows (if n is odd, then

Algorithm 3: Polynomial evaluation using the Paterson and Stockmeyer algorithm

Input: a polynomial $P(X)$ of degree n and an input x.
Note that X designates the symbolic indeterminate variable of the polynomial P and x designates the value over which $P(X)$ is evaluated.
Output: $P(x)$.
1   $k \leftarrow \lfloor \sqrt{n/2} \rfloor$
2   $m \leftarrow \lceil \log_2(n/k + 1) \rceil$
3   $bs \leftarrow (x^2, x^3, \ldots, x^k)$
4   $gs \leftarrow (x^k, x^{2k}, x^{4k}, x^{8k} \ldots, x^{k2^{m-1}})$
5   $P'(x) := P(x) + X^{k(2^m-1)}$
6   $value = ps(P', x, bs, gs, k, m)$
7   **return** $value - x^{k(2^m-1)}$

Input: a polynomial $P(X)$; an input x, the small ("baby step") powers bs and large ("giant step") powers gs as computed in Steps 3 and 4 above, and integers k and m as computed in Steps 1 and 2 above
Output: $P(x)$.
8   **Function** ps(P, x, bs, gs, k, m):
9   **if** *P is a polynomial of degree 0 of the form* $P(X) = a_0$ **then**
10   **return** a_0
11   **end**
12
13   Divide $P(X)$ by $X^{k2^{m-1}}$ using polynomial long division to get the quotient $q(X)$ and the remainder $r(X)$ so that $P(X) = X^{k2^{m-1}} q(X) + r(X)$
14     $r'(X) := r(X) - X^{k(2^{m-1}-1)}$
15   Divide $r'(X)$ by $q(X)$ using polynomial long division to get the quotient $c(X)$ and the remainder $s(X)$ so that $r'(X) = c(X)q(X) + s(X)$
16     $G \leftarrow x^{k2^{m-1}}$
17     $C \leftarrow c(x)$
18     $Q \leftarrow ps(q, x, bs, gs, k, m-1)$
19     $s'(X) := s(X) + X^{k(2^{m-1}-1)}$
20     $S \leftarrow ps(s', x, bs, gs, k, m-1)$
21     $result \leftarrow (G + C)Q + S$
22   **return** $result$
23 **end**

an unpaired term is also included):

$$P(x) = c(X^2 - (r_1 + r_2)X + r_1 r_2)(X^2 - (r_3 + r_4)X + r_3 r_4)$$
$$\cdots (X^2 - (r_{n-1} + r_n)X + r_{n-1} r_n) \quad (4.13)$$

The X^2 term here can be reused in all the pairs, so that each pair will essentially involve one scalar product like $(r_1 + r_2)X$ and one non-scalar product of the pair with the next pair. The total number of products is thus $\frac{n}{2}$ non-scalar products and

$\frac{n}{2} + 1$ scalar products (including the multiplication by c), and the product depth is $\lceil \log_2(n+1) \rceil$ as before.

We can continue to reduce the number of non-scalar products by working with larger groups of root terms, where the optimum is reached if we can divide the n root terms to $\sqrt{n}$ groups of $\sqrt{n}$ root terms. If n is not a square number, then an additional smaller group must also be used. In the optimal case (when n is square), $\sim \sqrt{n}$ non-scalar products are needed to compute all the powers shared by all the groups of root terms and then $\sim \sqrt{n}$ more non-scalar products to multiply all the groups. Each group also involves $\sim \sqrt{n}$ different scalar products, so the overall count is $\sim n$ scalar products and $\sim 2\sqrt{n}$ non-scalar products, at a product depth of $\lceil \log_2(n+1) \rceil$.

This can be compared with the lower number of $\sim \sqrt{2n} + log_2 n$ non-scalar products of the Paterson-Stockmeyer method, which also has the comparable product depth of $\sim \log_2 n$ (Sect. 4.8.4). However, the evaluation by roots method is more numerically stable for FHE than the Paterson-Stockmeyer method, as will be described in the next section.

4.8.6 Considerations of Accuracy and Numerical Stability

Consider the term with the highest degree in an n degree polynomial, namely, $a_n X^n$. For the values of x that are of relatively high magnitude, whether positive or negative, the power x^n might reach very large magnitudes indeed. For example, the highest power in a 10-degree polynomial that serves to estimate a function at X=100 would have a value of 10^{20}. Presumably, the function to be estimated by the polynomial does not reach such high magnitudes and so it follows that the coefficient a_n must be extremely minute in order to balance the magnitude of the huge value of x^n. This places a difficult demand upon the FHE scheme which is now required to encode and multiply an extremely small value with an extremely large one while maintaining a reasonable accuracy in the result.

A similar difficulty arises when we wish to evaluate the polynomial on a very small value of x. In this case, if the computed term $a_n X^n$ is to contribute anything to the final sum, then a_n would need to be huge in order to balance the very small magnitude of x^n, and again the FHE scheme would be required to encode and multiply an extremely large value with an extremely small one.

One way to ease this difficulty is to compute the term $a_n X^n$ as $(\sqrt[n]{a_n} X)^n$, thus avoiding extremely large or small intermediate values during the computation of the product. See Eq. 4.14. The problem with this solution is that now we can no longer reuse the computed powers of x among the different terms of the polynomial:

$$P(X) = (\sqrt[n]{a_n} X)^n + \ldots + (\sqrt[3]{a_3} X)^3 + (\sqrt[2]{a_2} X)^2 + a_1 X + a_0 \qquad (4.14)$$

In the method of evaluation by roots (see Sect. 4.8.5) when applied to large values of x, the minute constant c can be distributed among the multiplied root terms, e.g. by multiplying each of the n root terms by $|c|^{1/n}$ and correcting the overall sign if needed—see Eq. 4.15 (and Eq. 6.1 for the definition of the Sign function used in this equation). Thus, we can again reuse the $|c|^{1/n}X$ value that occurs in all the root terms. This is more numerically stable than X when raised to the power n:

$$P(X) = \text{Sign}(c) \prod_{i=0}^{n-1} |c|^{1/n}(X - r_i) \qquad (4.15)$$

If we divide the n root terms into pairs in order to reduce the non-scalar product count, as in Eq. 4.13, then $|c|^{2/n}$ can be distributed among the pairs as follows:

$$P(X) = \text{Sign}(c) \prod_{i=0}^{\frac{n}{2}-1} |c|^{2/n}(X^2 - (r_{2i} + r_{2i+1})X + r_{2i}r_{2i+1})$$

$$= \text{Sign}(c) \prod_{i=0}^{\frac{n}{2}-1} (|c|^{2/n}X^2 - (r_{2i} + r_{2i+1})|c|^{2/n}X + |c|^{2/n}r_{2i}r_{2i+1})$$

(4.16)

With this equation, we can again compute $|c|^{2/n}X^2$ just once for all the terms, with just one non-scalar product and one scalar product. In addition, we need a scalar product for each of the $n/2$ pairs and $n/2 - 1$ non-scalar numerically stable products between the $n/2$ pairs.

As mentioned in the previous section, the optimal number and size of the groups of root terms that minimize the number of non-scalar products is $m = \sqrt{n}$. This form requires $m - 1$ non-scalar products to compute the powers common to all the groups and another $m - 1$ non-scalar products to multiply the m groups themselves. Each of the m groups involves m scalar products, and therefore, the total is $m^2 = n$ scalar products and $\sim 2\sqrt{n}$ non-scalar products. The product depth of each group is $\sim log_2(\sqrt{n})$, and the product of the m groups adds $\sim log_2(\sqrt{n})$ more to the depth, giving a total product depth of $\sim log_2(n)$

Thus, the number of non-scalar products of this method ($\sim 2\sqrt{n}$) is somewhat higher than the number of non-scalar products of the Paterson-Stockmeyer method ($\sim\sqrt{2n} + log_2(n)$), which also has the comparable product depth of $\sim log_2(n)$. However, the gradual computation of the high power of x, together with the distribution of the small coefficient $a_n = c$ among the groups as described above, leads to a much better numerical stability of the computation.

4.9 FHE Code as Circuits

As mentioned in Sect. 4.1.1, circuit model fits FHE code with some resemblance to hardware circuits. It is only natural that we also represent FHE code as circuits. In what follows, we discuss the benefits of representing FHE as a circuit.

4.9.1 Circuit Representation

As is often mentioned, with FHE we can implement an arithmetic circuit, i.e., a DAG. In such a graph, a gate (or a node) represents an operation, where each gate has one outgoing edge and 0,1 or 2 input edges. To be exact, since a single output can be the input on multiple gates, these are hyper-edges, but for short we call them edges. The edges here are directed and they have exactly one node for whom the edge is an output edge. An edge in the circuit represents a ciphered message, either an input, an output, or an intermediate value in the computation. The edges in the circuit representation resemble very much the wires in hardware circuits where each wire has exactly one gate setting the voltage on that wire and multiple gates whose wire is their input. Obviously the voltage is the same across the entire wire.

When having a circuit representation, we can apply a more holistic view of the code. We list some of the advantages the circuit representation has:

- Scheduled bootstrapping
- Caching and prefetching.
- Parallel and distributed computations

4.9.1.1 Scheduled Bootstrapping

The problem of where to apply bootstrapping is not trivial. For example, sometimes it may be better to push bootstrapping forward to decrease the number of bootstraps. In fact, it was shown in [13] that the problem of determining the best locations to apply bootstrap is an NP-hard question. They showed an L-approximation for this problem, where L is the highest chain index the key supports. For their approximation they had a variable x_i (with constraints $0 \leq x_i \leq 1$) indicating whether there should be a bootstrap after the i-th gate. The L factor of approximation comes from moving from the continuous LP solution to the discrete solution where $x_i = 0, 1$.

Using a circuit representation allows us to identify places where redundant operations (e.g., bootstraps) are performed. Consider a gate whose output requires a bootstrap. Ideally, the output of this gate is the input of only one bootstrapping gate. In practice, compilers may defer the decision to bootstrap until the output value is being used. However, treating the output value as the input to multiple gates can eventually lead to multiple bootstraps that apply to the same gate (in parallel). Static analyzers that process circuits can identify these redundant bootstraps' computation and merge them into only one.

4.9.1.2 Caching and Prefetching

Ciphertexts in FHE have a large overhead and they take up a large amount of memory. To run large FHE code on systems with small random access memory (RAM), a traditional approach (e.g., one that is used by operating systems) is to cache memory to the disk and fetch them again when needed. The online caching problem received a lot of attention since it has a large effect on performance. In [36] the authors took advantage of FHE being input independent, thus running in the same way and described a system that in a preliminary step runs the FHE code and records the memory access pattern. In consequent runs, they described how the recorded access pattern can be used to make better decisions when caching to the disk and prefetching them from the disk before they are needed.

Using circuit representation completely turns caching into an offline problem. Knowing the circuit and when each ciphertext is going to be accessed, we can devise an optimal caching policy for writing and prefetching ciphertexts.

4.9.1.3 Parallel and Distributed Computations

The engineering task of running a code on multiple CPUs (that can communicate over a shared RAM) and multiple servers (that communicate through a network) is sometimes challenging, specifically synchronizing dependent tasks and also writing code that efficiently uses all available computing resources. Given circuit representation, we can treat each gate as a micro-task that needs to be assigned to a computing resource taking into account dependencies, caching, and networking time. This opens the door to a generic execution tool that efficiently executes circuits in different environments.

For example, suppose that a circuit needs to run on two servers. An offline analysis can partition the nodes into two subsets, such that each subset is intended to run on a different server. An edge going from a node in one subset into a node in the other means a ciphertext that needs to be communicated from one server to the other. An efficient partition would be such that the total execution time is minimal (or close to it). Again, we see here resemblance to hardware design, specifically to circuit partitioning, although the goals in hardware design are somewhat different.

4.9.2 Compiling HE Code into a Circuit

After having discussed its advantages, we need to show that it is easy to move from FHE code into circuit representation. Indeed, since the execution of FHE code does not depend on the input, we can run the FHE code while recording each operation we execute with its inputs and outputs. Following our previous observation, each intermediate value during the execution is an edge. It is an output of one operation and an input to others.

4.9.3 Running a Circuit

To run a circuit, we iteratively find a gate that has not been computed yet and whose inputs are already computed and then compute it. In fact, this process is easily parallelizable and we can have multiple processes scanning the circuit to run gates that are ready to be run. Given an infinite number of processors, a circuit with depth d can be executed in d steps. With a single processor that can execute one gate at a time, the order of execution affects the memory required for the running. If we allow the same gate to be executed multiple times, a circuit can be executed while using a storage of d ciphertexts, where d is the circuit depth.

4.10 Lab Exercises

4.10.1 Polynomial Evaluation Methods

In the following exercises, consider the polynomial

$$P(X) = 15x^8 + 0.5x^7 - 38x^6 - 2x^5 + 32x^4 - 10x^2 + x + 10$$

and create a CKKS ciphertext c_x that encrypts a vector of sequential real values rising from 0 to 1:

1. Evaluate P over the encrypted values in c_x using the naïve method under FHE.

    A Report the mean.
    B Report the maximal. errors
    C Report the number of non-scalar and scalar products as well as the product depth.

2. Repeat Exercise 1, only this time use the exponentiation by squaring method under FHE.
3. Implement a non-FHE version of the Paterson-Stockmeyer algorithm using, e.g., Python's SymPy library [44] to perform the necessary polynomial divisions and other polynomial operations that are part of the algorithm.
4. Use the Paterson-Stockmeyer algorithm to evaluate $P(x)$ for $x = 0.8$. Report the number of non-scalar and scalar products as well as the product depth.
5. Use the Paterson-Stockmeyer algorithm to evaluate a polynomial $P(x)$ of degree 100 with all coefficients equaling 1.0 over $x = 0.8$, and report the number of non-scalar and scalar products as well as the product depth. Compare this to the expected number of non-scalar products for such a polynomial when evaluated using Horner's rule or the naïve method using exponentiation by squaring.

4.10.1.1 Polynomial Evaluation by Roots

In the next exercises, consider the 16 degree polynomial $P(x)$:

$$\begin{aligned}
P(x) = &- 2.1225348045163603e{-}11 x^{16} - 6.817121637851839e{-}11 x^{15} \\
&+ 5.051692241028642e{-}09 x^{14} + 7.433305644755716e{-}09 x^{13} \\
&- 4.5659833073111e{-}07 x^{12} - 2.9933873233888765e{-}07 x^{11} \\
&+ 2.1075147674815448e{-}05 x^{10} + 5.259833568629501e{-}06 x^{9} \\
&- 0.0005472744425741752 x^{8} - 3.167560745397922e{-}05 x^{7} \\
&+ 0.008188257999608619 x^{6} - 0.00011643195078502193 x^{5} \\
&- 0.07012299945361615 x^{4} + 0.0015778800829984456 x^{3} \\
&+ 0.40393641355042276 x^{2} + 0.4973694343718088 x \\
&+ 0.12477746137138203
\end{aligned}$$

The 16 complex roots of $P(x)$ are

$$-12.167505663283148$$
$$-5.932146327049738,$$
$$-5.632508528122107,$$
$$-5.108243271469325,$$
$$-4.3831346534677325,$$
$$-3.489764198347683,$$
$$-2.46991567692673,$$
$$-1.3816913713845256,$$
$$-0.34595415937565827,$$
$$7.5886990604867375,$$
$$6.9901856637753 + 1.1479313265787225i,$$
$$6.9901856637753 - 1.1479313265787225i,$$
$$5.300857713592042 + 2.009898099473336i,$$
$$5.300857713592042 - 2.009898099473336i,$$
$$2.7641474144428306 + 2.2731417864173022i,$$
$$2.7641474144428306 - 2.2731417864173022i$$

4.10 Lab Exercises

7. Create a CKKS ciphertext C_X that encrypts a vector of sequential real values rising from -2 to 2.

   Evaluate $P(X)$ over the values in C_X using exponentiation by squaring under FHE.

   A Report the mean and maximal errors compared with the precise non-FHE evaluation of the polynomial.
   B Report the number of non-scalar and scalar products as well as the product depth.

8. Evaluate $P(X)$ as a product of the differences from the roots according to the form of Eq. 4.15. Report the mean and maximal errors compared with the precise non-FHE evaluation of the polynomial, the number of non-scalar and scalar products, and the product depth.

   Hint: First, identify the coefficient c of the equation and compute $|c|^{1/16}$. c is not encrypted, so its sign is known and the correction of the sign of the final result can be managed by an FHE-negate operation. In order to optimize the product depth, it is best to compute the product of the 16 root terms in a tree computation, i.e., first, compute all pairs of terms, then multiply pairs of pairs, etc.

9. Equation 4.16 shows how the root terms can be grouped in pairs in order to reduce the non-scalar product count. Construct a similar equation where the root terms are grouped in groups of size 4 and each group is multiplied by $|C|^{4/n}$.

10. Evaluate $P(X)$ as a product of the differences from the roots using the equation constructed above. In the case of $P(X)$, there are 16 roots, so there are 4 groups of size 4. Report the mean and maximal errors compared with the precise non-FHE evaluation of the polynomial, the number of non-scalar and scalar products, and the product depth.

    Hint: First, compute $|C|^{4/n}x^4$ and x^2 by squaring, and then use these terms in all the four groups. In order to optimize the product depth, it is best to compute the product of the four groups in a tree computation, i.e., multiply two pairs of groups and then multiply the two results.

4.10.1.2 Sort an Array of Numbers

In the following exercises, you will need to implement a circuit that sorts an array of eight numbers. The numbers are all integers in the range $1, \ldots, 128$ and are encrypted with CKKS. You can assume you are given a function isGreaterThan(Enc(x), Enc(y)) that gets two ciphertexts Enc(x) and Enc(y) and returns a ciphertext Enc(r) where $r = 1$ if $x > y$ and otherwise 0. Such functions are discussed in Chap. 6, but in this question, you can implement $isGreater$(Enc(x), Enc(y)) by decrypting Enc(x) and Enc(y), comparing x and y and then returning Enc(1) or Enc(0).

11. Implement merge sort in plaintext: split the array into two halves; merge each half in recursion; and merge the two sorted halves. Does your code have the same behavior (e.g., memory access pattern) for different inputs?
12. Use the general recipe in Sect. 4.1.2 to transform your merge sort to a circuit and implement it under FHE.
13. Write code that implements a Batcher sort [9] circuit without using SIMD.
14. Change your Batcher sort implementation to use SIMD.
15. List your different FHE implementations from slowest to fastest.
16. Run your different FHE implementations on arrays of sizes $2^3, 2^4, 2^5, 2^6, \ldots$ and plot the running times of each FHE implementation. What is the asymptotic runtime of each implementation?
17. Repeat Exercises 11–15, but this time use BGV. Compare your results with the CKKS-based implementation.
18. Repeat Exercises 11–12, but this time use TFHE. Compare your results with your CKKS-based and BGV-based implementations.

4.10.1.3 Private Information Retrieval (PIR)

Consider an array A of size $n = 2^{20}$. Implement PIR, i.e., given $1 \le i \le n$ as an encrypted input, you should output $A[i]$. Use CKKS and choose parameters that yield 2^{16} slots in a ciphertext:

1. Assume the input is a single ciphertext C where the value i is broadcasted over all slots of C. In your answer, you can use a function $Eq(C_1, C_2)$ that returns a ciphertext C_3 where each slot indicates whether the respective slots of C_1 and C_2 are equal. For your tests, implement $Eq(C_1, C_2)$ by decrypting C_1 and C_2, comparing their slots in plaintext, and encrypting the result of the comparison.
2. Assume the input is $\log n$ ciphertexts $C_1, \ldots, C_{\log n}$, where the b-th bit of i (in its binary representation) is broadcasted along the slots of C_b.
3. For which values of n each method is faster? How would this answer change if Eq was ×16 slower?

4.10.1.4 Copy and Recurse: BTree

Implement BTree [10] search under FHE using copy and recurse:

- The BTree has four levels.
- Each inner node v has $4 < n_v < 8$ elements (and therefore $n_v + 1$ children, where n_v are known.
- The elements in the tree are integers in the range $0, \ldots 128$.
- Each node in the tree is given in a single ciphertext C. For a node v, let $x_1, x_2, \ldots$ be the values v holds and then slots $1, \ldots 7$ of C encode the bits of x_1; slots $8, \ldots 14$ encode the bits of x_2 and so on (you can decide how the bits are ordered).

4.10 Lab Exercises

Given an encrypted input in (given in binary), report (under FHE) whether in appears in the tree:

- Use the generic recipe to traverse the tree.
- Use copy and recurse to traverse the tree. Hint: you will need to add empty nodes to make the tree full.

4.10.1.5 Implement $argmax$

Let A be an input array of size $n = 2^{20}$ encrypted with CKKS, where the elements of A are given as the slots of multiple ciphertexts, where each slot holds a single value. Write code that find the index of the maximal value (assume the maximal value is unique). You may assume you are given a function $isGreater(C_1, C_2)$ that gets two ciphertexts C_1 and C_2 and returns an encrypted ciphertext where each slot indicates whether the respective slot of C_1 is bigger than the slot of C_2. In your tests, implement $isGreater$ by decrypting C_1 and C_2, comparing their plaintext slots, and encrypting the result:

- Write your code efficiently (using SIMD).
- How many times (not in O-notation) was $isGreater$ invoked? Write this number as a function of n.

Appendix

Table 4.2 A comparison of various polynomial evaluation methods. We consider a method to have better numerical stability when the value of the term of the highest degree is reached gradually both when the value of the highest degree is very high (e.g., when $X > 1$) and the corresponding coefficient is very small and also when the value of the highest degree is very small (e.g., when $X \in (0, 1)$) and the corresponding coefficient is very high

Method	Non-scalar products	Scalar products	Depth	Numerical stability
The naïve method	$n(n-1)/2$	n	n	Low
Horner's method	$n-1$	1	n	Moderate
Exponentiation by squaring	$n-1$	n	$\lceil \log_2 n \rceil + 1$	Low
Paterson and Stockmeyer	$\sim\sqrt{2n} + log_2(n)$	$\sim n - \sqrt{2n}$	$\sim log_2(n)$	Moderate
Eval. by poly. roots: Eq. 4.15	$n-1$	2	$\lceil \log_2 (n+1) \rceil$	High
Eval. by poly. roots: $\sqrt{n}$ groups	$2\sqrt{n}$	n	$\sim log_2(n)$	Moderate

Algorithm 4: Polynomial evaluation using the Paterson and Stockmeyer algorithm: an annotated version of Algorithm 3

Input: a polynomial $P(X)$ of degree n and an input x.
Note that X designates the symbolic indeterminate variable of the polynomial P and x designates the value over which $P(X)$ is evaluated.
Output: $P(x)$.

// k and m set so that $k(2^m - 1) \geq n$
1   $k \leftarrow \lfloor \sqrt{n/2} \rfloor$
2   $m \leftarrow \lceil log_2(n/k + 1) \rceil$

// The "baby-step" powers in bs are computed using exponentiation by squaring as described in Sect. 4.8.3, taking $k-1$ non-scalar products, and a product depth of $\lceil log_2(k) \rceil$ (for the largest power).
3   $bs \leftarrow (x^2, x^3, \ldots, x^k)$

// The "giant-step" powers in gs are computed by repeated squaring starting from x^k taken from bs, and taking $m-1$ additional non-scalar products, and an added product depth of $m-1$.
4   $gs \leftarrow (x^k, x^{2k}, x^{4k}, x^{8k} \ldots, x^{k2^{m-1}})$

// The term $X^{k(2^m-1)}$ is added to ensure that the function evaluates a polynomial of degree $k(2^m - 1)$, even if originally $k(2^m - 1) > n$. Note that the term is not evaluated at this point, but merely added to the definition of $P'(x)$. It will be evaluated as part of step 6, and then removed in step 7 below.
5   Define the polynomial $P'(x) = P(x) + X^{k(2^m-1)}$
6   $value = ps(P', x, bs, gs, k, m)$

// $x^{k(2^m-1)}$ can be computed here as the product of the m values in gs with $m-1$ products and a product depth of $\lceil log_2(m) \rceil$
7   **return** $value - x^{k(2^m-1)}$

Input: a polynomial P(x); an input x, baby step powers (bs) and giant step powers (gs) as computed in steps 3 and 4 above, and integers k and m as computed in steps 1 and 2 above

Output: $P(x)$.

```
8  Function ps(P, x, bs, gs, k, m):
9      if P is a polynomial of degree 0 of the form P(X) = a₀ then
10         return a₀    // This is where the recursion of this function ends
11     end
```
12   Divide $P(X)$ by $X^{k2^{m-1}}$ using polynomial long division to get the quotient $q(X)$ and the remainder $r(X)$ so that $P(X) = X^{k2^{m-1}} q(X) + r(X)$

13   Define the polynomial $r'(X) = r(X) - X^{k(2^{m-1}-1)}$

14   Divide $r'(X)$ by $q(X)$ using polynomial long division to get the quotient $c(X)$ and the remainder $s(X)$ so that $r'(X) = c(X)q(X) + s(X)$

```
    // It can be seen at this point that
    // P(X) = (X^(k2^(m-1)) + c(X))q(X) + s(X) + X^(k(2^(m-1)-1))
    // and steps 15-20 below evaluate this expression over x.
```

15      $G \leftarrow x^{k2^{m-1}}$ // This term can be taken from gs

16      $C \leftarrow c(x)$ // It can be shown that the degree of c(x) is no more than k and thus this polynomial can be evaluated directly using the powers already computed in bs, with no additional non-scalar products.

17      $Q \leftarrow ps(q, x, bs, gs, k, m-1)$

18   Define the polynomial $s'(X) = s(X) + X^{k(2^{m-1}-1)}$

19      $S \leftarrow ps(s', x, bs, gs, k, m-1)$

         // The final step involves one non-scalar product. Counting all the recursive calls, this adds up to $2^{m-1} - 1 \approx n/2k$ non-scalar products, and an extra product depth of m (the depth of the recursion).

20      $result \leftarrow (G+C)Q + S$

21   **return** $result$

22 **end**

References

1. Aharoni, E., Drucker, N., Ezov, G., Kushnir, E., Shaul, H., Soceanu, O.: E2e near-standard and practical authenticated transciphering. Cryptology ePrint Archive, Paper 2023/1040 (2023). https://eprint.iacr.org/2023/1040
2. Aharoni, E., Drucker, N., Ezov, G., Kushnir, E., Shaul, H., Soceanu, O.: E2E near-standard hybrid encryption. Poster session at 6th HomomorphicEncryption.org Standards Meeting (2023). https://homomorphicencryption.org/6th-homomorphicencryption-org-standards-meeting/
3. Ajtai, M., Komlós, J., Szemerédi, E.: An 0(n log n) sorting network. In: Proceedings of the Fifteenth Annual ACM Symposium on Theory of Computing, STOC '83, pp. 1–9. Association for Computing Machinery, New York (1983). https://doi.org/10.1145/800061.808726

4. Akavia, A., Feldman, D., Shaul, H.: Secure search on encrypted data via multi-ring sketch. In: Proceedings of the 2018 ACM SIGSAC Conference on Computer and Communications Security, CCS '18, pp. 985–1001. Association for Computing Machinery, New York (2018). https://doi.org/10.1145/3243734.3243810
5. Akavia, A., Feldman, D., Shaul, H.: Secure data retrieval on the cloud: homomorphic encryption meets coresets. IACR Trans. Cryptogr. Hardw. Embed. Syst. **2019**(2), 80–106 (2019). https://doi.org/10.13154/tches.v2019.i2.80-106
6. Akavia, A., Gentry, C., Halevi, S., Vald, M.: Achievable cca2 relaxation for homomorphic encryption. In: Kiltz, E., Vaikuntanathan, V. (eds.) Theory of Cryptography, pp. 70–99. Springer, Cham (2022)
7. Akavia, A., Shaul, H., Weiss, M., Yakhini, Z.: Linear-regression on packed encrypted data in the two-server model. In: Proceedings of the 7th ACM Workshop on Encrypted Computing & Applied Homomorphic Cryptography, WAHC'19, pp. 21–32. Association for Computing Machinery, New York (2019). https://doi.org/10.1145/3338469.3358942
8. Akl, S.G.: Bitonic Sort, pp. 139–146. Springer, Boston (2011). https://doi.org/10.1007/978-0-387-09766-4_124
9. Batcher, K.E.: Sorting networks and their applications. In: Proceedings of the April 30–May 2, 1968, Spring Joint Computer Conference, AFIPS '68 (Spring), pp. 307–314. Association for Computing Machinery, New York (1968). https://doi.org/10.1145/1468075.1468121
10. Bayer, R., McCreight, E.M.: Organization and maintenance of large ordered indexes. Acta Inform. **1**(3), 173–189 (1972). https://doi.org/10.1007/BF00288683
11. Beimel, A., Ishai, Y., Malkin, T.: Reducing the servers computation in private information retrieval: PIR with preprocessing. In: Annual International Cryptology Conference, pp. 55–73. Springer, Berlin (2000)
12. Benhamouda, F., Lepoint, T., Mathieu, C., Zhou, H.: Optimization of bootstrapping in circuits. In: Proceedings of the Twenty-Eighth Annual ACM-SIAM Symposium on Discrete Algorithms, SODA '17, pp. 2423–2433. Society for Industrial and Applied Mathematics, USA (2017)
13. Benhamouda, F., Lepoint, T., Mathieu, C., Zhou, H.: Optimization of bootstrapping in circuits. In: Proceedings of the Twenty-Eighth Annual ACM-SIAM Symposium on Discrete Algorithms, SODA '17, p. 2423–2433. Society for Industrial and Applied Mathematics, USA (2017)
14. Bergerat, L., Boudi, A., Bourgerie, Q., Chillotti, I., Ligier, D., Orfila, J.B., Tap, S.: Parameter optimization and larger precision for (t) fhe. J. Cryptol. **36**(3), 28 (2023). https://doi.org/10.1007/s00145-023-09463-5
15. Brakerski, Z., Gentry, C., Vaikuntanathan, V.: (leveled) fully homomorphic encryption without bootstrapping. ACM Trans. Comput. Theory **6**(3) (2014). https://doi.org/10.1145/2633600
16. Cheon, J., Kim, A., Kim, M., Song, Y.: Homomorphic encryption for arithmetic of approximate numbers. In: Proceedings of Advances in Cryptology - ASIACRYPT 2017, pp. 409–437. Springer Cham (2017). https://doi.org/10.1007/978-3-319-70694-8_15
17. Cheon, J.H., Choe, H., Park, J.H.: Tree-based lookup table on batched encrypted queries using homomorphic encryption. Cryptology ePrint Archive, Paper 2024/087 (2024). https://eprint.iacr.org/2024/087
18. Cheon, J.H., Han, K., Kim, A., Kim, M., Song, Y.: Bootstrapping for approximate homomorphic encryption. In: Annual International Conference on the Theory and Applications of Cryptographic Techniques, pp. 360–384. Springer, Berlin (2018)
19. Chillotti, I., Gama, N., Georgieva, M., Izabachène, M.: TFHE: fast fully homomorphic encryption over the torus. J. Cryptol. **33**(1), 34–91 (2020). https://doi.org/10.1007/s00145-019-09319-x
20. Chillotti, I., Joye, M., Paillier, P.: Programmable bootstrapping enables efficient homomorphic inference of deep neural networks. In: Dolev, S., Margalit, O., Pinkas, B., Schwarzmann, A. (eds.) Cyber Security Cryptography and Machine Learning, pp. 1–19. Springer, Cham (2021). https://doi.org/10.1007/978-3-030-78086-9_1

21. Cong, K., Das, D., Park, J., Pereira, H.V.: SortingHat: efficient private decision tree evaluation via homomorphic encryption and transciphering. In: Proceedings of the 2022 ACM SIGSAC Conference on Computer and Communications Security, CCS '22, pp. 563–577. Association for Computing Machinery, New York (2022). https://doi.org/10.1145/3548606.3560702
22. Cormen, T.H., Leiserson, C.E., Rivest, R.L., Stein, C.: Introduction to Algorithms, 3rd edn. PHI Learning Pvt. Ltd. (Originally MIT Press) (2010)
23. Cover, T.M., Thomas, J.A.: Elements of Information Theory. John Wiley & Sons, London (1991)
24. Cramer, R., Shoup, V.: Design and analysis of practical public-key encryption schemes secure against adaptive chosen ciphertext attack. SIAM J. Comput. **33**(1), 167–226 (2003). https://doi.org/10.1137/S0097539702403773
25. Crawford, J.L.H., Gentry, C., Halevi, S., Platt, D., Shoup, V.: Doing real work with FHE: the case of logistic regression. In: Proceedings of the 6th Workshop on Encrypted Computing & Applied Homomorphic Cryptography, WAHC '18, pp. 1–12. Association for Computing Machinery, New York (2018). https://doi.org/10.1145/3267973.3267974
26. Drucker, N., Moshkowich, G., Pelleg, T., Shaul, H.: BLEACH: cleaning errors in discrete computations over CKKS. J. Cryptol. **37**(1), 3 (2023). https://doi.org/10.1007/s00145-023-09483-1
27. Fleischhacker, N., Larsen, K.G., Simkin, M.: How to compress encrypted data. In: Hazay, C., Stam, M. (eds.) Advances in Cryptology – EUROCRYPT 2023, pp. 551–577. Springer, Cham (2023)
28. Games, M.: The fantastic combinations of John Conway's new solitaire game "life" by Martin Gardner. Sci. Am. **223**, 120–123 (1970)
29. Giacomelli, I., Jha, S., Joye, M., Page, C.D., Yoon, K.: Privacy-preserving ridge regression with only linearly-homomorphic encryption. In: Preneel, B., Vercauteren, F. (eds.) Applied Cryptography and Network Security - 16th International Conference, ACNS 2018, Leuven, Belgium, July 2-4, 2018, Proceedings, Lecture Notes in Computer Science, vol. 10892, pp. 243–261. Springer, Berlin (2018). https://doi.org/10.1007/978-3-319-93387-0_13
30. Giacomelli, I., Jha, S., Joye, M., Page, C.D., Yoon, K.: Privacy-preserving ridge regression with only linearly-homomorphic encryption. In: Preneel, B., Vercauteren, F. (eds.) Applied Cryptography and Network Security - 16th International Conference, ACNS 2018, Leuven, Belgium, July 2-4, 2018, Proceedings, Lecture Notes in Computer Science, vol. 10892, pp. 243–261. Springer, Berlin (2018). https://doi.org/10.1007/978-3-319-93387-0_13
31. Guimarães, A., Borin, E., Aranha, D.F.: Revisiting the functional bootstrap in tfhe. IACR Trans. Cryptogr. Hardw. Embedd. Syst. **2021**(2), 229–253 (2021). https://doi.org/10.46586/tches.v2021.i2.229-253
32. Hong, S., Kim, S., Choi, J., Lee, Y., Cheon, J.H.: Efficient sorting of homomorphic encrypted data with k-way sorting network. IEEE Trans. Inform. Forensics Secur. **16**, 4389–4404 (2021). https://doi.org/10.1109/TIFS.2021.3106167
33. Iliashenko, I., Zucca, V.: Faster homomorphic comparison operations for BGV and BFV. Proc. Privacy Enhancing Technol. **2021**(3), 246–264 (2021). https://doi.org/10.2478/popets-2021-0046
34. Kluczniak, K., Schild, L.: FDFB: full domain functional bootstrapping towards practical fully homomorphic encryption. IACR Trans. Cryptogr. Hardw. Embedd. Syst. **2023**(1), 501–537 (2022). https://doi.org/10.46586/tches.v2023.i1.501-537
35. Knuth, D.E.: The Art of Computer Programming, vol. 2, Seminumerical Algorithms, vol. 2, 2nd edn. Addison-Wesley Pub (Sd) (1981)
36. Kumar, S., Culler, D.E., Popa, R.A.: MAGE: nearly zero-cost virtual memory for secure computation. In: 15th USENIX Symposium on Operating Systems Design and Implementation (OSDI 21), pp. 367–385. USENIX Association (2021). https://www.usenix.org/conference/osdi21/presentation/kumar
37. Kushnir, E., Moshkowich, G., Shaul, H.: Secure range-searching using copy-and-recurse (2024). To appear in Proceedings on Privacy Enhancing Technologies (PETS) 2024

38. Lyubashevsky, V., Micciancio, D., Peikert, C., Rosen, A.: SWIFFT: a modest proposal for FFT hashing. In: Nyberg, K. (ed.) Fast Software Encryption. Springer, Berlin (2008)
39. MacKay, D.J.: Information Theory, Inference, and Learning Algorithms. Cambridge University Press, Cambridge (2003)
40. Nandakumar, K., Ratha, N., Pankanti, S., Halevi, S.: Towards deep neural network training on encrypted data. In: Proceedings of the IEEE/CVF Conference on Computer Vision and Pattern Recognition (CVPR) Workshops (2019). https://openaccess.thecvf.com/content_CVPRW_2019/html/CV-COPS/Nandakumar_Towards_Deep_Neural_Network_Training_on_Encrypted_Data_CVPRW_2019_paper.html
41. Paterson, M.S., Stockmeyer, L.J.: On the number of nonscalar multiplications necessary to evaluate polynomials. SIAM J. Comput. **2**(1), 60–66 (1973). https://doi.org/10.1137/0202007
42. Rendell, P.: Turing Universality of the Game of Life, pp. 513–539. Springer, London (2002). https://doi.org/10.1007/978-1-4471-0129-1_18
43. Shaul, H., Feldman, D., Rus, D.: Secure k-ish nearest neighbors classifier. Proc. Privacy Enhancing Technol. **2020**(3), 42–61 (2020). https://doi.org/10.2478/popets-2020-0045
44. SymPy: A python library for symbolic mathematics (2023). https://www.sympy.org/en/index.html
45. Trama, D., Clet, P.E., Boudguiga, A., Sirdey, R.: A homomorphic AES evaluation in less than 30 seconds by means of TFHE. In: Proceedings of the 11th Workshop on Encrypted Computing & Applied Homomorphic Cryptography, WAHC '23, pp. 79–90. Association for Computing Machinery, New York (2023). https://doi.org/10.1145/3605759.3625260
46. Zhou, L., Wang, Z., Zhang, X., Yu, Y.: HEAD: an FHE-based outsourced computation protocol with compact storage and efficient computation. IACR Cryptol. ePrint Arch. p. 238 (2022). https://eprint.iacr.org/2022/238

Part II
Approximations

Approximation theory is a wide and well-established branch of mathematics which studies the process of approximating given functions by using other functions that are simpler to compute for various reasons. The following chapters will focus on the considerations and techniques used when applying various approximation methods in fully homomorphic encryption (FHE). Chapter 5 describes general approximation methods, and Chap. 6 applies these methods and other more targeted methods to approximate specific functions that are commonly used by modern analytics.

Chapter 5
Approximation Methods Part I: A General Overview

Abstract The arithmetic of modern fully homomorphic encryption (FHE) schemes is basically limited to addition, subtraction, and multiplication, so that most computations are actually evaluations of some finite (possibly multivariate) polynomial over the input ciphertexts. Thus, FHE computations that involve functions that cannot be expressed with a finite polynomial (e.g., division, square roots, or trigonometric functions) must be approximated with some finite polynomial, and finding the "best" estimating polynomial and the "best" way to evaluate it is part of the art of developing FHE applications. Section 4.8 described and compared different methods for evaluating polynomials under FHE. This section describes some general aspects that should be considered when looking for the best polynomial to approximate a given targeted function under FHE and then proceeds to describe some specific methods used for finding such polynomial approximations for general functions. The next chapter (Chap. 6) will then show how to use these and other more targeted techniques to approximate specific useful functions, such as comparisons, reciprocals, trigonometric functions, and activation functions.

5.1 Approximating Functions with Finite Polynomials Under FHE

This section deals with the problem of finding a polynomial that accurately approximates a function in some required domain, where the function cannot be directly computed under FHE. The quality of a polynomial approximation can be measured in terms of how closely it approximates the targeted function (e.g., the maximal or mean absolute approximation error in the target domain) and in terms of the performance of their evaluation process.

When the form of the approximated function is complex or includes many details over a wide approximation domain, the polynomial may need to be of a high degree, because usually a polynomial of higher degree has more potential for capturing such complexity. However, in some cases, raising the degree of the polynomial actually increases the estimation error. For example, as will be seen in the next section,

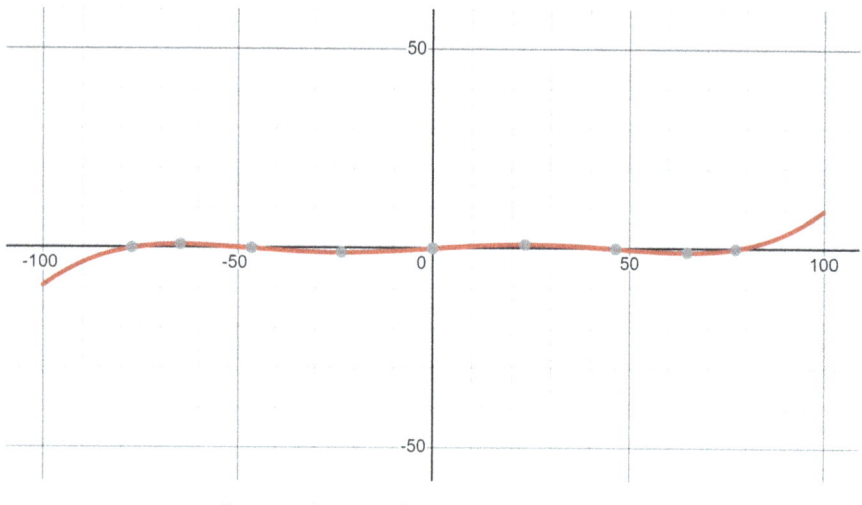

Fig. 5.1 $P(X) = -\dfrac{X^{11}}{3 \cdot 10^{20}} + \dfrac{X^9}{10^{16}} - \dfrac{X^7}{8 \cdot 10^{11}} + \dfrac{X^5}{9 \cdot 10^7} - \dfrac{X^3}{2 \cdot 10^4} + \dfrac{X}{15}$

the estimation accuracy of Taylor polynomials may deteriorate with higher degrees when estimating outside the "radius of convergence" (see Sect. 5.2 and Fig. 5.2b).

In addition, high-degree terms raise numerical stability considerations (see Sect. 4.8.6). If the polynomial aims to estimate the function for high x values, then the power of x in the polynomial term of the highest degree may rise to very high values indeed.

Example 5.1 Consider the polynomial $P(X)$ that roughly approximates the function $f(x) = \sin\left(\frac{x}{15}\right)$ in the domain $x \in [-100, 100]$ that is illustrated in Fig. 5.1. Its highest term degree is 11 and when evaluated at the point $x = 100$ this term would have a value of 10^{22} (a 72-bit number that does not fit in standard 64-bit containers).

Presumably, the approximated function does not reach such high magnitudes, and so it follows that the coefficient must be extremely minute in order to balance the magnitude of the huge value of the last term. In the example above, the coefficient of X^{11} is $\frac{-1}{3 \cdot 10^{20}}$. Direct computation under the IEEE standard for floating-point arithmetic, of such a product between a very small and a very large value, may thus fail to give accurate results. This difficulty is exacerbated when such a product is carried out under FHE which is less flexible in the scales of the represented values. Moreover, homomorphic noise is also accumulated in these high-valued, high-degree terms. When $x < 1$, the high-order terms merely add some small correction to the estimate, and at some point, this correction becomes small enough to be completely masked by the numeric or homomorphic errors. Finally, high degrees imply a high product depth which may require costly FHE bootstrap operations. So as a rule of thumb, for FHE applications, the degree of the estimating polynomials should generally be kept low if possible.

For x values going to infinity or to negative infinity, the corresponding values of a finite polynomial also go to infinity or to negative infinity (depending on the sign and parity of the high-order term). This means that a finite polynomial can usually only estimate a function over a limited domain of x values, which is not too far from the origin. If one wants to cover a wider domain of inputs and still keep the degree of the estimating polynomial low, then it is possible to use a "piecewise polynomial" based on a sequence of different polynomials that estimate different input domains, and decide which polynomial to use at evaluation time, depending on the input. However, such a solution is usually not possible in FHE applications because during the homomorphic computation, we do not know the value of x, and so we cannot choose the correct estimating polynomial.

In some cases, the domain of expected inputs can be determined ahead of time from the nature of the input data (say water temperature), but otherwise one approach could be to ask the user to provide the expected domain of input values to the computing server, either during setup or at every invocation. However, such a solution may violate the privacy of the user's data. A better approach could be to ask the user to pre-process the input data and normalize it to fit the given supported input domain of the estimating polynomial. This would be fine in the common case where the unit of the input feature is not important. In any case, extending the estimation domain (whether via direct approximation in the wider domain or via input normalization) usually adds more details to the form that needs to be approximated, which implies a need for a higher degree in the approximating polynomial, and a corresponding rise in the performance costs.

Example 5.2 Suppose that we directly approximate $f(x) = \sin(x)$ in $[0, k]$ where $k > 1$ using the Remez Algorithm (as will be described in Sect. 5.3). The shape of $f(x) = \sin(x)$ in the domain $[0, k]$ becomes more complex as k rises, in the sense that there are more periods of sin to estimate in the targeted domain, which would require a higher degree in the approximating polynomial. We may alternatively approximate $f(x) = \sin(kx)$ by the method of interpolating via Chebyshev nodes in the domain $[0, 1]$ (as will be described in Sect. 6.4.2), and then use the resulting polynomial to approximate $f(x) = \sin(x)$ in $[0, k]$ by first dividing x by k. However, $f(x) = \sin(kx)$ has the same high number of details that need to be approximated in $[0, 1]$ as $f(x) = \sin(x)$ has in $[0, k]$, which would require more interpolation points and again result in a similarly high degree in the estimating polynomial.

5.2 Estimating a Function with a Taylor Polynomial

Perhaps the best known method to estimate a function with a polynomial is to use a Taylor polynomial of a desired degree n. This method considers a polynomial $P(X)$ of degree n to be a good estimate of the function $f(x)$ if P and f share the same value and the same values of the first n derivatives at some given point a. For example, the following Taylor polynomial has the same value and the same first, second, and third derivatives as f at the point a:

$$P(X) = f(a) + \frac{f'(a)}{1!}(X-a) + \frac{f''(a)}{2!}(X-a)^2 + \frac{f'''(a)}{3!}(X-a)^3 \qquad (5.1)$$

A Taylor polynomial is the finite prefix of the infinite Taylor series or Taylor expansion of a function $f(x)$ at the point a. Thus, a Taylor series can only be defined for functions that are infinitely differentiable at that point.

The nice thing about estimating with a Taylor polynomial is that for many useful functions $f(x)$, a Taylor polynomial $P(X)$ with a high enough degree estimates $f(x)$ well not only at the point a where it shares f's derivatives but also at a whole continuous domain of x values around a (i.e., at an open interval containing a).

Example 5.3 The function $f(x) = \frac{1}{1-x}$ can be approximated in the domain $(-1, 1)$ with a finite prefix $P(X)$ of the Taylor series expanded at the point $x = 0$:

$$\sum_{n=1}^{\infty} X^n$$

The approximation continues to improve in this domain when $P(X)$ is expanded to higher and higher degrees; see Fig. 5.2a. In fact, the limit of this Taylor series converges to $f(x)$ in this domain when n goes to infinity. A function that equals the limit of its Taylor series in a given domain like $\frac{1}{1-x}$ is called analytic function in that domain.

However, the Taylor polynomials for $f(x) = \frac{1}{1-x}$ actually diverge further away from $f(x)$ to infinity beyond the "radius of convergence," and the divergence increases when the Taylor polynomials are raised to higher degrees. See Fig. 5.2b.

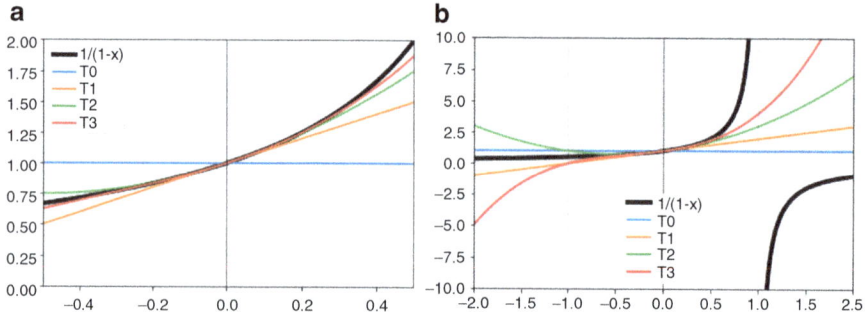

Fig. 5.2 Taylor polynomials for $\frac{1}{1-x}$. (**a**) Taylor polynomials of degrees 0, 1, 2, and 3 approximating $\frac{1}{1-x}$ with improving accuracy when inside the radius of convergence ($x \in (-1, 1)$). (**b**) Taylor polynomials of degrees 0, 1, 2, and 3 approximating $\frac{1}{1-x}$ with deteriorating accuracy when outside the radius of convergence ($x \notin (-1, 1)$)

In fact, for some functions, the radius of convergence is 0, i.e., the Taylor series does not converge to the function anywhere except at the point a itself. Nevertheless, for many useful functions such as e^x, $\sin(x)$, and $\cos(x)$, the Taylor series converges to the approximated function for every x, and the accuracy of the estimating Taylor polynomial always improves with higher degrees, at least until numeric and homomorphic noise start taking over.

Even when the Taylor series converges everywhere, the finite Taylor polynomials do not provide uniform accuracy for all inputs, and the accuracy usually deteriorates as the distance between the input and the Taylor series' expansion point increases. Furthermore, a Taylor polynomial is usually not the most accurate approximation possible given the required input domain and polynomial degree. The next section describes how such an optimal approximation can be found.

5.3 Minimax Polynomial Approximation and the Remez Algorithm

A minimax polynomial approximation $P(X)$ with degree at most n of a function $f(x)$ in the domain $[a, b]$ is the polynomial that minimizes the maximal approximation error across the entire domain among all polynomials of degree at most n, i.e.,

$$\arg\min_{\substack{P(X) \\ \deg(P) \leq n}} \max_{a \leq x \leq b} |f(x) - P(X = x)| \tag{5.2}$$

Interestingly, there exists a unique minimax polynomial of degree at most n for every continuous function in a closed domain [4].

For some $P(x)$ approximating $f(x)$ in the domain $[a, b]$, let $r(x)$ be defined as the estimation error, i.e., $r(x) = f(x) - P(x)$, and let E be the maximal estimation error in the domain $E = \max_{a \leq x \leq b} |r(x)|$. Chebyshev's Equioscillation theorem claims that $P(X)$ is a minimax polynomial approximation of $f(x)$ in this domain among all polynomials of degree at most n if and only if there exist $n+2$ "extreme" points $a \leq x_0 < x_1 < \ldots < x_{n+1} \leq b$ such that $r(x_i) = -r(x_{i-1}) = \pm E$ for $1 \leq i \leq n+1$. In other words, the minimax polynomial $P(X)$ oscillates back and forth along the curve of $f(x)$ such that the maximal distances in all the oscillations are the same. See an illustration in Fig. 5.3.

The Remez algorithm [5] is an iterative algorithm that searches for the minimax polynomial $P(X)$ given the function $f(x)$, the maximal degree n of $P(X)$, and the input domain in which the estimation is required $[a, b]$. The algorithm starts by selecting $n+2$ points in the domain $[a, b]$, and at every iteration, the algorithm adjusts these points so that they converge closer and closer to the $n+2$ extreme points of the equioscillation theorem. At every iteration, the algorithm finds a polynomial P that has a fixed oscillating difference from f at the current set of

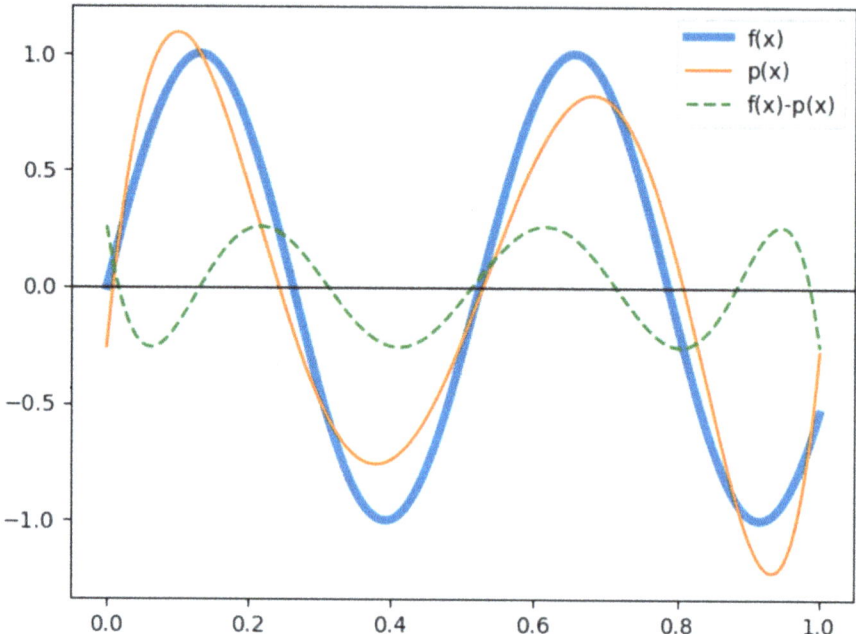

Fig. 5.3 $P(X)$ is the minimax polynomial of degree $n = 6$ for $f(x) = \sin(12x)$ in the domain $[0, 1]$. $P(X)$ oscillates around $f(x)$ so that the difference $r(x) = f(x) - P(x)$ reaches the maximal error *eight* times in this domain, including at $x = 0$ and $x = 1$, and with alternating error signs

$n + 2$ points. P is similar to the targeted minimax polynomial, which according to the equioscillation theorem also has such fixed oscillating differences, except for the crucial difference, that the fixed difference in the case of the minimax polynomial is not arbitrary, but is equal to the polynomial's maximal error.

Both f and P are continuous functions, so the difference between them (r) must cross the x axis $n + 1$ times (i.e., between each pair of points in which r oscillates). These $n + 1$ crossing points divide the domain $[a, b]$ into $n + 2$ segments. The algorithm now finds the $n + 2$ extreme points of r in these $n + 2$ segments (i.e., the points where r is maximal in the positive segments or where r is minimal in the negative segments). Subsequently, these $n + 2$ extreme points are used as the basis for the next iteration, etc. The algorithm stops when the magnitudes of the extreme values of r at the current set of $n + 2$ extreme points are close enough to each other to justify the conclusion that Chebyshev's equioscillation condition holds for these points and that the minimax polynomial approximation has been found. See further details in Algorithm 5 that includes the Remez algorithm.

5.3 Minimax Polynomial Approximation and the Remez Algorithm

Algorithm 5: Remez algorithm

Input: A domain $[a, b]$, a function $f(x)$ that is continuous in this domain, a maximal degree n for the required estimating polynomial, and an approximation parameter γ.

Output: The minimax approximation polynomial $P(X)$.

1 Select points $x_1, x_2, \ldots, x_{n+2} \in [a, b]$ in strictly increasing order.
2 Find the n degree polynomial $P(X) = \sum_{i=0}^{n} c_i X^i$ for which $P(x_i) - f(x_i) = -1^i E$ for some E.
3 In each of the $n+1$ intervals (x_i, x_{i+1}), find the point z_i for which $P(z_i) = f(z_i)$.
   // z_i must exist because $P(x_i) - f(x_i)$ is continuous and crosses the
   $x - axis$ in the interval (x_i, x_{i+1}) due to the way that $P(X)$ was
   constructed in step 2.
4 Divide the interval $[a, b]$ into the $n+2$ sections $[a, z_1], [z_1, z_2], \ldots, [z_{n+1}, b]$ and for every section S_i find a point y_i such that $y_i = \max_{x \in S_i} |P(X = x) - f(x)|$. That is, y_i is the point in which the magnitude of the error is the largest in the section.
5 $\epsilon_{max} \leftarrow \max_i |P(y_i) - f(y_i)|$
6 $\epsilon_{min} \leftarrow \min_i |P(y_i) - f(y_i)|$
7 **if** $(\epsilon_{max} - \epsilon_{min})/\epsilon_{min} < \gamma$ **then**
8 $\quad$ return $P(x)$
9 **else**
10 $\quad$ Replace x_i with y_i for all i and go to step 2 above.
11 **end**

Some notes on the Remez algorithm:

- There are several heuristics for selecting the initial points in Step 1, which expedite the convergence. However, even with an arbitrary initialization, the Remez algorithm is guaranteed to converge to the minimax estimation polynomial $P*$ according to the following inequality [4]:

$$\|P_k - P * \|_\infty \leq A\theta^k, \qquad (5.3)$$

where P_k is the polynomial of the k'th iteration, A is a nonnegative constant, and $0 < \theta < 1$.

- The Remez algorithm can usually be executed un-encrypted and in advance of any use of the resulting minimax polynomial under FHE, so performance considerations such as the rate of convergence are much less important than, for example, the quality of the resulting polynomial.
- Step 2 involves solving a system of $n+2$ linear equations (one for every point x_i) with $n+2$ variables (the $n+1$ coefficients c_i and E). This system of equations is guaranteed not to be singular by Haar's condition [4], and thus a solution exists and is unique.
- There are several variants for the above basic Remez algorithm, including the following:
  - Different polynomial basis: The above discussion assumed that the polynomial is defined on a *power basis*, in the sense that it is a linear combination of

powers of x. The fact that such a basis satisfies Haar's condition was used in the above bullet to conclude the success of Step 2. There are other polynomial bases that also satisfy Haar's condition and that can thus also be used by the minimax polynomials for which the algorithm searches. For example, the set of Chebyshev polynomials of degrees 0 to n satisfy Haar's condition, and a polynomial defined on top of such a basis has the benefit that it can often be computed with better numerical stability. See the definition of Chebyshev polynomials and their relation to numerical stability in Sect. 6.4.

- Complex domain: This variant searches for polynomial approximations in the complex domain, i.e., where both the approximated function and the approximating polynomial take complex inputs [8].
- Multi-interval: This variant optimizes the polynomial in multiple disjoint intervals rather than in a single continuous interval $[a, b]$. In this multi-interval variant [2], there may be more than $n + 2$ extreme points over the multiple intervals at each iteration, and the algorithm shows how to select $n + 2$ points from among them so that the convergence continues. Section 6.1.1 discusses the multi-interval version in the context of the Sign function in the domain $[-1, -\epsilon] \cup [\epsilon, 1]$.

It may first seem that with the Remez algorithm, the search for the best estimating polynomial for any function would be over. After all, the algorithm is guaranteed to converge at the unique polynomial that has the least maximal error in the domain of interest and with a maximal degree that can be specified according to the maximal product count and depth that the user is willing to tolerate. See, for example, Fig. 5.4 for a comparison of the accuracy of the minimax polynomials of degree 6 estimating $\sin(12x)$ at various domains as returned by the Remez algorithm, with the corresponding accuracy of the Taylor polynomial of the same degree and in the same domains.

However, as mentioned above, accuracy is not the only quality measure for polynomial estimations. Polynomials that are not the minimax polynomial may, for example, be much faster to compute, or may incur fewer issues of numerical stability in their evaluation, or may lend themselves more easily to be expressible in special forms that are faster or more numerically stable to compute. For example, as will be seen in the next section, the number of non-scalar products can be dramatically lowered if it is possible to express the estimating polynomial as a composition of polynomials, possibly minimax polynomials, rather than as a single minimax polynomial of the same degree. The minimax polynomial is probably not itself expressible in the form of a composed polynomial. The composed polynomial will be much faster to compute than the minimax polynomial, but its maximal error would not be as low as the optimal error of the single minimax polynomial returned by Remez for the same degree. However, the difference in error can be made negligible or completely masked by the homomorphic error introduced by the FHE scheme. Another example is Goldschmidt's division algorithm for estimating the function $f(x) = \frac{1}{x}$, which will be discussed in Sect. 6.3.2. This method is able to compute efficiently and with good numerical stability expressions that are equivalent to very-high-degree polynomials, which would be impracticable to

5.4 Estimating a Function by Composing Multiple Polynomials

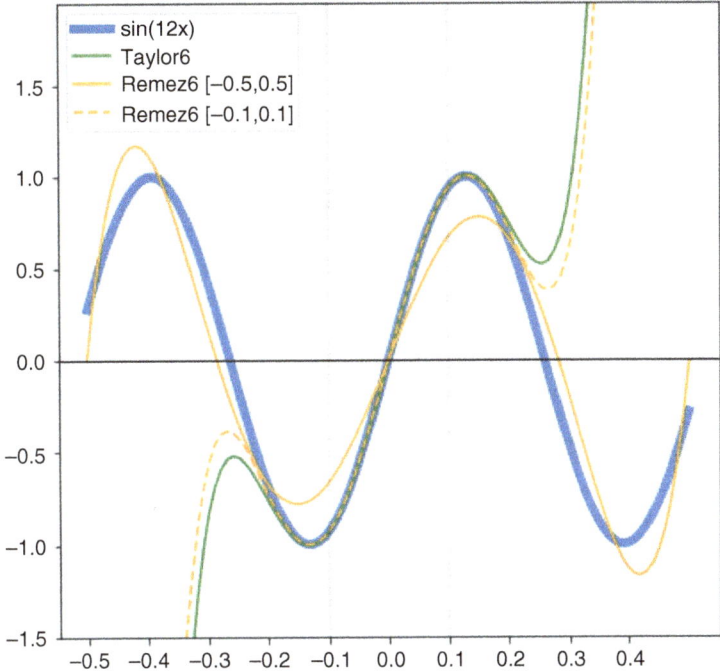

Fig. 5.4 The Taylor polynomial of degree 6 for sin(12x) expanded around $x = 0$ compared to the minimax polynomials of degree 6 returned by the Remez algorithm for the domains $[-0.5, 0.5]$ and $[-0.1, 0.1]$. The mean/max error of the minimax polynomial for the domain $[-0.5, 0.5]$ is 0.170/0.266 compared to a mean/max error of 5.104/35.079 for the Taylor series at the same domain. The mean/max error of the minimax polynomial for the domain $[-0.1, 0.1]$ is $6.586e^{-6}/1.036e^{-5}$ compared to a mean/max error of $9.166e^{-5}/5.23e^{-4}$ for the Taylor series at the same domain

estimate more accurately with the minimax polynomial of that high degree. Finally, recall that even if we decide to use the minimax polynomial, we could still improve its practical accuracy by expressing it in the form of a product of differences from the roots as presented in Sect. 4.8.5, which can improve the numerical stability of the evaluation process.

5.4 Estimating a Function by Composing Multiple Polynomials

Consider the polynomial function

$$f(X) = 256X^9 + 576X^8 + 816X^7 + 924X^6 + 768X^5 \qquad (5.4)$$
$$+ 519X^4 + 292X^3 + 120X^2 + 40X + 10$$

which is decomposable as $f(X) = P(P(X))$ for

$$P(X) = 4X^3 + 3X^2 + 2X + 1 \qquad (5.5)$$

If, for example, the polynomial evaluation method uses exponentiation by squaring (see Sect. 4.8.3), then evaluating Eq. 5.4 would take eight non-scalar products, nine scalar products, and a product depth of $\lceil log_2(9) \rceil = 4$. In contrast, evaluating $P(X)$ would take only two non-scalar products, three scalar products, and the product depth of 2. So, evaluating $f(X)$ as $P(P(X))$ would take only four non-scalar products, six scalar products, and the same product depth of 4.

In general if $f(X)$ can be decomposed as $f(X) = P_1(P_2(\ldots P_m(X)\ldots))$ where every P_i is a polynomial of degree n, then $f(X)$ is thus equal to a polynomial of degree n^m. If the polynomial evaluation method uses exponentiation by squaring, then evaluating $f(X)$ directly would require $n^m - 1$ non-scalar products and n^m scalar products and a product depth of $\lceil log_2(n^m) \rceil = \lceil m log_2 n \rceil$. On the other hand, evaluating the composed polynomials would only require $(n-1)m$ non-scalar products, nm scalar products, and the comparable product depth of $m \cdot \lceil log_2 n \rceil$.

Evaluating $f(X)$ via its decomposition requires many fewer multiplications, even when we evaluate all the polynomials using Paterson and Stockmeyer's method (Sect. 4.8.4). Recall that evaluating a polynomial of degree n with Paterson and Stockmeyer's method requires $\sim \sqrt{2n} + log_2 n$ non-scalar products, $n - \sqrt{\frac{n}{2}}$ scalar products, and a product depth of $\lceil log_2 n \rceil$. Thus, using Paterson and Stockmeyer's method directly on $f(X)$ represented as a polynomial of degree n^m would require $\sim \sqrt{2n^m} + log_2 n^m$ non-scalar products, $n^m - \sqrt{\frac{n^m}{2}}$ scalar products, and the same product depth as before of $\lceil m log_2 n \rceil$. On the other hand, the number of non-scalar products required when evaluating the composed polynomials would only be $\sim (\sqrt{2n} + log_2 n)m = \sqrt{2nm^2} + log_2 n^m$. The number of scalar products also drops to $nm - \sqrt{\frac{n \cdot m^2}{2}}$, but the depth remains as before $(m \cdot \lceil log_2 n \rceil)$.

Unfortunately, most polynomials cannot be decomposed in this way and so cannot benefit from the improvement in the number of multiplications [6]. Even when a polynomial is in fact decomposable, it may be difficult to actually find a decomposition. However, in the case of polynomial approximations, we generally do not need to decompose a specific given polynomial $g(X)$, but rather we try to find a polynomial $g(X)$ that approximates another function $f(X)$. Thus, it may make sense to find such an approximation $g(X)$ that is also decomposable. Luckily, this happens naturally when $g(X)$ is found using an approximation algorithm that includes repeated iterations.

Example 5.4 The work of [3] shows how to estimate $f(x) = \frac{1}{\sqrt{x}}$ under FHE using a fixed number of iterations of the Newton-Raphson method. The latter is a method for searching for the roots of a function when the function is differentiable in the search domain. The process starts with an initial guess for the root, say r_0, and then at each subsequent iteration, it computes an improved guess of $r_{n+1} = r_n - \frac{g(r_n)}{g'(r_n)}$.

5.4 Estimating a Function by Composing Multiple Polynomials

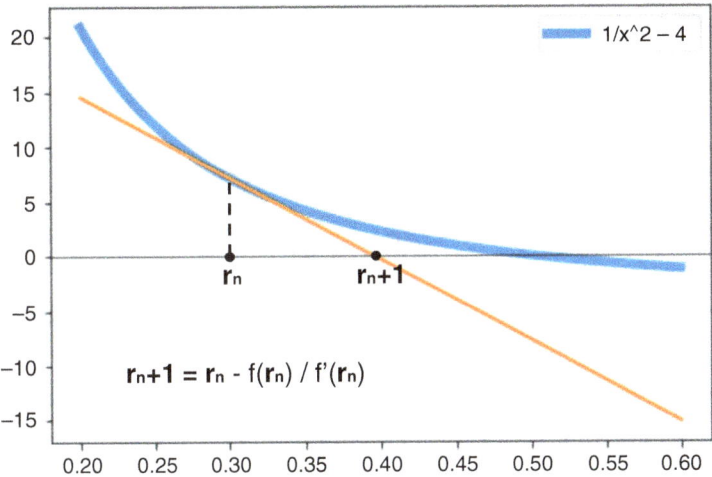

Fig. 5.5 Estimating $\frac{1}{\sqrt{4}}$ by searching for the root of $f(x) = \frac{1}{x^2} - 4$ with the Newton-Raphson method. The derivative at $x = r_n$ is $f'(r_n) = \frac{f(r_n)}{(r_n - r_{n+1})}$ so $r_{n+1} = r_n - f(r_n)/f'(r_n)$, where r_{n+1} is closer to the root than r_n. If we start with an initial guess of $r_0 = 0.3$, then the next estimates are $r_1 = 0.396$, $r_2 = 0.4698$, and $r_3 = 0.4973$, which is very close to the actual root at $x = 0.5$, which is also the value of $\frac{1}{\sqrt{4}}$

Under certain conditions, and a good initial guess, this process is guaranteed to quickly converge on a root of $f(x)$ [1].

Suppose we wish to estimate $\frac{1}{\sqrt{v}}$ where v is the (unknown) value encrypted by a given ciphertext. We then consider the function $g(x) = \frac{1}{x^2} - v$. If r is an estimate for the positive root of $g(x)$, then r is also an estimate for $\frac{1}{\sqrt{v}}$. The Newton-Raphson method can now be used to estimate r by iterations starting with an initial encrypted guess r_0, and then each iteration computes the next estimate as $r_{n+1} = r_n - \frac{g(r_n)}{g'(r_n)} = \frac{r_n}{2}(-v r_n^2 + 3)$; see Fig. 5.5. This expression can be easily computed with three multiplications under FHE where both v and r_n are given as ciphertexts.

Note that the Newton-Raphson iteration formula $\frac{g(r_n)}{g'(r_n)}$ does not always produce such an FHE-friendly expression for every function. In any case, if we decide to run, say, three iterations of Newton-Raphson, then $f(x) = \frac{1}{\sqrt{x}}$ can be approximated as the composition $P(x, P(x, P(x, r_0)))$ where $P(X, r) = \frac{r}{2}(-Xr^2 + 3)$. Note that coming up with an initial guess r_0 that would converge quickly enough for an unknown v is not a trivial matter, and a proposed method is described in detail in the original paper [3]. Another example for a function estimate computed by iterations represented as polynomial compositions is described in Sect. 6.1.1 which discusses estimations of the Sign function.

5.5 Lab Exercise: Implement and Test the Remez Algorithm

5.5.1 Taylor Series

1. Find the infinite Taylor series for $f(x) = \dfrac{1}{1+x}$ expanded at $x = 0$.
2. Let $T_n(x)$ be the prefix of the above Taylor series up to degree n. Compute the maximal error of $T_3(x)$, $T_4(x)$ and $T_5(x)$ in the domain $[-0.5, 0.5]$ and in the domain $[1.1, 1.5]$.
3. Plot the errors of T_{10}, T_{20}, and T_{30} in the domain $[-0.9, 1.1]$. What appears to be the radius of convergence of this Taylor series?

5.5.2 Remez

4. Implement the Remez algorithm. Note that the algorithm is not run under FHE, so any software environment can be used (e.g., Python).
   [Hint:] To simplify the implementation, you may assume that the algorithm will only be applied to the function e^x, as required by the next exercise.
5. Use your implementation to find the minimax polynomial of degree 3 estimating e^x in the domain $[0,1]$.
6. Generalize your Remez implementation to support all continuous functions. To do this, you will need to find the derivative of the given function in order to find its extreme points in the intervals defined in Step 4 of the algorithm.
7. Plot the function $\sin(x)$ and the approximation generated by the Remez algorithm in the domain $[-2, 2]$. Compare it with other approximation methods that standard libraries provide.
8. Train an neural network (NN) to approximate the $\sin(x)$ function in the domain $[-2, 2]$ using a polynomial approximation $P(x)$ of degree n. The NN should receive pairs $(x, \sin(x))$ as input and should output the coefficients of $P(x)$, where the loss function minimizes the maximal error $|P(x) - \sin(x)|$. Did you get the same polynomial you received using the Remez algorithm?
9. Repeat the above experiment but this time approximate the function $\text{ReLU}(x) = \max\{0, x\}$.
10. [*Hard*]: Generalize the Remez algorithm above to multi-interval Remez—see [2].

5.5.3 Polynomial Composition

11. Find the polynomial $P_3(x)$ that is the Taylor series of degree 3 for the function e^x expanded at x=0.

12. Expand $P_3(P_3(x))$ into a polynomial $P_9(x)$ of degree 9 that estimates $e^{(e^x)}$ at $x = 0$.
    [Hint:] You may use the compose operation from Python's *SymPy* library [7] for symbolic polynomial manipulation.
13. Evaluate $P_9(0.8)$ under FHE using Horner's rule. Count the product count (scalar and non-scalar) and depth of this evaluation.
14. Evaluate $P_3(P_3(x))$ under FHE where each evaluation of $P_3(x)$ uses Horner's rule. Count the total number of products (scalar and non-scalar) and the product depth of this evaluation.

References

1. Encyclopedia of mathematics: Newton method (2020). https://encyclopediaofmath.org/index.php?title=Newton_method
2. Lee, J.W., Lee, E., Lee, Y., Kim, Y.S., No, J.S.: High-precision bootstrapping of RNS-CKKS homomorphic encryption using optimal minimax polynomial approximation and inverse sine function. In: Canteaut, A., Standaert, F.X. (eds.) Advances in Cryptology – EUROCRYPT 2021, pp. 618–647. Springer International Publishing, Cham (2021). https://doi.org/10.1007/978-3-030-77870-5_22
3. Panda, S.: Polynomial approximation of inverse sqrt function for fhe. In: Dolev, S., Katz, J., Meisels, A. (eds.) Cyber Security, Cryptology, and Machine Learning, pp. 366–376. Springer International Publishing, Cham (2022). https://doi.org/10.1007/978-3-031-07689-3_27
4. Powell, M.J.D., et al.: Approximation Theory and Methods. Cambridge University Press, Cambridge (1981). https://doi.org/10.1017/CBO9781139171502
5. Remez, E.Y.: Sur la détermination des polynômes d'approximation de degré donnée. Comm. Soc. Math. Kharkov **10**(196), 41–63 (1934)
6. Rickards, J.: When is a polynomial a composition of other polynomials? Am. Math. Mon. **118**(4), 358–363 (2011). https://doi.org/10.4169/amer.math.monthly.118.04.358
7. SymPy: A python library for symbolic mathematics (2023). https://www.sympy.org/en/index.html
8. Tang, P.T.P.: A fast algorithm for linear complex chebyshev approximations. Math. Comput. **51**(184), 721–739 (1988)

Chapter 6
Approximation Methods Part II: Approximations of Standard Functions

Abstract Chapter 5 described some commonly used general approximation methods and their trade-offs when used under fully homomorphic encryption (FHE). The reader is now ready to learn how to apply these and more targeted methods for estimating specific functions that are commonly used by modern analytics. We will describe how these functions can be estimated both in integer and approximate FHE contexts and then discuss the performance of the described methods in terms of time and accuracy. The chapter will start with a description of various methods for comparing encrypted values under FHE, as required when computing conditional code, sorting, or finding the maximum among a set of values under FHE. Section 6.2 builds upon these methods to compute the ReLU activation function used in neural networks (NNs). Section 6.3 describes how to compute reciprocals, both in the approximate and integer cases. Finally, Sect. 6.4 describes Chebyshev polynomials and how they can be used in the context of interpolation and function estimation, specifically focusing on trigonometric functions.

6.1 Approximating Comparisons: Equality and Inequality and General and Tailored Versions

Comparison of two values is a very common operation in any application, including machine learning (ML) applications. For example, as will be shown below, comparisons can be used to compute max pooling and activation functions such as ReLU used in NNs. In addition, comparisons are used in any data science application that needs to query data from an encrypted private database where the record fields are compared to the value in the query.

Computations often split to different paths based on conditionals that compare values. Consider, for example, the following FHE computation, where a, b, and c are ciphertexts:

```
if (a>b) then
    c = a
else
    c = b
```

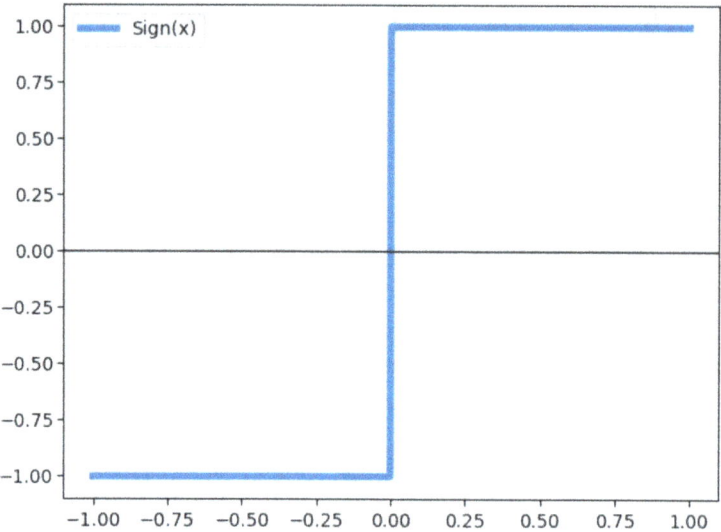

Fig. 6.1 A Sign function in the range $[-1, 1]$

Recall that Sect. 4.1.1 explained why conditional branches, such as the one implied by the above code, cannot be securely implemented under FHE using a process that only computes the correct branch. Instead, one must compute some intermediate control value based on the condition and use that to compute the function implied by both branches of the conditional code. For that the function

$$\text{Sign}(x) = \begin{cases} \frac{x}{|x|} & x \neq 0 \\ 0 & otherwise \end{cases} = \begin{cases} 1 & \text{if } x > 0 \\ 0 & \text{if } x = 0 \\ -1 & \text{if } x < 0 \end{cases} \quad (6.1)$$

is often used. It is illustrated in Fig. 6.1.

If we can compute Sign under Cheon, Kim, Kim, and Song (CKKS) (as we shall see in the next section), then we can also implement the above conditional code using the following equivalent code for computing $\max(a, b)$:

$$\max(a, b) = \frac{(a+b) + (a-b)\text{Sign}(a-b)}{2} \quad (6.2)$$

Similarly, $\min(a, b)$ can be computed as

$$\min(a, b) = \frac{(a+b) - (a-b)\text{Sign}(a-b)}{2} \quad (6.3)$$

The above expression can also be used to implement the code of Sect. 4.1.1 as $\min(x, y)^2$. Similarly, $\text{Sign}(x)$ can be used in other cases of conditional compu-

6.1 Approximating Comparisons: Equality and Inequality and General and...

tations. For example, to compute the compare function:

$$\text{Eq}(x, y) = \begin{cases} 1 & \text{if } x > y \\ 1/2 & \text{if } x = y \\ 0 & \text{if } x < y \end{cases} = \frac{\text{Sign}(x - y) + 1}{2}$$

A different approach for comparing two values is discussed in Sect. 6.1.2 where the two values are represented in a bitwise form (see Example 2.5) and then compared bit by bit.

6.1.1 Estimating the Sign Function by Composing Polynomials

From Chap. 5 we know that Sign can be estimated by using the Remez algorithm to find the minimax approximation polynomial $P(X)$ in the range $[-1, 1]$ at some required maximal degree. If some other input domain is required, say $[-B, B]$, then $\text{Sign}(x)$ can be estimated by scaling x down with $P(\frac{x}{B})$. Note though that P is bound to be inaccurate in some hopefully small region around 0 where Sign is discontinuous, and down-scaling x in the above manner would make this inaccurate region B times wider.

As the Sign function is discontinuous at 0, it makes sense to search for a minimax polynomial in a split domain $[-1, -\epsilon] \cup [\epsilon, 1]$ that avoids a small area around 0 (as in [11]). The objective then is to find a polynomial $P(X)$ that is (α, ϵ)-close to $f(x)$, that is, $|P(x) - f(x)| \leq 2^{-\alpha}$ for all $x \in [-1, -\epsilon] \cup [\epsilon, 1]$.

Figure 6.2a shows the minimax polynomial $P_0(X)$ of degree 7 for $\text{Sign}(x)$ in the range $[-1, -0.1] \cup [0.1, 1]$. The maximum error of $P_0(X)$ relative to $\text{Sign}(x)$ in this range is 0.34, so $P_0(X)$ is $(\alpha = -\log_2(0.34) \approx 1.55, \epsilon = 0.1)$-close to $\text{Sign}(x)$. P_0 is not a very good approximation of Sign as seen by the figure and by the values α and ϵ. We can improve the estimation accuracy by raising the maximal allowed polynomial degree specified to the Remez algorithm. However, using a more accurate minimax polynomial of higher degree would come at a cost of more non-scalar products and at a higher product depth.

In Sect. 5.4 we saw that the number of non-scalar products can be reduced dramatically if we are able to express the polynomial as a composition of lower degree polynomials. The methods reported in [5] and [9] use such composed polynomials for estimating the Sign function, and the following discussion follows the method introduced in the latter work.

As mentioned above, instead of improving the accuracy of P_0 by raising its degree, it would be preferable to compose one or more polynomials to it. Thus, we are looking for another polynomial $P_1(X)$ such that $P_1(P_0(X))$ will more accurately estimate $\text{Sign}(x)$, In other words, it will be (α, ϵ)-close to Sign with a higher α.

We can again use the Remez algorithm to find P_1. Note that P_0 needs to be the minimax polynomial on the original domain $D_0 = [-1, -\epsilon] \cup [\epsilon, 1]$, whereas P_1

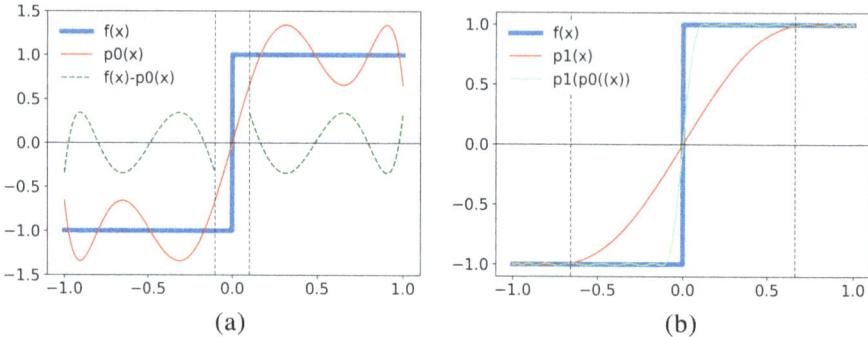

Fig. 6.2 A composite of two polynomials estimating Sign(x). (**a**) $P_0(X)$ (orange solid line) is the 7 degree minimax estimation polynomial for Sign(x) in the split domain $[-1, -0.1] \cup [0.1, 1]$. Its range (on the Y axis) in this domain is $[-1.34, -0.66] \cup [0.66, 1.34]$. The figure also shows the equi-oscillating error (green dashed line) of this estimation typical of a minimax polynomial. (**b**) $P_1(X)$ (orange solid line) is the 7 degree minimax estimation polynomial for Sign(x) in the split domain $[-1.34, -0.66] \cup [0.66, 1.34]$. The figure also shows the composite polynomial $P_1(P_0(X))$ (green line), which is already a pretty close approximation of the Sign function

needs to be the minimax polynomial on the function range of P_0 when applied to the function domain D_0.

Example 6.1 In Fig. 6.2a P_0 is the minimax polynomial on the domain $D_0 = [-1, -0.1] \cup [0.1, 1]$ and results in the range $D_1 \approx [-1.34, -0.66] \cup [0.66, 1.34]$, so P_1 just needs to approximate Sign in D_1. This is an easier domain to estimate in because it is further away from the problematic discontinuous area of 0. As seen in Fig. 6.2b, the minimax 7-degree polynomial $P_1(X)$ estimates Sign quite accurately in D_1. Moreover, the figure also shows that the composed polynomial $P(X) = P_1(P_0(X))$ also estimates Sign quite accurately in the original range D_0. $P(X)$ is equivalent to a polynomial of degree $7 \times 7 = 49$ and indeed requires a similar product depth of $\lceil \log_2(49) \rceil = 6$ but with many less non-scalar products, as seen in Sect. 5.4.

We can continue to compose more minimax polynomials in the same manner. That is, each new polynomial is a minimax polynomial that can be found using the Remez algorithm on the range that results from the previous polynomial in the composition. At each iteration, the composition has a smaller output range around 1, and the next polynomial in the composition only needs to give a good estimate in this reduced range, until a satisfactory accuracy is reached, i.e., the composed polynomial is (α, ϵ)-close to Sign with a high enough α for the original domain. This is guaranteed to happen after enough compositions, but the question is, how many compositions are needed and what should their polynomial degrees be in order for the resulting composed polynomial to be (α, ϵ)-close to Sign with the optimal number of non-scalar products or with the lowest multiplication depth required.

The work of [9] indeed proves that a composition of minimax polynomials achieves the required accuracy with an optimal number of non-scalar products. The paper then proceeds to use dynamic programming to search for the optimal number of compositions and corresponding polynomial degrees for the optimal composition for various α parameters (and taking $\epsilon = 2^{-\alpha}$).

Example 6.2 It turns out that to get an accuracy with $\alpha = 7$, one would need to compose two minimax polynomials of degree 3 followed by three minimax polynomials of degree 5. If the target is to optimize the product depth rather than the product count, then one would need to compose two polynomials of degree 7 followed by a minimax polynomial of degree 13. The full table of optimal compositions for optimizing product count or depth for a range of α values is given in the original paper [9].

6.1.2 Comparisons: Equality and Inequality of Integers

The methods for comparing two values using estimating polynomials for Sign as described in the above section assume an FHE scheme that supports approximate arithmetic such as CKKS. When the compared values are integers, other approaches can be used to perform the comparison as described below.

- **Computing equality of two bitwise values with arithmetic mod p = 2**: FHE schemes such as FHE over the torus (TFHE) or Brakerski, Gentry, Vaikuntanathan (BGV) with arithmetic mod $p = 2$ support the bitwise representation of integer values—see Example 2.5. Assume that the compared n bit values a and b are represented as two arrays of n ciphertexts $c_a[0 : n − 1]$ and $c_b[0 : n − 1]$ each encrypting one bit of the value a, b, respectively, where array indices run from 0 to $n − 1$ and bit 0 is the most-significant bit (MSB).

    The following expression equals 1 if the two bit arrays are equal and is 0 otherwise. The expression can be computed with $n − 1$ multiplications at a product depth of $\lceil \log_2(n) \rceil$:

$$\text{Eq}(c_a, c_b) = \prod_{i \in [0, n-1]} (c_a[i] \oplus c_b[i] + 1) \bmod 2 \quad (6.4)$$

- **Computing equality of two hybrid bitwise values with arithmetic mod $p > 2$ or with approximate arithmetic:** Assume as before that the compared n bit values a and b are represented as arrays of n ciphertexts. The values are hybrid bitwise. That is, when the arithmetic is mod $p > 2$ (as possible for example with BGV), each bit is a ciphertext holding either 0 or 1. When the arithmetic is approximate floating-point arithmetic (e.g., as possible with CKKS), each bit is a ciphertext holding either ~ 0.0 or ~ 1.0. In both these cases, Eq. 6.4 would no longer work because the arithmetic is no longer mod 2.

The following expression equals 1 if the two bit arrays are equal and is 0 otherwise. The expression can be computed with n multiplications at a product depth of $\lceil \log_2(n) \rceil + 1$:

$$\text{Eq}(c_a, c_b) = \left(\prod_{i \in [0, n-1]} (c_a[i] \oplus c_b[i] - 1) \right)^2 \tag{6.5}$$

- **Computing equality of two integer values with arithmetic mod $p > 2$ where p is prime:** Assume we are working with arithmetic mod $p > 2$, where p is prime, as possible, for example, with BGV. The following expression equals 1 if the two integer values are equal and is 0 otherwise. It can be computed using exponentiation by squaring with $\lceil \log_2(p-1) \rceil$ multiplications at a product depth of $\lceil \log_2(p-1) \rceil$:

$$\text{Eq}(a, b) = 1 - (a-b)^{p-1} \bmod p \tag{6.6}$$

The formula relies on Fermat's little theorem, which implies that if p is a prime and x is any integer not divisible by p, then

$$x^{p-1} = 1 \bmod p. \tag{6.7}$$

- **Computing whether one bitwise or hybrid bitwise integer is greater than another:** Assume as before that the compared n bit values a and b are represented as arrays of n ciphertexts $c_a[0:n-1]$ and $c_b[0:n-1]$, respectively. $c_a[i:j]$ is the sub-array of bits i to j inclusive. The values are either bitwise with arithmetic mod $p = 2$ (as possible in schemes such as TFHE or BGV) or hybrid bitwise with arithmetic mod $p > 2$ (as possible with BGV) or hybrid bitwise with approximate floating-point arithmetic (as with CKKS).

In all these cases, the following expression equals 1 if $a > b$ and 0 otherwise:

$$\text{isGreaterThan}(c_a, c_b) = c_a[0](1 - c_b[0]) \oplus$$

$$\sum_{i=1}^{n-1} c_a[i](1 - c_b[i]) \text{Eq}(c_a[0:i-1], c_b[0:i-1]) \tag{6.8}$$

The $\text{Eq}(c_a, c_b)$ part of the above expression is taken from Eq. 6.4 if a and b are bitwise values with arithmetic mod $p = 2$ or from Eq. 6.5 if they are hybrid bitwise values with either arithmetic mod $p > 2$ or with approximate floating-point arithmetic. For the case of arithmetic mod $p = 2$, the expression for $a > b$ can be computed with $(n^2 - 1)/2$ multiplications at a product depth of $\lceil \log_2(n-1) \rceil$. For the case of arithmetic mod $p > 2$ or approximate arithmetic, the expression for $a > b$ can be computed with $(n^2 + 3n - 2)/2$ multiplications at a product depth of $\lceil \log_2 n \rceil + 2$. By clever use of caching intermediate products, one can further reduce the number of multiplications while maintaining the product depth.

Note that in all these possible representations and arithmetics, $a[i](1 - b[i])$ equals 1 if bit $a[i] > b[i]$ and is 0 otherwise. Let j be the smallest index for which $a[j] \neq b[j]$. If $j > 0$, then all the elements of the sum are 0 except the element where $i = j$, which equals 1 if $a > b$ and 0 otherwise. If $j = 0$, then the entire sum is 0 and $a[0](1 - b[0])$ equals 1 if $a > b$ and 0 otherwise. If $a \leq b$, then $a[0](1 - b[0])$ and the entire sum are all 0.

6.1.3 Approximating Max

The ability to select the maximal among a set of two or more values is useful for a variety of ML applications, including the following:

- **Max pooling.** Max pooling layers in NNs are used to downsample the data flowing through the network by replacing sets of values with their maximum. See detailed treatment of this use case in Sect. 11.1.4.
- **XGBoost.** When training XGBoost [2] and other tree-based models, a node in the trained tree is split according to the split that has the best score among many candidate splits.
- **Transformers.** A Transformer NN typically outputs many possible words along with their corresponding probability, and the most probable word is selected as the final prediction. If this selection is made as the final step of the network, then one acceptable solution could be to return the entire set of output words along with their encrypted probabilities to the user, who would proceed to decrypt the probabilities and select the most probable word. However, if this output serves as an input to the next generative step, then such reliance on the user may be problematic (see Sect. 3.3), and one would need to select the word with the maximal probability under FHE.

The maximum of two bitwise integer values a and b may be computed by using the method described in 6.1 to compute $GT = \text{isGreaterThan}(c_a, c_b)$ and then computing the maximum as $GT * c_a + (1 - GT) * c_b$. To compute the maximum of two real values, one can refer to the methods reported in [5, 9], and [10] that describe various ways of estimating $\max(a, b)$ based on $\text{Sign}(x)$ as in the following equation, where $\text{Sign}(x)$ is in turn estimated using composed polynomials (see Sect. 6.1.1):

$$\max(a, b) = \frac{(a + b) + (a - b) * \text{Sign}(a - b)}{2} \tag{6.9}$$

The above expression requires one additional non-scalar product with $1/2$ and the consequent increase of the product depth beyond the corresponding costs of the Sign computation.

Note that Eq. 6.9 can also be rewritten as

$$\max(a, b) = b + \frac{(a - b) * (1 + Sign(a - b))}{2} \quad (6.10)$$

This form suggests that the additional costs due to the multiplication with $1/2$ can be avoided if we modify the final composed polynomial of the $Sign(x)$ function by adding 1 to its free coefficient and then dividing all the coefficients by 2. If we call this modified composed polynomial $ModifiedSign(x)$, then Eq. 6.10 becomes

$$\max(a, b) = b + (a - b) * \text{ModifiedSign}(a - b) \quad (6.11)$$

Note that the modification of $Sign(x)$ described above is carried out in plaintext, ahead of this more efficient FHE computation.

Assume that computing Eq. 6.11 involves $maxM$ multiplications and a product depth of $maxD$. To compute the maximum of a set of more than two values, one can use repeated applications of the above methods for comparing pairs of values. Thus, to compute the maximum of four values would require two applications of Eq. 6.11 for each pair followed by a third final application among the two results. In general, to compute the maximal value among a set of n values where n is a power of 2, one could run a *Tournament* involving $n - 1$ applications of Eq. 6.11, which would involve $maxM * (n - 1)$ multiplications and a product depth of $maxD * \log_2 n$.

If the n values are all packed inside a single ciphertext, then one could exploit the single instruction multiple data (SIMD) character of the FHE operations in order to compute the maximal slot value more efficiently by using a method similar to the *rotate-and-sum* technique described in Sect. 7.2. The only difference is that instead of applying the SIMD $\oplus$ operation between the rotated ciphertexts (in Step 3 of Algorithm 7), we now need to apply Eq. 6.11 to the pair of ciphertexts. If the ciphertext includes $n = 2^m$ values in its slots, then after m iterations of the algorithm, the first slot will contain the maximal value among them. So SIMD helps us reduce the number of multiplications from $maxM * (n - 1)$ to $maxM * \log_2 n$, at the same product depth of $maxD * \log_2 n$.

Another method for computing the maximal value among a set of values is to conduct a *League* where each of the n values is separately compared with each of the other values, and only a value that comes out the larger in all these comparisons wins the League. This is again carried out using repeated applications of Eq. 6.11 and appropriate manipulation of the resulting masks. This method would involve $maxM * n^2 + n$ multiplications at the product depth of $maxD + \log_2 n$ (including multiplications coming from n^2 comparisons at a depth of one comparison, plus the number of products and depth needed for the final mask manipulation). The work of [8] also describes an efficient way to combine Tournaments and Leagues to compute the maximal value of a set.

6.2 ReLU and Other Activation Functions

The most popular activation functions used in NNs, such as ReLU and its variants, Sigmoid and Softmax, are unfortunately not directly computable with the basic FHE operations and must therefore be handled by some other methods. Section 3.3 describes the approach of computing the activation function via communication with the client and how the final Softmax activation in a NN can also be handled with the assistance of the client. Here we focus on the approach of estimating the activation function with a polynomial.

The activation function can be approximated with a polynomial using one of the methods described above, such as using Taylor polynomials, minimax polynomials, or composed polynomials. See Fig. 6.3.

$\text{ReLU}(x) = \max(x, 0)$ and so $\text{ReLU}(x)$ can also be computed using the methods described in Sect. 6.1.3 for computing $\max(a, b)$, with the simplifying factor that b is a plain scalar 0 and not a ciphertext. For example, the work of [10] describes how to estimate ReLU via an optimal composed polynomial for the $\text{Sign}(x)$ function. First, an (α, ϵ)-close composed polynomial is found for the required α and ϵ parameters and within the allowed degree. The result is guaranteed to be the optimal composed polynomial in terms of the number of non-scalar products or product depth (see Sect. 6.1.1 above). Then the ReLU function can be estimated according to the following equation, involving two additional products—one non scalar (with x) and one scalar (with $\frac{1}{2}$), beyond the corresponding costs of the Sign computation.

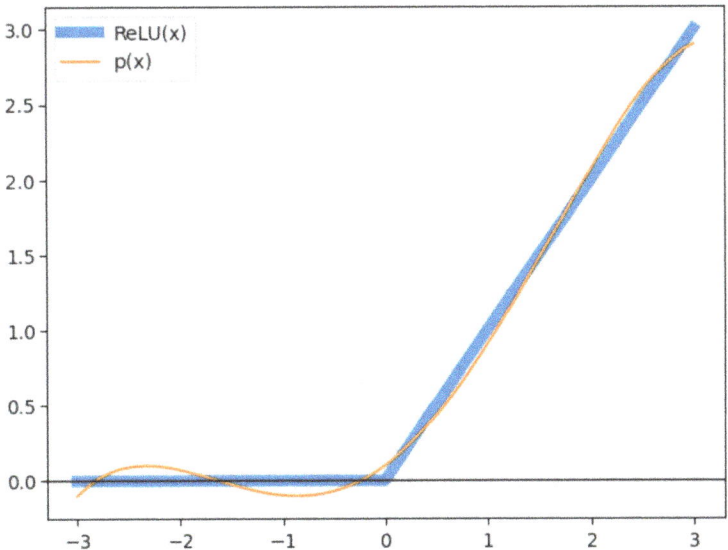

Fig. 6.3 The ReLU function and its approximation with a minimax polynomial of degree 4 for the domain $[-3, 3]$

The product depth also becomes deeper by two:

$$\text{ReLU}(x) = \frac{x + x * \text{Sign}(x)}{2} \qquad (6.12)$$

The additional scalar product with $\frac{1}{2}$ and the consequent increase of the product depth can be avoided if we modify the final composed polynomial of the Sign(x) function using a similar approach to that used in Eq. 6.11 for the case of max(a, b). Here again we can use the function ModifiedSign(x) in which we modify the final composed polynomial of Sign(x) by adding 1 to its free coefficient and then dividing all the coefficients by 2. Equation 6.12 becomes

$$\text{ReLU}(x) = x * \text{ModifiedSign}(x) \qquad (6.13)$$

Again, as with Eq. 6.11, the modification of Sign(x) described above is carried out in plaintext, ahead of this more efficient FHE computation.

The main problem with approximating an activation function with a polynomial is that such approximations are only good for limited input domains. If the polynomial $P(X)$ is a minimax polynomial for ReLU(x) in the domain $[-1, 1]$, then its accuracy may deteriorate quickly even for x values that are just outside this range. It is not always easy to know in advance what would be the input domain of an activation function that is placed deep inside a NN. If we wish to extend the domain of the approximation of $P(x)$ to $[-B, B]$ for some $B > 1$, then we can use the formula $\text{ReLU}(x) = B * P(x/B)$. However, such a transformation comes at a cost of a larger approximation error and is also not possible for most activation functions. Section 11 describes an alternative approach where instead of using a polynomial that tries to approximate ReLU in a limited domain, we use alternative more FHE-friendly activation functions, or use a polynomial that is trained to directly improve the loss function of the NN.

6.3 Computing Reciprocals: $\frac{1}{X}$

6.3.1 Computing Reciprocals When Working with Integers

FHE schemes that rely on ring LWE (R-LWE) perform the arithmetic operations under polynomial rings, which include the operations of addition, subtraction, and multiplication, but do not necessarily include the division operation. BGV is also based on R-LWE, but, when the plaintext modulus is a prime, the plaintext space is a finite field, which includes division. This happens because of Fermat's little theorem (Eq. 6.7), where for any integer x not divisible by p, $x * x^{p-2} = 1 \mod p$. In other words, the multiplicative inverse, (a.k.a. the *reciprocal*) of any x is $x^{p-2} \mod p$.

Example 6.3 Working in arithmetic modulo the prime $p = 97$, the reciprocal of 5 is $5^{95} \mod 97 = 39$. So, if we are working with BGV with plaintext space mod 97

6.3 Computing Reciprocals: $\frac{1}{x}$

and we happen to know that the ciphertext c encrypts some multiple of 5, then we can compute $c/5$ as $c * 39$. For example, $55 * 39 \mod 97 = 11$. This only works if indeed we know that c encrypts a multiple of 5, because we can only represent integers when working with integers mod 97. For example, if we try to divide 56 by 5 by multiplying 56 by the reciprocal of 5, we get $56 * 39 \mod 97 = 50$.

We can also use the above method to divide one ciphertext by another in SIMD fashion. So, if c_1 and c_2 encrypt two vectors of integers, where each slot of c_1 is divisible by the corresponding slot of c_2, then we can compute $c_3 = \frac{c_1}{c_2} = c_1 \odot (c_2{}^{p-2})$ under FHE where every slot of c_3 contains the division of the corresponding slots of c_1 and c_2. The power $c_2{}^{p-2}$ can be computed using exponentiation by squaring with a product depth of $O(\log_2 p)$ as described in Sect. 4.8.3. However, the requirement that the numerator is divisible by the denominator usually cannot be met, and one may need to estimate the division by computing polynomials in an FHE scheme that supports approximate arithmetic such as CKKS, as described in the next section.

6.3.2 Estimating Reciprocals when Working with Approximate Arithmetic

FHE schemes that support approximate arithmetic, such as CKKS, do not support the division operation directly. In order to divide the values encrypted in the slots of a ciphertext c by a known plaintext scalar value b, one can simply compute $1/b$ in plaintext and then multiply c by the scalar result. However, if we wish to compute $\frac{a}{b}$ for the respective values in the slots of ciphertexts c_a and c_b, then one must estimate $\frac{1}{b}$ for the values in c_b using a polynomial estimation of the function $f(x) = \frac{1}{x}$ under FHE and then multiply the resulting ciphertext by c_a.

The paper presenting CKKS [3] also showed that Goldschmidt's division algorithm [7] can be used to produce accurate approximations of reciprocals with CKKS. This algorithm relies on the fact that for $x \in (0, 2)$, the following polynomials converge to $\frac{1}{x}$ as $d \to \infty$:

$$G_d(x) = \prod_{i=0}^{d} \left(1 + (1-x)^{2^i}\right) \quad (6.14)$$

Note that $G_d(x)$ is a polynomial of degree $n = 2^{d+1} - 1$, which is a very large degree when d is large. However, in spite of this high degree, the polynomial can be computed with just $2d \approx 2 \log_2 n$ multiplications at a product depth of $d + 1 \approx \log_2 n$. This is much better than, say, the Paterson-Stockmeyer method for evaluating general polynomials of degree n, which requires $\sqrt{2n} + \log_2 n$ non-scalar products at a similar product depth. The numerical stability of the form of $G_d(x)$ is also good because there is no product of a small coefficient with a large power as generally

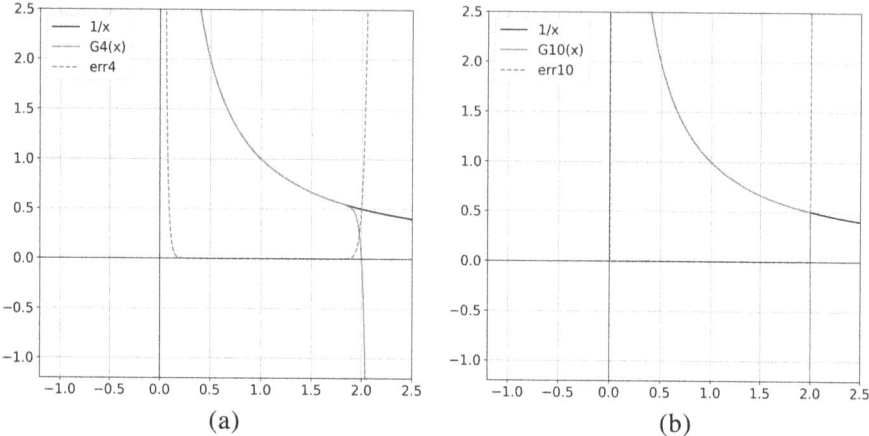

Fig. 6.4 Approximating $\frac{1}{x}$ with Goldschmidt's polynomials. (**a**) A Goldschmidt polynomial with $d = 4$. (**b**) A Goldschmidt polynomial with $d = 10$

occurs with n degree approximations, like minimax approximations resulting from the Remez algorithm.

Example 6.4 For $d = 4$, we get a polynomial of degree 31 with an error less than $1e−04$ for $x \in [0.2794, 1.7633]$. We can raise d either in order to reduce the error in this domain or to extend the domain of x where this error limit still applies. For example, if we raise d to 10, we get $G_d(x)$ of degree 2047 with an error less than $1e−04$ for $x \in [0.0069, 1.9958]$. Note that we are still limited to within the radius of convergence of $[0, 2]$. See Fig. 6.4.

When the above computations are done under FHE, homomorphic noise will also naturally be added to the error. Testing on a secure 2^{15} slot CKKS configuration from HEaaN's library [6], we were able to preserve the $1e−04$ limit on the error for $d = 4$ in the corresponding range given above ($[0.2794, 1.7633]$). However, for $d = 10$ the range of x preserving the $1e−04$ limit under FHE was reduced a little from $[0.0069, 1.9958]$ to $[0.1011, 1.9958]$.

6.4 Chebyshev Polynomials and Approximation of Trigonometric Functions

6.4.1 Chebyshev Polynomials

Simple trigonometric equalities can be used to show that the cosine of a multiple of some angle can be expressed as a polynomial over the cosine of that angle. The corresponding polynomials are named Chebyshev polynomials, after the

6.4 Chebyshev Polynomials and Approximation of Trigonometric Functions

$$\cos(0\alpha) = 1$$
$$\cos(1\alpha) = \cos(\alpha)$$
$$\cos(2\alpha) = 2\cos^2(\alpha) - 1$$
$$\cos(3\alpha) = 4\cos^3(\alpha) - 3\cos(\alpha)$$
$$\vdots$$
$$\cos(2n\alpha) = 2\cos^2(n\alpha) - 1$$
$$\cos((2n+1)\alpha) =$$
$$2\cos((n+1)\alpha)\cos(n\alpha) \quad \cos(\alpha)$$

(a)

$$T_0(x) = 1$$
$$T_1(x) = x$$
$$T_2(x) = 2x^2 - 1$$
$$T_3(x) = 4x^3 - 3x$$
$$\vdots$$
$$T_{2n}(x) = 2T_n^2(x) - 1$$
$$T_{2n+1}(x) = 2T_{n+1}(x)T_n(x) - x$$

(b)

Fig. 6.5 $\cos(n\alpha) = T_n(\cos(\alpha))$. (**a**) Cosine of a multiple of an angle can be expressed as a polynomial over the cosine of that angle. (**b**) Chebyshev polynomials

Russian mathematician Pafnuty Chebyshev. The nth Chebyshev polynomial $T_n(x)$ corresponds to $\cos(n\alpha)$, so that $\cos(n\alpha) = T_n(\cos(\alpha))$ as can be seen in Fig. 6.5. The final two relations given in the figure enable the recursive computation of $T_n(x)$ with a computation of depth $\log_2(n)$. Chebyshev polynomials have several useful applications when approximating functions by polynomials as will be described below.

6.4.2 Interpolation via Chebyshev Nodes

A common method for approximating a known function $f(x)$ with a polynomial is to use a polynomial $P(X)$ that shares the same value as $f(x)$ on some set of *interpolation* points $x_i \in I = x_0, x_1, x_2, \ldots x_n$. The hope is that if $P(X)$ is identical to $f(x)$ at the interpolation points in I, then it also approximates $f(x)$ well on a wider range of x values that includes I.

A polynomial $P(X)$ of degree n has $n + 1$ coefficients. The coefficients can be uniquely determined by solving a set of $n + 1$ linear equations corresponding to $n + 1$ interpolation points, where the coefficients c_i are the unknowns and the values of $f(x_i)$ as well as the powers of x_i are known, for every $i \in [0, n]$:

$$f(x_0) = c_n x_0^n \ldots + c_2 x_0^2 + c_1 x_0 + c_0$$
$$f(x_1) = c_n x_1^n \ldots + c_2 x_1^2 + c_1 x_1 + c_0$$
$$\vdots$$
$$f(x_n) = c_n x_n^n \ldots + c_2 x_n^2 + c_1 x_n + c_0$$

(6.15)

The interpolating polynomial is thus completely determined from the choice of interpolation points, and the question arises—which choice of interpolation points would result in a polynomial that gives the best approximation in the required approximation range? A natural choice may be to interpolate via equidistant points

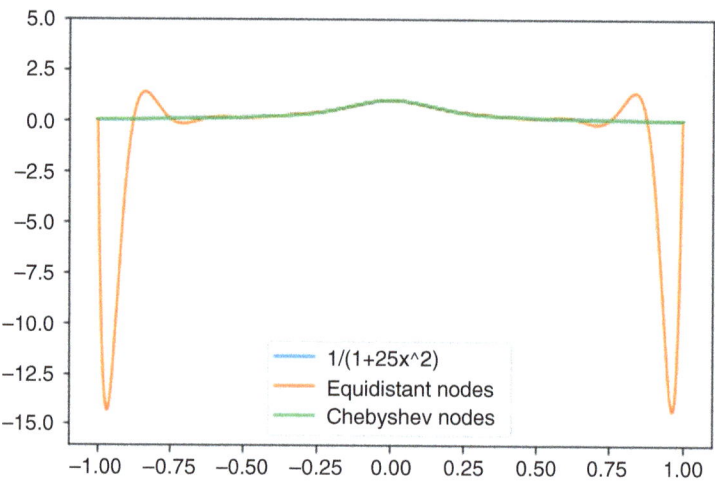

Fig. 6.6 Interpolating via equidistant points may suffer from Runge's phenomenon near the ends of the range. Interpolating via the same number Chebyshev nodes results in smaller more uniform errors

along the target range, but this is usually not the best choice and may suffer from Runge's phenomenon where the approximation deteriorates near the ends of the range, and the error may even worsen when higher polynomial degrees are used. See an example in Fig. 6.6 for the result of interpolating $f(x) = 1/(1 + 25x^2)$ with a 16-degree polynomial via equidistant points along the $[-1, 1]$ range.

It turns out that if we want to interpolate in the range $[-1, 1]$ with a polynomial of degree n, then interpolating via the n roots of the $n'th$ Chebyshev polynomial $T_n(x)$ gives a very good approximation, which is also more uniformly distributed along the range. This is because it can be shown that the approximation error includes the term

$$\max_{x \in [-1,1]} \left| \prod_{i=1}^{n} (x - x_i) \right|, \tag{6.16}$$

which is minimized when the interpolation points x_i are the n roots of $T_n(x)$. Figure 6.6 shows two interpolations of $f(x) = 1/(1 + 25x^2)$ using 16-degree polynomials. The polynomial that interpolates via equidistant points along the range suffers from Runge's phenomenon near the ends, whereas the interpolation via the Chebyshev nodes has a very small approximation error along the entire range (with a mean error of ~ 0.014 and a maximal error of ~ 0.0367).

Recall that $T_n(\cos(\alpha)) = \cos(n\alpha)$, and therefore the $n + 1$ roots of the degree $n + 1$ polynomial T_{n+1} are at

$$x_i = \cos(\frac{2i + 1}{2(n + 1)}\pi) \tag{6.17}$$

6.4 Chebyshev Polynomials and Approximation of Trigonometric Functions

for $i = 0, 1, \ldots, n$. Note that the points x_i are strictly inside the range $[-1, 1]$, and the polynomial that passes via these $n + 1$ points is of degree n. There is also an alternative set of $n + 1$ Chebyshev nodes that includes the two ends of the $[-1, 1]$ range, namely, $x_i = \cos(\frac{i}{n}\pi)$ for $i = 0, 1, \ldots, n$.

6.4.3 Function Estimation with Chebyshev Expansions

$T_0(x), T_1(x), \ldots, T_n(x)$, being polynomials, are linear combinations of $1, x, x^2, \ldots, x^n$, as seen, for example, in Fig. 6.5b. These $n + 1$ linear combinations can be expressed in the form of a matrix which can be shown to be invertible, and thus any power x^n can also be expressed as a linear combination of the Chebyshev polynomials $T_0(x), T_1(x), \ldots, T_n(x)$.

This implies that any polynomial $P(X) = \sum_{i=0}^{n} a_i X^i$ of degree n estimating some function $f(x)$ in the range $[-1, 1]$ can also be expressed as a linear combination of Chebyshev polynomials $\sum_{i=0}^{n} b_i T_i(x)$. This form is called a Chebyshev series or the Chebyshev expansion of $f(x)$, and the conversion of $P(X)$ to this form is always possible (because the Chebyshev polynomials $T_n(x)$ form an orthonormal basis). For example,

$$P(X) = a_5 X^5 + a_4 X^4 + a_3 X^3 + a_2 X^2 + a_1 X + a_0$$
$$= \left(\frac{a_5}{16}\right) T_5(X) + \left(\frac{a_4}{8}\right) T_4(X) + \left(\frac{a_3}{4} + \frac{5a_5}{16}\right) T_3(X) + \left(\frac{a_2}{2} + \frac{a_4}{2}\right) T_2(X) +$$
$$\left(a_1 + \frac{3a_3}{4} + \frac{5a_5}{8}\right) T_1(X) + \left(a_0 + \frac{a_2}{2} + \frac{3a_4}{8}\right) T_0(X)$$
(6.18)

A Chebyshev expansion of degree n can be evaluated by first computing all terms $T_0(x) \ldots T_n(x)$ and then using scalar products to multiply these with the corresponding coefficients. The $T_i(x)$ terms can be computed recursively using the recurrence relations

$$T_{2n}(x) = 2T_n^2(x) - 1 \tag{6.19}$$

or

$$T_{2n+1}(x) = 2T_{n+1}(x)T_n(x) - x \tag{6.20}$$

in the case of Chebyshev terms of an even or odd index, respectively (see Fig. 6.5). Some of the methods described in Sect. 4.8 also have variants for evaluation over Chebyshev rather than power terms. For example, [1] describes a modified version of the Paterson-Stockmeyer algorithm for fast evaluation of Chebyshev series.

Table 6.1 Errors of a Taylor expansion and of its economized version

Estimation of $\sin(\pi x)$ in $[-1, 1]$	Mean error	Max error
Taylor series of degree 5	0.07196	0.52404
Taylor series of degree 3	0.37474	2.02612
Taylor series of degree 5 economized to a Chebyshev series of degree 3	0.16095	0.37377
Chebyshev series of degree 3 interpolating on 4 Chebyshev nodes	0.08244	0.18062

Example 6.5 The Taylor expansion of degree 5 for $f(x) = \sin(\pi x)$ (showing rounded coefficients) is

$$P(X) = 2.55016X^5 - 5.16771X^3 + 3.14159X \qquad (6.21)$$

and the same polynomial in the form of a Chebyshev expansion using Eq. 6.18 is

$$P(X) = 0.15939 T_5(X) - 0.495 T_3(X) + 0.85966 T_1(X) \qquad (6.22)$$

Note that these are simply two forms of the same function so they share the same mean and maximal error with respect to the estimated function $f(x)$ in the range $[-1, 1]$. However, as seen in Table 6.1, if we drop the last (degree 5) term from the Taylor series, we get a polynomial of degree 3 with deteriorating mean and maximal errors, whereas if we drop the last term from its Chebyshev expansion, we get a polynomial of degree 3 with a lower maximal error than the original Taylor series of degree 5, though with some deterioration of the mean error. In any case, both the mean and maximal errors of the shortened Chebyshev expansion of degree 3 are better than in a Taylor series of the same degree. This suggests that the degree of the Taylor polynomial can be reduced in this manner while maintaining its accuracy. This process is called an *economization* of a power series with a Chebyshev series and is particularly useful for function estimation under FHE, where the motivation for reducing the product depth is high. Figure 6.7 shows the function $f(x) = \sin(\pi x)$ against its estimation with a Taylor series of degree 5 and the economized Chebyshev series of degree 3.

Instead of economizing an existing polynomial into the form of a Chebyshev series of a lower degree, one can also directly estimate the target function with a Chebyshev series $\sum_{i=0}^{n} c_i T_i(x)$ of the same low degree by interpolation, preferably via Chebyshev nodes (see Sect. 6.4.2). Table 6.1 shows that the interpolating Chebyshev series can give an even better estimate than the Chebyshev series that economizes a Taylor expansion. The coefficients c_i for $i \in [0, n]$ can be found, for example, by solving a set of $n + 1$ linear equations corresponding to the $n + 1$ interpolation points, similar to those shown in Eq. 6.15, except that $T_i(x)$ replace x^i for every i in all the equations.

6.4 Chebyshev Polynomials and Approximation of Trigonometric Functions

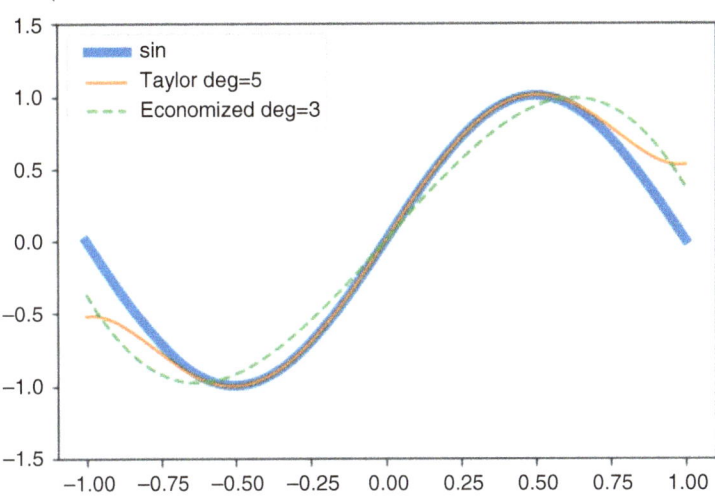

Fig. 6.7 A Taylor series of degree 5 for $\sin(\pi x)$ is economized into a Chebyshev series of degree 3 by converting to Chebyshev form and then dropping the last Chebyshev term. The maximal error of the economized series is lower but with some rise in the mean error, when compared with the original Taylor series

A minimax polynomial approximation of $f(x)$ in the range $[-1, 1]$, such as the one which can be found with the Remez algorithm (see Sect. 5.3), would give an even better approximation than an interpolation with a Chebyshev series. However, with a high enough degree, the difference between these two estimation methods becomes negligible, while the method of Chebyshev expansion is easier to compute.

It is possible to convert the interpolating Chebyshev expansion to the form of a power series and use any of the methods described in Sect. 4.8 in order to evaluate it, rather than using the variants for Chebyshev series. However, it is usually more advisable to evaluate the Chebyshev series directly, rather than its corresponding power series form, because the matrix used to perform the transformation of the coefficients from Chebyshev to power form is ill-conditioned, meaning that small errors in the source coefficients may magnify in the target coefficients. In addition, the coefficients of the resulting power series would also differ in many orders of magnitude, inducing large numerical errors in the polynomial evaluation [1].

6.4.4 Estimating Sin and Cos Beyond the [−1, 1] Range

We saw in the above sections that we can get a good approximation of functions like $\sin(k\pi x)$ and $\cos(k\pi x)$ for $x \in [-1, 1]$ using a Chebyshev series that interpolates via Chebyshev nodes in this range. However, we may naturally wish to extend the approximation beyond this limited range. The following methods can be used for this purpose:

1. **By Euler's formula** [4]: Euler's formula states that for any real α,

$$e^{i\alpha} = \cos(\alpha) + i \sin(\alpha) \tag{6.23}$$

Therefore,

$$(\cos(\alpha) + i \sin(\alpha))^n = e^{in\alpha} = \cos(n\alpha) + i \sin(n\alpha) \tag{6.24}$$

Algorithm 6 uses the above to estimate $\cos(k\pi x)$ and $\sin(k\pi x)$ for x in the range $[-n, n]$, where n is an integer greater than 1, preferably a power of 2. Algorithm 6 involves complex operations and thus mostly fit schemes such as CKKS that operate over the complex plaintext space. Specifically, operations on complex values as in Steps 3 and 5 are natively supported by CKKS. Step 4 can use an exponentiation by squaring algorithm, and when n is a power of 2, this step can be efficiently carried out using repeated squaring.

Algorithm 6: Large range approximation

Input: A value $x \in [-n, n]$, where $n > 1$, preferably a power of 2.
Output: An approximation of $\cos(k\pi x)$ and $\sin(k\pi x)$.
1 $x' \leftarrow \frac{x}{n}$; // $x' \in [-1, 1]$
2 Approximate $c = \cos(k\pi x')$ and $s = \sin(k\pi x')$ using a Chebyshev series that interpolates via Chebyshev nodes in the range $[-1, 1]$.;
3 $e = c + i \cdot s$;
4 $e' = e^n$;
5 $Re(e') = \frac{e' + \text{Conj}(e')}{2}$; // Eq. 6.24: $\cos(nk\pi x)$, the real part of e'.
6 $Im(e') = \frac{e' - \text{Conj}(e')}{2}$; // Eq. 6.24: $\sin(nk\pi x)$, the imaginary part of e'.
7 return $Re(e'), Im(e')$

---

2. **By the formula for Cos of a double angle** [4]:
The following double-angle formulas can be used to compute cos/sin of large values from cos/sin of values in the range $[-1, 1]$:

$$\cos(2x) = 2\cos^2(x) - 1$$
$$\cos(2x) = \cos^2(x) - \sin^2(x) \tag{6.25}$$
$$\sin(2x) = 2\sin(x)\cos(x)$$

The first of these formulas can be used if one just needs to compute the cosine of a large angle, and the last two formulas can be used to compute the cosine and sine together from the cosine and sine of the smaller angle. Other trigonometric equalities can be used if the ratio between the large and small angles is not

6.4 Chebyshev Polynomials and Approximation of Trigonometric Functions

a power of 2. In any case, one starts as in the above procedure using Euler's formula, by computing the sin/cos of the divided angle, but now the angle is raised by repeated application of the above formulas.

3. **By scaling** [1]:
   Suppose we wish to estimate $\cos(k\pi x)$ for $x \in [-n, n]$ where n is an integer greater than 1. If $x' = x/n$ then by change of variables, we can compute $\cos(kn\pi x')$ for $x' \in [-1, 1]$. $\cos(kn\pi x')$ can again be estimated using a Chebyshev series that interpolates via Chebyshev nodes in this range. However, the curve of $\cos(kn\pi x')$ in the range $[-1, 1]$ is more complex than that of $\cos(k\pi x')$, so it would require more interpolation points in order to reach a similar level of accuracy and a corresponding higher degree for the estimating polynomial.

Figure 6.8 shows the result of the comparison of the above three methods using the CKKS FHE scheme of HEaaN's library [6]. We set the parameters of the keys to have 128 bits of security with 2^{14} slots, chain length 11, integer precision of 20 bits, and fractional precision of 40 bits. We tested the methods by evaluating $\cos(2\pi x)$ on four ranges of x: $[-1, 1], [-2, 2], [-4, 4], [-8, 8]$. In all the methods, we start by dividing x with a value that would ensure to bring it into the $[-1, 1]$ range (e.g., by 8 if we start from the $[-8, 8]$ range) and then proceed to evaluate a Chebyshev series that interpolates via Chebyshev nodes in $[-1, 1]$. For the Euler and double-angle method, we estimated $\cos(2\pi x)$ with a Chebyshev series of degree 16, and for the scaling method, we used Chebyshev series of degrees 16, 16∗2, 16∗4, 16∗8 to

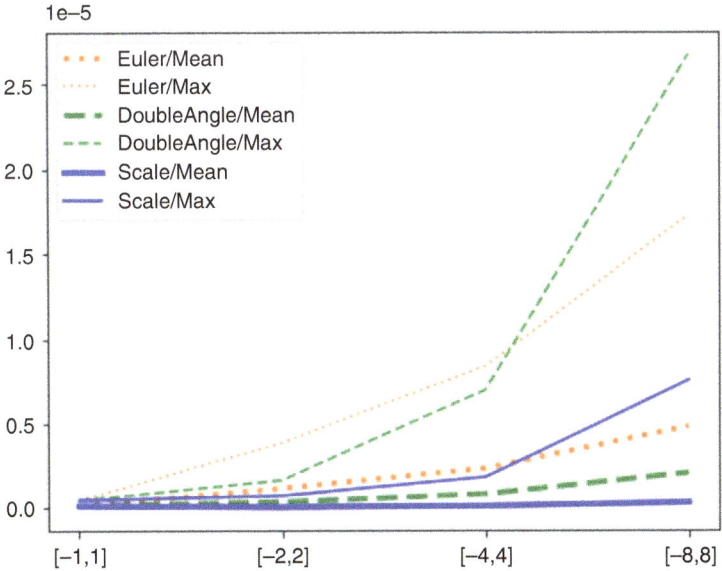

Fig. 6.8 Estimation errors of different estimation methods for $\cos(2\pi x)$ at large angles

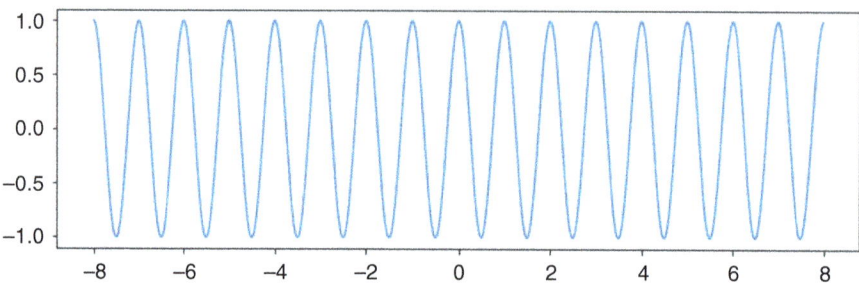

Fig. 6.9 $\cos(2\pi x)$ evaluated under FHE in the range $[-8, 8]$ using the scaling method

estimate $\cos(2\pi x)$, $\cos(2*2\pi x)$, $\cos(2*4\pi x)$, $\cos(2*8\pi x)$ for the corresponding input ranges of x. The product depth needed for the double-angle and scaling methods was the same (5, 7, 8, and 9 for the four ranges) and the Euler method requires one extra product. Figure 6.8 shows the mean and maximal estimation errors of the three methods in the four ranges, which include both numeric and FHE errors. The scaling method is shown to produce the least errors, with a mean error of $\sim 3.2e-7$ and a maximal error of $\sim 7.5e-7$ in the range $[-8, 8]$. Figure 6.9 shows the resulting estimation of $\cos(2\pi x)$ in this range using the scaling method.

6.5 Lab Exercises

6.5.1 Estimate the Sign Function

For the next exercises, consider the minimax polynomial P_0 of degree 7 estimating Sign(x) over the domain $R_0 = [-1, -0.1] \cup [0.1, 1]$:

$$\begin{aligned} P_0(X) = &- 28.3725X^7 + 5.2677e{-}6X^6 + 53.6501X^5 \\ &- 2.9007e{-}6X^4 - 31.5068X^3 + 2.6783e{-}7X^2 \\ &+ 6.8869X - 2.3935e{-}9 \end{aligned} \quad (6.26)$$

The range of P_0 for $x \in R_0$ is $R_1 = [-1.3423, -0.6577] \cup [0.6577, 1.3423]$. Consider also the minimax polynomial P_1 of degree 7 estimating Sign(x) over the domain R_1:

$$\begin{aligned} P_1(X) = &- 0.3263X^7 + 1.6666e{-}11X^6 + 1.4879X^5 \\ &- 2.7850e{-}11X^4 - 2.5119X^3 + 1.3339e{-}11X^2 \\ &+ 2.3431X - 1.9079e{-}12 \end{aligned} \quad (6.27)$$

6.5 Lab Exercises

1. Use P_0 and P_1 to estimate Sign(x) in the range $[-1, 1]$.
   Run this estimation under FHE and measure the number of multiplications and the product depth.
2. Use the above estimation of Sign to compute max(a, b) according to Eq. 6.11 where a and b are in separate ciphertexts.
3. Use the above estimation of max(a, b) to compute the maximal value among 16 values represented by 16 separate ciphertexts. Note that the computation would now involve a product depth that may require bootstraps.
4. Repeat the above where the 16 values are represented by 16 slots of the same ciphertext. Use the rotate-and-sum-like technique described in Sect. 6.1.3 where the max(a, b) operation is applied between the rotated versions of the ciphertext and the overall maximal value ends up in the first slot.
   Compare the number of multiplications and product depth of this technique and the technique used by the answer to the previous question.
5. Use the above estimation of Sign to estimate ReLU as in Eq. 6.12, on $x = 0.8$.
   Run this estimation under FHE and measure the number of multiplications and the product depth.
6. Use the above estimation of Sign to estimate ReLU as in Eq. 6.13, on $x = 0.8$.
   Again measure the number of multiplications and the product depth, and compare with the result of the previous question.
7. Measure the mean and maximal error of the above estimate when the domain is $[-1, 1]$ and when the domain is $[-1.1, 1.1]$.
   Hint: Create a ciphertext where the slots include a sequence of values in the range to be measured, and compute the ReLU function on the entire ciphertext.

6.5.2 Compare Two Hybrid Bitwise Values with CKKS

8. Compare two 8-bit values with CKKS using hybrid bitwise representation and arithmetic, with the following formula:

$$\text{Eq}(a, b) = \left(\prod_{i \in [0, n-1]} (a[i] + b[i] - 1) \right)^2$$

9. Repeat the same comparison using the following alternative formula:

$$\text{Eq}(a[], b[]) = \prod_{i \in [0, n-1]} (1 - (a[i] - b[i])^2)$$

10. Compare the number of multiplications and product depth required by the above two methods.
11. Use the first of the above implementations of Eq to compute isGreaterThan (a, b) for two 8-bit values using Eq. 6.8.

12. Use the above implementation to compute the following piecewise function under CKKS:

$$f(a,b) = \begin{cases} 5 & a > b \\ 9 & a \leq b \end{cases}$$

6.5.3 Computing Reciprocals Under FHE

13. Use BGV with arithmetic modulo 53 to divide by 2 all the values in a ciphertext that encrypts only even numbers.
14. Use $G_{d=4}$ to compute the reciprocals of different values in the range [0.1, 1.9] encrypted in a CKKS ciphertext.

Estimate $\sin(\pi x)$ in the range $x \in [-1, 1]$

15. Implement the recursive function $T(n, x)$ that computes the Chebyshev polynomial $T_n(x)$.
    Use the recursive formulas in Fig. 6.6b. Test that indeed $\cos(5 * 0.1) = T_5(\cos(0.1))$.
16. Find the polynomial of degree 6 that interpolates a given function $f(x)$ via seven equidistant points along $[-1, 1]$. Test this on $f(x) = \sin(\pi x)$.
17. Find the polynomial of degree 6 that interpolates a given function $f(x)$ via the seven Chebyshev nodes: $\cos(\pi i/6)$ for $i \in [0, 6]$. Test this on $f(x) = \sin(\pi x)$.
18. Find the Chebyshev series of degree 6 that interpolates a function $f(x)$ via the seven Chebyshev nodes: $\cos(\pi i/6)$ for $i \in [0, 6]$. Test this on $f(x) = \sin(\pi x)$.
19. Compare the mean and maximal errors of the above three estimations of $\sin(\pi x)$ in the range $[-1, 1]$.
    Compare these errors to the mean/max errors of the Taylor series of the higher degree of 8 for $\sin(\pi x)$ expanded at $x = 0$.
20. Use the above Chebyshev series method to estimate $f(x) = \sin(4\pi x)$ in the range $[-1, 1]$ using polynomials of degrees 6, 12, and 24. Plot the three estimating polynomials in this range and compare their mean and maximal errors relative to $f(x)$.
21. Use the above polynomial of degree 24 estimating $f(x) = \sin(4\pi x)$ in the range $[-1, 1]$ to estimate $f(x) = \sin(\pi x)$ in the range $[-4, 4]$ using the scaling method. Plot the result and compute the mean and maximal errors relative to $\sin(\pi x)$ in this range.
22. In your Remez implementation of Sect. 5.5, Exercise 4, replace the power basis with a Chebyshev basis. Did you observe any difference with the generated polynomial approximations?

References

1. Chen, H., Chillotti, I., Song, Y.: Improved bootstrapping for approximate homomorphic encryption. In: Advances in Cryptology–EUROCRYPT 2019: 38th Annual International Conference on the Theory and Applications of Cryptographic Techniques, Darmstadt, Germany, May 19–23, 2019, Proceedings, Part II, pp. 34–54. Springer (2019). https://doi.org/10.1007/978-3-030-17656-3_2
2. Chen, T., Guestrin, C.: XGBoost: a scalable tree boosting system. In: Proceedings of the 22nd ACM SIGKDD International Conference on Knowledge Discovery and Data Mining, KDD '16, p. 785–794. Association for Computing Machinery, New York (2016). https://doi.org/10.1145/2939672.2939785
3. Cheon, J., Kim, A., Kim, M., Song, Y.: Homomorphic encryption for arithmetic of approximate numbers. In: Proceedings of Advances in Cryptology - ASIACRYPT 2017, pp. 409–437. Springer, Cham (2017). https://doi.org/10.1007/978-3-319-70694-8_15
4. Cheon, J.H., Han, K., Kim, A., Kim, M., Song, Y.: Bootstrapping for approximate homomorphic encryption. In: Nielsen, J.B., Rijmen, V. (eds.) Advances in Cryptology – EUROCRYPT 2018, pp. 360–384. Springer, Cham (2018)
5. Cheon, J.H., Kim, D., Kim, D.: Efficient homomorphic comparison methods with optimal complexity. In: Moriai, S., Wang, H. (eds.) Advances in Cryptology – ASIACRYPT 2020, pp. 221–256. Springer, Cham (2020). https://doi.org/10.1007/978-3-030-64834-3_8
6. CryptoLab: HEaaN: Homomorphic encryption for arithmetic of approximate numbers, version 3.1.4 (2022). https://www.cryptolab.co.kr/eng/product/heaan.php
7. Goldschmidt, R.E.: Applications of division by convergence. Ph.D. Thesis, Massachusetts Institute of Technology (1964). https://dspace.mit.edu/bitstream/handle/1721.1/11113/34136725-MIT.pdf
8. Iliashenko, I., Zucca, V.: Faster homomorphic comparison operations for BGV and BFV. Proc. Privacy Enhancing Technol. **2021**, 246–264 (2021). https://doi.org/10.2478/popets-2021-0046
9. Lee, E., Lee, J.W., No, J.S., Kim, Y.S.: Minimax approximation of sign function by composite polynomial for homomorphic comparison. IEEE Trans. Dependable Secure Comput. **19**(6), 3711–3727 (2022). https://doi.org/10.1109/TDSC.2021.3105111
10. Lee, J., Lee, E., Lee, J.W., Kim, Y., Kim, Y.S., No, J.S.: Precise approximation of convolutional neural networks for homomorphically encrypted data. IEEE Access **11**, 62062–62076 (2023). https://doi.org/10.1109/ACCESS.2023.3287564
11. Lee, J.W., Lee, E., Lee, Y., Kim, Y.S., No, J.S.: High-precision bootstrapping of RNS-CKKS homomorphic encryption using optimal minimax polynomial approximation and inverse sine function. In: Canteaut, A., Standaert, F.X. (eds.) Advances in Cryptology – EUROCRYPT 2021, pp. 618–647. Springer, Cham (2021). https://doi.org/10.1007/978-3-030-77870-5_22

Part III
Packing Methods

By now the reader should be familiar with basic fully homomorphic encryption (FHE) concepts such as the FHE computation model, the type of supported operations, and various ways to generate some of the missing functionalities, e.g., by using approximations. Nevertheless, most of the prior discussion was focused on ciphertexts that encrypt only one value at a time. This approach is implemented, for example, in FHE schemes such as TFHE or FHEW, that show good results in terms of latency and memory for various applications. This approach does not possess any limitation, i.e., it is possible to use these schemes to evaluate every function by, e.g., converting the functions to Boolean circuits. However, it may miss some interesting opportunities for optimizations.

This part of the book considers ciphertexts that can encrypt more than one value. For that, a technique called *packing* (a.k.a packing scheme) is explored, which presents ways to arrange multiple pieces of data inside an encoded plaintext which is subsequently encrypted. This part contains three chapters. Chapter 7 presents basic packing concepts and methods. Chapter 8 builds on these methods and describes a data structure named tile tensors that leverages the packing methods from Chap. 7 to achieve more general solutions. Finally, in Chap. 9 we discuss more advanced ways to utilize the power of tile tensors.

Chapter 7
SIMD Packing Part I: Basic Packing Techniques

Abstract This chapter describes basic packing techniques to efficiently utilize single instruction multiple data (SIMD) and operations over SIMD-based elements. The primitives described are used to implement a simple matrix-vector multiplication use case. Subsequent chapters build upon these techniques to create more sophisticated and general packing schemes.

7.1 Why Pack?

Most modern fully homomorphic encryption (FHE) schemes, such as Cheon, Kim, Kim, and Song (CKKS), Brakerski, Gentry, Vaikuntanathan (BGV), and Brakerski / Fan and Vercauteren (B/FV), support SIMD operations, where plaintexts are vectors of numbers and ciphertext operations apply in a parallel fashion to all the plaintext vector elements. These schemes can be configured to operate over plaintext vectors with a fixed *number of slots*. For example, if a scheme is configured for 16 slots, then each ciphertext has the capacity to contain 16 encrypted numbers. Using a larger number of slots has the additional side effect that it enables the scheme to be more resistant to cryptographic attacks, because it uses polynomials of larger degrees. For that reason, modern applications configure the scheme for a large number of slots, typically in the order of thousands. It may seem that using the largest supported number of slots is always better, but this is not always the case. Increasing the ciphertext size also increases the latency of the computations in a way that is more than linear in the number of slots. The exact number should be defined per application.

To understand the concept of SIMD, we consider the following toy example. Assume a scheme that is configured with a low number of slots, e.g., $s = 16$. We are given two vectors $V = (v_0, v_1, \ldots, v_{15})$ and $U = (u_0, u_1, \ldots, u_{15})$ over some arbitrary vector space. Using homomorphic encryption (HE), we can encrypt them into two ciphertexts $c_1 = \text{Enc}(V)$ and $c_2 = \text{Enc}(U)$, and add the two ciphertexts as follows: $c_3 = c_1 \oplus c_2$. Decrypting c_3 yields the vector $(v_0 + u_0, v_1 + u_1, \ldots, v_{15} + u_{15})$, which is the sum of the 16 pairs of numbers. The above required only a single (encrypted) *add* operation. That's SIMD.

What can we do if V and U have only four elements each, while the FHE scheme is still configured for 16 slots? Recall that the number of slots is mostly determined by security and other constraints and we cannot arbitrarily reduce it. The solution is to pad U and V with zeros, i.e., c_1 is the encryption of the vector $V||0^{12} = (v_0, v_1, v_2, v_3, 0, 0, \ldots, 0)$, and similarly $c_2 = \text{Enc}(U||0^{12})$. When computing $c_1 \oplus c_2$, we still add 16 pairs of numbers, but only *four* pairs are actually meaningful, which is wasteful. We paid for adding 16 numbers, but used only *four*. That is like renting a large bus for a trip, but utilizing only a small part of the seats.

Let us consider a more complex example. V and U each has *four* elements and we want to compute the 16 products: $(v_0 u_0, v_0 u_1, v_0 u_2, \ldots, v_3 u_3)$. Here is an efficient way to do so. Set c_1, c_2 to be the ciphertexts:

$$c_1 = \text{Enc}\,((v_0, v_0, v_0, v_0, v_1, v_1, v_1, v_1, v_2, v_2, v_2, v_2, v_3, v_3, v_3, v_3))$$

$$c_2 = \text{Enc}\,((u_0, u_1, u_2, u_3, u_0, u_1, u_2, u_3, u_0, u_1, u_2, u_3, u_0, u_1, u_2, u_3))$$

Computing $c_1 \odot c_2$ will compute the 16 required products with a single (encrypted) multiplication operation. We see that by cleverly choosing how to place elements in the ciphertext slots, we managed to perform the outer product with optimal efficiency.

The above examples are small and simple. In modern scenarios, we need to perform more complex circuits of operations on data, possibly higher-ranking tensors, such as matrices or 3D arrays. Such circuits can implement neural networks (NNs) or decision trees, and the operators may involve matrix multiplications, convolutions, pooling, transposing, etc. Deciding how to arrange the values inside the ciphertexts for a given computation is called the packing scheme. Different packing schemes may have a dramatic impact on several different aspects of the computation: memory consumption, latency, throughput, network bandwidth, and the ability to parallelize the computation. Thus, we require the following goals from a packing scheme:

- **Slot utilization:** It should leverage the maximal number of slots possible in every step of the computation, since unused slots can increase the waste of time and memory.
- **Operation minimization:** It should allow minimizing the number of operations performed.
- **Parallelization:** It should increase the parallelizability of the circuit. This property depends, of course, on the environment, e.g., hardware parallelization capabilities.
- **Easy-transition between steps:** The output of each step should be properly packed for the next step or can be made so with as little effort as possible.
- **Efficient memory use:** The packing scheme should be memory efficient and use as few ciphertexts as possible.

7.2 The Rotate-and-Sum Algorithm

Remark 7.1 These goals are sometimes conflicting. A packing scheme that is more parallelizable might require more operations, or a packing scheme that requires less operations might use more memory.

The following sections describe basic packing techniques and simple operations that can be performed on data packed using these techniques. In Chaps. 8 and 9, we will use these basic methods as building blocks to construct a more versatile and powerful packing scheme called tile tensors.

Remark 7.2 In some scenarios, a computation has both encrypted and unencrypted inputs, for example, a scenario where we need to compute the outer product of an encrypted vector and a plaintext vector. This is usually a faster operation than computing the outer product when both inputs are encrypted, but for packing purposes, it makes little difference. The plaintext vector should still undergo encoding, and the encoded plaintext object has the same number of slots as the ciphertext, so we still need to design how we pack the vector inside it. Encrypted versus unencrypted packing is discussed further in Sect. 8.13.

7.2 The Rotate-and-Sum Algorithm

Consider the following problem: given a ciphertext c with s slots, compute the sum of the first k elements, $\sum_{j=0}^{j=k-1} c[j]$.

It will actually be easy to compute not just the sum of the first k elements, but the sum of any consecutive k elements, as in the following definition.

Definition 7.3 (Partial Sum) For a ciphertext c, $c' = \text{PartSum}(c, k)$ is a ciphertext that encrypts all the sums of consecutive (cyclic) k elements, i.e.,

$$c'[i] = \sum_{j=0}^{k-1} c[(i+j) \bmod s]. \tag{7.1}$$

Example 7.4 Let c = Enc $((10, 11, 12, 13, 14, 15, 16, 17))$. Then c_1 = PartSum$(c, 2)$ = Enc $((21, 23, 25, 27, 29, 31, 33, 27))$. Every element in c_1 contains the sum of two consecutive elements in c. This includes the last element, which is the sum of the last and first elements in c. Similarly, in c_2 = PartSum$(c, 4)$ = Enc $((46, 50, 54, 58, 62, 58, 54, 50))$ every element is the sum of four consecutive elements of c. A degenerate case is PartSum$(c, 1)$, which is equal to c.

Computing the partial sum can be done using the rotate operator Rot defined in Definition 2.7. If $k = 2^m$, for some positive integer m, then summing the first k elements is easy, using a function called RotateAndSum [4]:

Algorithm 7: RotateAndSum—Sum the first 2^m elements of c

Input: A ciphertext c of s slots and an integer $0 \le m \le \lfloor \log_2(s) \rfloor$.
Output: A ciphertext r, where $r = \text{PartSum}(c, 2^m)$.
1 **Function** RotateAndSum(c, m):
2      $r = c$
3   **for** $i \leftarrow 0$ **to** $m - 1$ **do**
4          $r = r \oplus \text{Rot}_{2^i}(r)$
5   **end**
6   **return** r

To see why this works, let us prove the following lemma.

Lemma 7.5 $\text{PartSum}(c, k) \oplus \text{Rot}_k(\text{PartSum}(c, l)) = \text{PartSum}(c, k + l)$

Proof By definition 7.3, for $0 \le i < s$,

$$\text{PartSum}(c, k)[i] + \text{Rot}_k(\text{PartSum}(c, l))[i] =$$

$$= \sum_{j=0}^{k-1} c[(i+j) \bmod s] + \sum_{j=0}^{l-1} c[(i+j+k) \bmod s]$$

$$= \sum_{j=0}^{k-1} c[(i+j) \bmod s] + \sum_{j=k}^{k+l-1} c[(i+j) \bmod s]$$

$$= \sum_{j=0}^{k+l-1} c[(i+j) \bmod s]$$

$$= \text{PartSum}(c, k+l)[i]$$

□

Using Lemma 7.5 it is easy to see why Algorithm 7 works. It starts with $c = \text{PartSum}(c, 1)$, and in the i'th iteration, it computes $\text{PartSum}(c, 2^i) \oplus \text{Rot}_{2^i}(\text{PartSum}(c, 2^i)) = \text{PartSum}(c, 2^{i+1})$.

Example 7.6 Continuing Example 7.4: $c_1 = \text{PartSum}(c, 2)$ is computed using $c_1 = c \oplus \text{Rot}_1(c)$. Then, $c_2 = \text{PartSum}(c, 4)$ is computed using $c_2 = c_1 \oplus \text{Rot}_2(c_1)$.

Remark 7.7 In Algorithm 7 the sum operator can be replaced with any associative binary operator. For example, we can compute partial products, partial max and min operations, etc.

Lemma 7.5 can also help us generalize the RotateAndSum algorithm to operate with any number of elements, not just powers of 2. For that, and only for this paragraph, consider the $\text{PartSum}(c, k)$ operation as a power operator on c, denoted by c^k. And further that $c^k \oplus \text{Rot}_k(c^m)$ is the product operation $c^k c^m$. Lemma 7.5 now takes the form of the usual law for exponents: $c^k c^m = c^{k+m}$. With this we can reuse efficient algorithms for evaluation of powers as in [3, 9] to compute c^k for arbitrary k.

One such exponentiation by squaring algorithm was already presented in Sect. 4.8, in the context of evaluation of polynomials. Below, we present two additional variants: Algorithm 8 and 9, already translated to deal with computing sums. Algorithm 9 has the advantage that it only uses rotations with offsets that are powers of 2. This works faster for common setups of FHE schemes that prepare rotations keys for these offsets. In contrast, Algorithm 8 has the advantage that it empirically introduces less FHE noise. In the algorithms, we use the following notation. Let k be an integer and $k = \sum_{i=0}^{\lceil \log_2(k) \rceil} 2^i (k)_j$ its binary decomposition, and then we refer to $(k)_j$ by the jth bit of k.

Algorithm 8: RotateAndSumLR—Summarize the first k elements of c, adapted for left-to-right exponentiation by squaring from [3, 9]

Input: A ciphertext c of s slots and an integer $0 \leq k < s$
Output: A ciphertext y such that $y = \text{PartSum}(c, k)$.
1 **Function** RotateAndSumLR(c, k):
2     $e \leftarrow 1$
3     $y \leftarrow c$
4   **for** $j \leftarrow \text{NumBits}(k) - 2$ **downTo** 0 **do**
5         $y \leftarrow y \oplus \text{Rot}_e(y)$
6         $e \leftarrow 2e$
7   **if** $(k)_j = 1$ **then**
8             $y \leftarrow c \oplus \text{Rot}_1(y)$
9             $e \leftarrow e + 1$
10   **end**
11   **end**
12   **return** y
13 **end**

The following observation shows that executing PartSum over all slots results in a single duplicated value. This property will be useful in later chapters.

Observation 7.8 *For a ciphertext c with s slots,* $\text{PartSum}(c, s)$ *is a ciphertext where all the slots are equal to the sum of the slots of c. This follows from Definition 7.3 and since the sum covers all the s elements.*

7.3 Flattening

Packing a vector inside a ciphertext is easy in general, because we assume a ciphertext natively encrypts a vector, as is the case in common FHE schemes. Of course, the vector may be larger than the number of slots in a ciphertext, but we will

Algorithm 9: RotateAndSumRL—Summarize the first k elements of c, adapted for right-to-left exponentiation by squaring from [3, 9]

Input: A ciphertext c of s slots and an integer $0 \leq k < s$
Output: A ciphertext y such that $y = \text{PartSum}(c, k)$.

```
1  Function RotateAndSumRL(c, k):
2      e ← 1
3      x ← c
4      y ← null
5      while true do
6          if k mod 2 = 1 then
7              if y = null then
8                  y ← x
9              else
10                 y ← x⊕Rot_e(y)
11             end
12             k ← (k − 1)/2
13         else
14             k ← k/2
15         end
16         if k = 0 then
17             return y
18         end
19         x ← x⊕Rot_e(x)
20         e ← 2e
21     end
```

deal with that in Chap. 8. Sometimes, however, we need to pack tensors of higher rank: matrices or three-dimensional arrays. In Chap. 9, for example, we discuss NN 2D convolutions that use four-dimensional arrays.

We start with defining the term tensor in the sense used within the artificial intelligence (AI) domain.

Definition 7.9 (Tensor) A tensor is a multidimensional array of numbers. A rank-k tensor is a k-dimensional array of shape $[n_0, \ldots, n_{k-1}]$, where n_i is the size of the i'th dimension.

For example, a rank-0 tensor is a scalar. A rank-1 tensor is a vector, and a rank-2 tensor is a matrix. A rank-k tensor is also sometimes referred to as a tensor of order k.

Consider, for example, the following matrix M whose shape is [2, 4] (meaning, *two* rows and *four* columns) (Fig. 7.1).

Fig. 7.1 An example matrix M of shape [2, 4]

$$M = \begin{array}{|cccc|} \hline 1 & 2 & 3 & 4 \\ 5 & 6 & 7 & 8 \\ \hline \end{array}$$

7.3 Flattening

```
c row  = | 1 | 2 | 3 | 4 | 5 | 6 | 7 | 8 |
major
                    (a)

c col  = | 1 | 5 | 2 | 6 | 3 | 7 | 4 | 8 |
major
                    (b)
```

Fig. 7.2 Flattening conventions of the matrix M of shape $[2, 4]$ in a ciphertext c of $s = 8$ slots. (**a**) M packed using row-major order. (**b**) M packed using column-major order

Say we want to pack it in a ciphertext c with $s = 8$ slots. There are two natural ways to do that: row-major or column-major, depending on whether consecutive elements are in the same row or the same column, respectively (Fig. 7.2).

The terms row-major and column-major generalize for higher-ranking tensors as well. Consider, for example, a tensor T of shape $[2, 4, 3]$. Flattening in column-major order, the elements are ordered as follows:

$$T[0, 0, 0], T[1, 0, 0], T[0, 1, 0], T[1, 1, 0], T[0, 2, 0], T[1, 2, 0], T[0, 3, 0], \ldots \tag{7.2}$$

That is, the first dimension (most left dimension) is moving first when moving to the next slot. In contrast, when using row-major order, the elements are ordered such that the last dimension (most right dimension) is moving first:

$$T[0, 0, 0], T[0, 0, 1], T[0, 0, 2], T[0, 1, 0], T[0, 1, 1], T[0, 1, 2], T[0, 2, 0], \ldots \tag{7.3}$$

Below are formal definitions for row-major and column-major flattening.

Definition 7.10 (Column-Major Order) Let T be a rank-k tensor of shape $[n_0, \ldots, n_{k-1}]$. A vector V of length $\prod_i n_i$ contains T flattened in column-major order if the element $T[i_0, \ldots, i_{k-1}]$ is placed in $V\left[\sum_l i_l s_l\right]$, where s_l are strides, given by $s_l = \prod_{j=1}^{l} n_{j-1}$.

Definition 7.11 (Row-Major Order) Let T be a tensor of shape $[n_0, \ldots, n_{k-1}]$. A vector V of length $\prod_i n_i$ contains T flattened in row-major order if the element $T[i_0, \ldots, i_{k-1}]$ is placed in $V\left[\sum_l i_l s_l\right]$, where s_l are strides, given by $s_l = \prod_{j=l+1}^{k-1} n_j$.

Example 7.12 If a tensor T of shape $[2, 4, 3]$ is flattened in row-major order, then the strides are $s_0 = 12, s_1 = 3, s_2 = 1$. Thus, the element at $(0, 1, 2)$ is placed at slot $s_1 + 2s_2 = 5$.

In this book, we adopt the row-major convention, and when we write "flattening," we in fact refer to "flattening in a row-major order." It might seem as though limiting ourselves to a single convention loses efficiency, since for some computations, one or the other conventions might be more useful. But that is not really the case. Packing a matrix M in row-major order is equivalent to packing M^T in column-major order. So even though we adopt a single convention, we can still transpose a matrix or reorder the dimensions of a tensor if it needs to be packed in a particular way. On the other hand, a single convention makes algorithms simpler and some properties more consistent. This is demonstrated, for example, in Lemma 7.18, where by assuming row-major order, we can deduce a certain property that the first dimension has.

Remark 7.13 "Row-major order" and "column-major order" are sometimes referred to as "last-order" and "first-order," because of the way indices advance while iterating over the tensor.

7.4 Rotations and Pseudo-Rotations

This section considers tensor rotations. In the context of FHE, rotation of ciphertexts means moving the encrypted elements cyclically by some offset. Here, we generalize the FHE rotation notion to tensors as well, and by rotation we mean moving (rolling) elements of a tensor cyclically.

Remark 7.14 Note that our use of the term "tensor rotation" is different from the usage in mathematics or physics, where it means changing the coordinate system to obtain different components, e.g., due to a change of frame of reference.

Definition 7.15 (Tensor Rotation) Let T be a tensor of shape $[n_1, n_2, \ldots, n_k]$. $\mathrm{Rot}_m^j(T)$ is a rotation of T along the j'th dimension by an offset of m, i.e.,

$$\mathrm{Rot}_m^j(T)[i_0, \ldots, i_{k-1}] = T[i_0, \ldots, (i_j + m) \bmod n_j, \ldots, i_{k-1}] \tag{7.4}$$

For example, consider again the matrix M of shape $[2, 4]$ from Fig. 7.1. Figure 7.3b shows rotating M along the first dimension an offset of 1, $\mathrm{Rot}_1^0(M)$. This means every element is moved one cell upward, with those at the top cycling back to the bottom. Figure 7.3c shows rotating M along its second dimension by the same offset, $\mathrm{Rot}_1^1(M)$, which means every element is moved once cell to the left, with those at the left edge cycling back to the right.

Rotating tensors is useful, for example, when computing convolutions, or for summing a tensor over a dimension. We will see the latter in the next section.

Assume that M is packed into a ciphertext c with $s = 8$ slots using the row-major order convention as in Fig. 7.4. We would like to rotate M, but we cannot do it

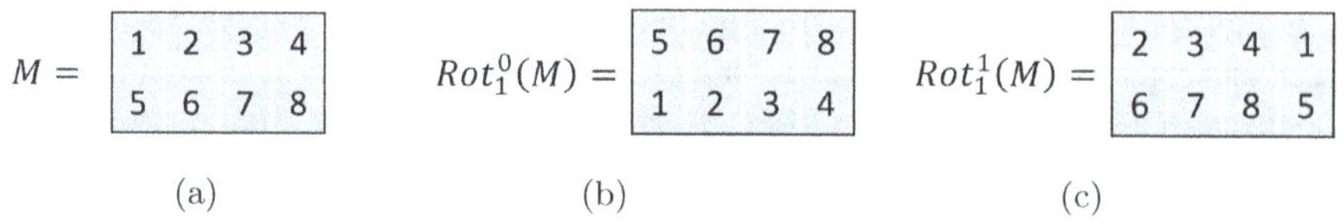

$$M = \begin{array}{|c|c|c|c|} \hline 1 & 2 & 3 & 4 \\ \hline 5 & 6 & 7 & 8 \\ \hline \end{array} \qquad \mathrm{Rot}_1^0(M) = \begin{array}{|c|c|c|c|} \hline 5 & 6 & 7 & 8 \\ \hline 1 & 2 & 3 & 4 \\ \hline \end{array} \qquad \mathrm{Rot}_1^1(M) = \begin{array}{|c|c|c|c|} \hline 2 & 3 & 4 & 1 \\ \hline 6 & 7 & 8 & 5 \\ \hline \end{array}$$

(a) (b) (c)

Fig. 7.3 Rotating the matrix M. (**a**) The original M. (**b**) Rotating M along its first dimension by 1. (**c**) Rotating M along its second dimension by 1

7.4 Rotations and Pseudo-Rotations

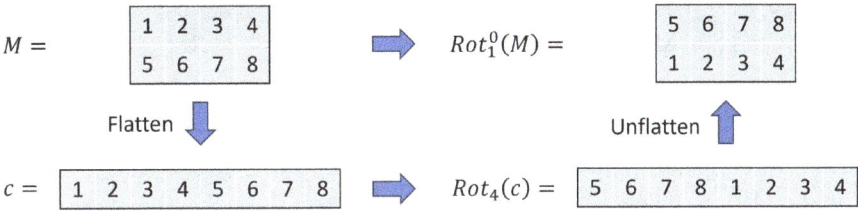

Fig. 7.4 A diagram showing flattening M to a ciphertext c, rotating it and unflattening the results. This is equivalent to rotating M along its first dimension

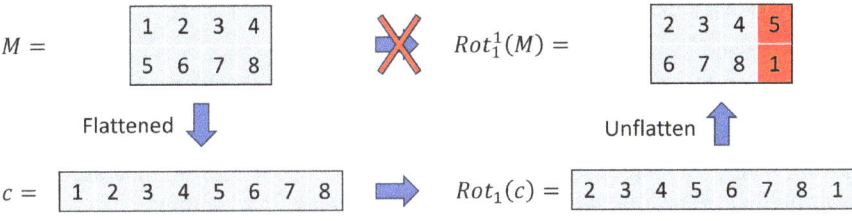

Fig. 7.5 A diagram of flattening M to a ciphertext c, rotating it and unflattening the results. Here the result is not equivalent to rotating M along its second dimension

directly. Our native FHE operators usually only include rotating an entire ciphertext as a flat vector. To rotate M's rows, we can rotate c by an offset of 4: $c' = \text{Rot}_4(c)$. We can then unflatten c' to get the correct result $\text{Rot}_1^0(M)$.

To rotate M's columns, we can try rotating c one position to the left: $c' = \text{Rot}_1(c)$. Unflattening c' will result in something similar to $\text{Rot}_1^1(M)$, but not exactly; see Fig. 7.5. The elements in the last column are not at their correct positions. Therefore, we call this a pseudo-rotation: we do not require elements to cycle back to their expected positions. Some elements shift out on one side, and arbitrary elements shift in on the other side. The concept of pseudo-rotation is useful since it is often easier to implement. As we proceed to discuss more advanced computations that require rotations, we will specify whether they require real rotations or whether pseudo-rotations are enough, making them simpler to implement.

Definition 7.16 (Tensor Pseudo-Rotation) Let T be a tensor of shape $[n_0, \ldots, n_{k-1}]$. $T' = \text{PseudoRot}_m^j(T)$ is a tensor of shape $[n_0, \ldots, n_{k-1}]$ such that $T'[i_0, \ldots, i_{k-1}] = T[i_0, \ldots, i_j + m, \ldots, i_{k-1}]$ for all indices in the range $\forall_l : 0 \leq i_l < n_l$ and $0 \leq i_j + m < n_j$.

The example above shows that when M is packed in a flat vector c, we can rotate it along its first dimension (rows), but along its second dimension, we can only pseudo-rotate it. This is true in general for any tensor: only the first dimension can be fully rotated, as can be seen by the following two lemmas.

We start by showing that any tensor flattened into a ciphertext c can be pseudo-rotated along any dimension.

Lemma 7.17 *Let T be a tensor of shape $[n_0, \ldots, n_{k-1}]$, flattened to a ciphertext c with $s \geq \prod_l n_l$ slots, and let $s_j = \prod_{l=j+1}^{k-1} n_l$. Then $c' = \text{Rot}_{m \cdot s_j}(c)$ contains $T' = \text{PseudoRot}_m^j(T)$ flattened.*

Proof By Definition 7.11, after flattening T, every element $T[i_0, \ldots, i_{k-1}]$ is located in $c[\sum_l i_l s_l]$, and after rotating c by ms_j slots, those elements are in $c'[(-ms_j + \sum_l i_l s_l) \bmod s]$.

To prove the lemma, we must show that all the elements of T that were not wrapped around due to the rotation of c are placed in the correct slot of c'. Particularly, we only care for T elements with their j'th index $0 \leq i_j - m \leq n_j$. These elements are placed in c' in slot index

$$-ms_j + \sum_l i_l s_l = (i_j - m)s_j + \sum_{l \neq j} i_l s_l,$$

and in T' they are placed in $T'[i_0, \ldots, i_j - m, \ldots, i_{k-1}]$. Hence, they are placed correctly with respect to the row-major order flattening of T' within c'. □

The second lemma shows that a tensor that is flattened into a ciphertext c, and fills it entirely, can be rotated along its first dimension.

Lemma 7.18 *Let T be a tensor of shape $[n_0, \ldots, n_{k-1}]$, flattened to a ciphertext c with $s = \prod_l n_l$ slots. Then $c' = \text{Rot}_{m \cdot s_0}(c)$, where $s_0 = \prod_{l=1}^{k-1} n_l$, contains $T' = \text{Rot}_m^0(T)$ flattened.*

Proof Let $T_{i_0=x}$ be a slice of T obtained by fixing the first dimension's index to x. Then the ciphertext c contains the first slice $T_{i_0=0}$ flattened, followed by the second slice $T_{i_0=1}$, and so on, which we denote here as $c = (T_{i_0=0}, T_{i_0=1}, \ldots, T_{i_0=n_0-1})$. Because $T_{i_0=x}$ has s_0 elements for every $0 \leq x < n_0$, we have that

$$c' = \text{Rot}_{ms_0}(c) = (T_{i_0=m}, T_{i_0=m+1}, \ldots T_{i_0=n_0-1}, T_{i_0=0}, \ldots, T_{i_0=m-1}),$$

which after unflattening is T'. □

Remark 7.19 (Native Tensor Representation) Some FHE schemes natively define a ciphertext as some tensor of rank higher than 1 and allow rotating it with a single FHE operation along any of its dimensions (see, e.g., properties of the BGV scheme in [5]). In such cases of course, we can rotate a tensor along dimensions other than its first, though we are still constrained as those rotatable dimensions are rigidly determined by the configuration.

Definition 7.20 summarizes the discussion so far by defining rotatable dimensions, and a follow-up theorem states which dimensions are rotatable and which can only be pseudo-rotated.

Definition 7.20 (Rotatable Dimension) Let T be a tensor of shape $[n_0, \ldots, n_{k-1}]$, flattened in row-major order to a ciphertext c with $s \geq \prod_i n_i$ slots. Dimension i is rotatable if for any offset m, there exists some offset l, such that $\text{Rot}_l(c)$ contains $\text{Rot}_m^i(T)$ packed flattened in row-major order.

Theorem 7.21 *Let T be a tensor of shape $[n_0, \ldots, n_{k-1}]$, flattened to a ciphertext c with $s = \prod_i n_i$ slots. Then dimension 0 is a rotatable dimension. Additional dimensions may be rotatable, depending on the native representation of a ciphertext. All other dimensions can be pseudo-rotated.*

Proof Follows from Lemmas 7.17, 7.18, and Remark 7.19. □

7.5 Simulating Small Ciphertexts

Consider the case in which we want to use some algorithm A for performing matrix-matrix multiplication under encryption. And say this algorithm assumes each of the input matrices fills all the slots of one ciphertext. For example, if each matrix is of shape $[n, n]$, then the algorithm assumes a ciphertext has $s = n^2$ slots, and it receives two ciphertexts as input.

To complicate this scenario, say that for security reasons, we use larger ciphertexts, with $s = kn^2$ slots for some positive k. Since we have more slots, we would like to use them, and compute the products of k pairs of matrices in batch. Can we still use algorithm A without modifications and without increasing the number of operations? The answer is yes, as the following theorem shows.

Theorem 7.22 *Let A be an algorithm that receives input data stored in one or more ciphertexts with s slots each. The algorithm computes a certain function using a sequence of FHE operations $Seq = op_1, op_2, \ldots, op_m$. Using larger ciphertexts with $s \cdot k$ slots, we can simulate k instances of A running in parallel on k inputs, using the same sequence of FHE operations Seq except adapting the offsets of the rotation operations.*

Proof Interpret each set of k ciphertexts as a matrix of shape $[s, k]$ flattened in row-major order into a large ciphertext of $s \cdot k$ slots. This allows us to simulate algorithm A on each column of the matrix as a separate ciphertext by performing the sequence Seq of operations.

1. When $op_i \in Seq$ is an elementwise operator such as addition, multiplication, squaring, and conjugation, perform it elementwise on the matrix.
2. When $op_i \in Seq$ is a rotation by l, rotate the first dimension of matrix. This dimension is rotatable by Lemma 7.18 and can be rotated by multiplying the original rotation offset by k, i.e., rotating the large ciphertext by $k \cdot l$.

□

Remark 7.23 Note that we lose efficiency by working with larger ciphertexts: the cost of certain operations such as rotations increases nonlinearly with the number of slots. However, we assume the ciphertext size was set to a large value due to security constraints; hence, this loss in efficiency cannot be avoided. In Sect. 9.2 we

7.6 Rotating Non-rotatable Dimensions

The term "non-rotatable dimension" does not mean this dimension cannot be rotated at all, but that it cannot be rotated with a single ciphertext rotation operation. This is because for such a dimension, a ciphertext rotation operation can only do pseudo-rotation. However, it can still be rotated using a quite simple chain of operations.

Let T be a tensor flattened to a ciphertext c. We want to compute $\text{Rot}_k^j(T)$, even though j is not a rotatable dimension. Let M be a mask such that the i'th slot has 1 if the corresponding slot in c contains an element of T whose index along dimension j is $i_j \geq k$ and 0 otherwise. Thus, $c_0 = c \odot M$ contains only those elements, and the rest are zeroed out. Similarly $c_1 = c - c \odot M$ contains only those elements where $i_j < k$.

This makes our task easier: c_0 contains all those elements that can be safely rotated k elements to the left without rolling back to the other side. Hence, a pseudo-rotation can perform this task. The elements remaining in c_1 are those that do roll back. So their overall change in position is k to the left, then n_j to the right, due to the rolling back, where n_j is the size of T along dimension j. A pseudo-rotation to the right by an offset of $n_j - k$ mimics this movement.

The final result is therefore $r = \text{PseudoRot}_k(c_0) + \text{PseudoRot}_{-n_j+k}(c_1)$, which contains $\text{Rot}_k^j T$.

The pseudo-rotations in the computation above correctly handle all the nonzero elements of c_0 and c_1. Regarding the zero elements, the pseudo-rotation does not guarantee they will rotate correctly, as some of them are in the range that roll back from one side to the other. However, note that in Fig. 7.5, the values that shifted in on the right are not entirely arbitrary; they were shifted from the left in a permuted way. Thus, if these values are all the same, a pseudo-rotation is the same as rotation. The following lemma proves this observation.

Lemma 7.24 *Let T be a tensor of shape $[n_0, \ldots, n_{k-1}]$, flattened in row-major order to a ciphertext c with $s = \prod_l n_l$ slots and let $s_j = \prod_{l=j+1}^{k-1} n_l$. For some positive m, let all the elements of T whose j'th index i_j satisfies $i_j < m$ be some constant value x. Then $= c' = \text{Rot}_{m \cdot s_j}(c)$ contains $T' = \text{Rot}_m^j(T)$ flattened.*

Proof For positive m, the elements of T such that $i_j \geq m$, are the elements of T' such that $i_j < n_j - m$. By Lemma 7.17 all these elements are placed in the appropriate slots of c'. Hence all the other slots of c' contain the elements of T such that $i_j < m$, which are the elements of T' such that $i_j \geq n_j - m$. They all have a constant value x, hence they are appropriately mapped as well. □

Corollary 7.25 *A similar argument holds for negative m and when the tensor has constant values at* $i_j \geq n_j + m$.

The method shown here for rotating a non-rotatable dimension costs *two* rotations, *one* mask multiplication, and *two* addition operations. This is a generalization of the method shown in [7] for rotating matrices along dimension 1. Since it requires a mask multiplication, it increases the multiplication depth. Hence, it is much preferred to pack tensor dimensions that should be rotated as rotatable dimensions where possible. For example, if a matrix needs to be rotated along dimension 1, it is preferred to pack it transposed.

7.7 Tensor Summation

Another useful operation on tensors is the summation.

Definition 7.26 (Tensor Summation) For a tensor A of shape $[n_0, \ldots, n_{k-1}]$, the operation $B = \text{Sum}^t(A)$ sums the elements of A along the t-th dimension, the resulting tensor B has the shape $[n_0, \ldots, n_{t-1}, 1, n_{t+1}, \ldots, n_{k-1}]$, and for all $j_i < n_i$ and $i \in \{0, 1, \ldots, k-1\} \setminus \{t\}$,

$$B[j_0, \ldots, j_{t-1}, 0, \ldots, j_{k-1}] = \sum_{l=0}^{n_t - 1} A[j_0, \ldots, j_{t-1}, l, \ldots, j_{k-1}], \qquad (7.5)$$

Given our example matrix M of shape $[2, 4]$, then $\text{Sum}^0(M)$ is a vector of shape $[1, 4]$ which is the sum of its two rows, and $sum^1(M)$ is a vector of shape $[2, 1]$ which is the sum of its four columns. See Fig. 7.6.

What happens when M is packed in a ciphertext c with *eight* slots? In that case, we can compute $\text{Sum}^0(M)$ as follows: $c' = c \oplus \text{Rot}_4(c)$ (Fig. 7.7).

This is in fact a special case of the RotateAndSum algorithm from Sect. 7.2. Instead of rotating the ciphertext c, we rotated the matrix M along its first dimension, $\text{Rot}_1^0(M)$, which is equivalent to rotating c by an offset of 4 (see Sect. 7.4). Since M only has two rows, one rotation was enough. If we unflatten c' to a $[2, 4]$ matrix M', we will get the values of $\text{Sum}^0(M)$ in all rows of M'. This is the same as in Observation 7.8.

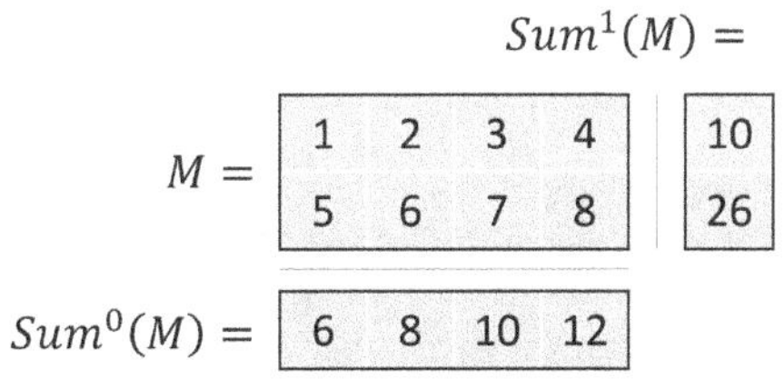

Fig. 7.6 Tensor summation across different dimensions of the matrix M

$$M = \begin{array}{|c|c|c|c|} \hline 1 & 2 & 3 & 4 \\ \hline 5 & 6 & 7 & 8 \\ \hline \end{array}$$

$$Sum^1(M) = \begin{array}{|c|} \hline 10 \\ \hline 26 \\ \hline \end{array}$$

$$Sum^0(M) = \begin{array}{|c|c|c|c|} \hline 6 & 8 & 10 & 12 \\ \hline \end{array}$$

Fig. 7.7 Tensor summation over the first dimension of M

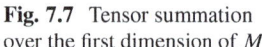

Fig. 7.8 Tensor summation over the second dimension of M

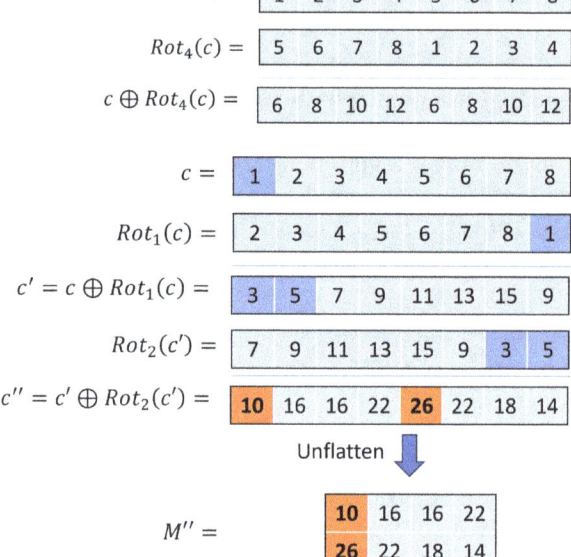

We can also compute $\text{Sum}^1(M)$. Because the size of dimension 1 of M is 4, the RotateAndSum algorithm will require two steps: $M' = M + \text{Rot}_1^1(M)$ and $M'' = M' + \text{Rot}_2^1(M')$. However, as explained in Sect. 7.2, we cannot rotate a packed tensor along an arbitrary dimension. In general, we can only do pseudo-rotations: $M' = M + \text{PseudoRot}_1^1(M)$ and $M'' = M' + \text{PseudoRot}_2^1(M')$, which is equivalent to computing $c' = c \oplus \text{Rot}_1(c)$ and $c'' = c' \oplus \text{Rot}_2(c')$ (Fig. 7.8).

For the RotateAndSum algorithm, pseudo-rotations are enough. We can see that by changing Definition 7.3 to exclude elements where the summation is cyclic. Moreover, Lemma 7.5 still holds, but Observation 7.8 breaks.

In general, we can summarize the elements of a packed tensor along any dimension, as in Algorithm 10. For generality we used pseudo-rotations, but we can use real rotations for rotatable dimensions (Definition 7.20) to get duplicated results, which is sometimes useful. We can similarly adapt the more general summation Algorithms 8 and 9 for cases where the number of slots in a given dimension is not a power of 2.

7.8 Complex Packing

Some FHE schemes such as CKKS natively support complex numbers. That means a ciphertext c with s slots can encrypt s complex numbers. If we happen to perform a computation involving complex numbers, then we can easily take advantage of that of course. But often a computation involves only real numbers.

7.8 Complex Packing

Algorithm 10: TensorRotateAndSum—Summarize the first 2^m elements of T along dimension i

Input: A tensor T, and dimension index i, and a number of elements m.
Output: A tensor containing at position 0 along dimension i the sum of the first 2^m elements of T along this dimension.

1 **Function** TensorRotateAndSum(T, i, m):
2   **for** $j \leftarrow 0$ **to** $m - 1$ **do**
3          $T = T + \text{PseudoRot}_{2^j}^i(T)$
4   **end**
5   **return** T
6 **end**

A naïve way to handle that is to simply treat these real numbers as a special case of complex numbers. That will work fine, but it basically ignores half of the ciphertext's capacity. The imaginary part of all s slots will be set to 0 and will be useless.

Complex packing is a more sophisticated way for handling this situation; see e.g., [1]. Given a vector of $n = 2s$ real numbers $a_0, a_1, \ldots, a_{n-1}$, we can convert it to s complex numbers $a_0 + ia_1, a_2 + ia_3, \ldots, a_{n-2} + ia_{n-1}$. This way we can fit twice as many numbers inside a ciphertext with s slots compared with the naïve method. We denote a complex-packed ciphertext c by $c_\mathbb{C}$.

Let $a_\mathbb{C}$ and $b_\mathbb{C}$ be two complex-packed ciphertexts with s slots encrypting vectors with $n = 2s$ values $A = (a_0, a_1, \ldots, a_{n-1})$ and $B = (b_0, \ldots, b_{n-1})$, respectively, and let c be a ciphertext encrypting in a naïve way the s values of the vector $C = (c_0, \ldots, c_{s-1})$. Then the following properties hold:

1. $a_\mathbb{C} \oplus b_\mathbb{C} = (A + B)_\mathbb{C}$.
2. $a_\mathbb{C} \odot c = (a_0 c_0, a_1 c_0, a_2 c_1, a_3 c_1, \ldots, a_{n-2} c_{s-1}, a_{n-1} c_{s-1})$.
3. $\text{Rot}_k(a_\mathbb{C}) = (a_{2k}, a_{2k+1}, \ldots, a_s, a_0, a_1, \ldots a_{2k-1})$, i.e., a complex-packed encryption of $a_0, \ldots, a_{n-1}$ rotated by $2k$.

Complex-packed ciphertexts can be used for accelerating many computations. For some computations, there is no overhead at all: additions, rotations with even offsets, and multiplying with a non-complex-packed ciphertext (see above). Other operations can be done with little overhead. For example, Algorithm 11 returns the rotation of a complex-packed ciphertext by an odd value k. Another example is Algorithm 12 that takes as input two complex-packed ciphertexts $a_\mathbb{C}$ and $b_\mathbb{C}$ and returns a ciphertext that encrypts the inner product $\sum a_i b_i$ in each of its slots. In the algorithm, we assume the existence of an operation Conj that returns the elementwise conjugate operation on a ciphertext c. Note that in schemes such as CKKS, conjugation operations are provided and share similar implementation to rotations.

Algorithm 11: CpRotate^k—Rotates a complex-packed ciphertext by an odd offset k to the left

Input: A complex-packed ciphertext $c_{\mathbb{C}}$ and an odd index k.
Output: $c_{\mathbb{C}}$ rotated by k to the left.

1 **Function** CpRotate1($c_{\mathbb{C}}$):
2     $c_{1\mathbb{C}} \leftarrow i \cdot c_{\mathbb{C}}$
3     $c_2 \leftarrow -\dfrac{c_{1\mathbb{C}} + \text{Conj}(c_{1\mathbb{C}})}{2}$
4     $c_3 \leftarrow \dfrac{c_{1\mathbb{C}} - \text{Conj}(c_{1\mathbb{C}})}{2}$
5     $c_{4\mathbb{C}} \leftarrow \text{Rot}_{\frac{k-1}{2}}(c_2) \oplus \text{Rot}_{\frac{k+1}{2}}(c_3)$
6   **return** $c_{4\mathbb{C}}$
7 **end**

Algorithm 12: CpInnerProduct—Computes the inner product of two complex-packed ciphertexts

Input: Complex-packed ciphertexts $a_{\mathbb{C}}$ and $b_{\mathbb{C}}$ of s slots.
Output: $c = \text{Enc}(\langle A, B \rangle, \ldots, \langle A, B \rangle)$

1 **Function** CpInnerProduct($a_{\mathbb{C}}, b_{\mathbb{C}}$):
2     $c_{1\mathbb{C}} \leftarrow a_{\mathbb{C}} \odot \text{Conj}(b_{\mathbb{C}})$
3     $c_2 \leftarrow \dfrac{c_{1\mathbb{C}} \oplus \text{Conj}(c_{1\mathbb{C}})}{2}$
4     $c_3 \leftarrow \text{RotateAndSum}(c_2, \log_2 s)$
5   **return** c_3
6 **end**

General computations may involve operators that cannot be directly applied to complex-packed ciphertexts, for example, evaluating the polynomial $x^2 + 1$ with a complex-packed ciphertext $a_{\mathbb{C}}$ as the input x. This computation requires multiplying $a_{\mathbb{C}}$ with itself elementwise, an operation that cannot be directly performed on complex-packed data. This is true in general when evaluating polynomials of degree 2 or more.

In such cases, we can temporarily separate $a_{\mathbb{C}}$ to two naïve-packed ciphertexts by extracting its real re and imaginary im parts as follows:

1. $re = \dfrac{a_{\mathbb{C}} + \text{Conj}(a_{\mathbb{C}})}{2}$ is a naïve packing of $a_0, a_2, \ldots, a_{n-2}$.

2. $im = \dfrac{-i(a_{\mathbb{C}} - \text{Conj}(a_{\mathbb{C}}))}{2}$ is a naïve packing of $a_1, a_3, \ldots, a_{n-1}$.

The inverse direction, i.e., combining two ciphertexts c_1 and c_2 to a complex-packed ciphertext $c_{\mathbb{C}}$ is done by simply computing $c = c_1 + ic_2$.

7.9 Toy Example: Matrix-Vector Multiplication

We are now ready to combine some of the techniques that we learned to perform a more complex task such as matrix-vector multiplication.

Let M be a matrix of shape [4, 4] and V a vector of shape [4, 1]. We would like to compute the matrix-vector product MV under encryption. To do so, let us define V' as a [4, 4] matrix where every row equals V. It is easy to see that $\mathrm{Sum}^1(M \odot V')$ is equal to MV (Fig. 7.9).

Consider now two ciphertexts c_M and $c_{V'}$ of $s = 16$ slots. We pack M flattened in c_M, and V' in $c_{V'}$. Computing MV is done as follows. First, compute the elementwise multiplication $c_{MV'} = c_M \odot c_{V'}$, which means that $c_{MV'}$ contains $M \odot V'$. All that is left to do is to sum over the second dimension using Algorithm 10. Specifically, we compute $c_1 = c_{MV'} \oplus \mathrm{Rot}_1(c_{MV'})$ and $c_{MV} = c_1 \oplus \mathrm{Rot}_2(c_1)$, where c_{MV} contains the resulting vector in every fourth element; see Fig. 7.10. This computation requires one homomorphic multiplication, two homomorphic rotations, and two homomorphic additions.

Remark 7.27 The resulting ciphertext contains unused slots. Furthermore, the used slots that actually contain the result are not consecutive. In Chap. 8 we introduce more advanced packing tools that simplify handling such cases.

$$M = \begin{bmatrix}1&2&3&4\\5&6&7&8\\9&10&11&12\\13&14&15&16\end{bmatrix} \quad V = \begin{bmatrix}1\\2\\3\\4\end{bmatrix} \quad V' = \begin{bmatrix}1&2&3&4\\1&2&3&4\\1&2&3&4\\1&2&3&4\end{bmatrix} \quad M \odot V' = \begin{bmatrix}1&4&9&16\\5&12&21&32\\9&20&33&48\\13&28&45&64\end{bmatrix} \quad Sum^1(M \odot V') = \begin{bmatrix}30\\70\\110\\150\end{bmatrix}$$

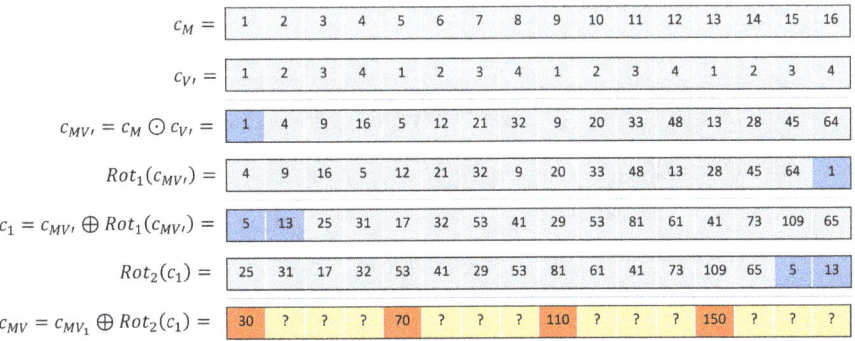

Fig. 7.9 A matrix M vector V multiplication example; the results are $\mathrm{Sum}^1(M \odot V') = MV$

$c_M =$	1	2	3	4	5	6	7	8	9	10	11	12	13	14	15	16
$c_{V'} =$	1	2	3	4	1	2	3	4	1	2	3	4	1	2	3	4
$c_{MV'} = c_M \odot c_{V'} =$	1	4	9	16	5	12	21	32	9	20	33	48	13	28	45	64
$Rot_1(c_{MV'}) =$	4	9	16	5	12	21	32	9	20	33	48	13	28	45	64	1
$c_1 = c_{MV'} \oplus Rot_1(c_{MV'}) =$	5	13	25	31	17	32	53	41	29	53	81	61	41	73	109	65
$Rot_2(c_1) =$	25	31	17	32	53	41	29	53	81	61	41	73	109	65	5	13
$c_{MV} = c_{MV_1} \oplus Rot_2(c_1) =$	30	?	?	?	70	?	?	?	110	?	?	?	150	?	?	?

Fig. 7.10 An encrypted matrix M and vector V multiplication example. Orange slots are the expected logical results, while the rest of elements of c_{mv} are meaningless info that can be ignored and thus marked with "?"

7.10 Diagonal Packing Techniques

This section shows an alternative method for matrix-vector multiplication referred to as the diagonal method [4, 8]. Let us return again to the example with the matrix M of shape [4, 4], only that this time we will assume each ciphertext has $s = 4$ slots. We will pack M into *four* ciphertexts $c_0, \ldots, c_3$, each contains a different (cyclic) diagonal of M, as in Fig. 7.11. The vector V naturally fits in one more ciphertext d, and Lemma 7.28 shows that

$$t = \sum_i c_i \mathrm{Rot}_i(d) \tag{7.6}$$

computes a ciphertext t that encrypts the matrix-vector multiplication MV.

A generalization of this approach referred to as the hybrid method [8] allows handling rectangular wide matrices (matrices with more columns than rows) as follows. Let M' be a matrix of shape [4, 8], and V' a vector of shape [8, 1], and let each ciphertext be of size $s = 8$. We pack M' in *four* ciphertexts $c_0, \ldots, c_3$ based on diagonals that wrap around, and V' is packed as before in one ciphertext d. See Fig. 7.12.

As before, the computation starts with $t = \sum_i c_i \mathrm{Rot}_i(d)$. The resulting ciphertext t does not hold the final outcome, $M'V'$. Interpreting t as a matrix of size [2, 4] flattened in row-major order, it contains two rows of *four* elements each, and the result is the sum of these rows. Thus, $t \oplus \mathrm{Rot}_4(t)$ will contain $M'V'$ in its first four slots.

After getting the intuition for the hybrid diagonal approach, we are ready to define it more formally for the general case. Let M be a matrix of shape $[n, kn]$ and V a vector of shape $[kn]$. Let the number of slots in a ciphertext be $s = kn$. Pack M in n ciphertexts $c_0, \ldots, c_{n-1}$, such that c_i encrypts a vector whose j'th

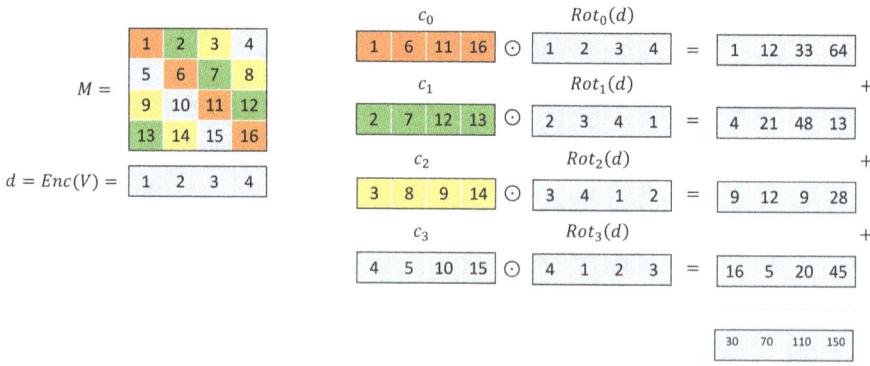

Fig. 7.11 A matrix-vector multiplication using the diagonal technique (see [4, 8])

7.10 Diagonal Packing Techniques

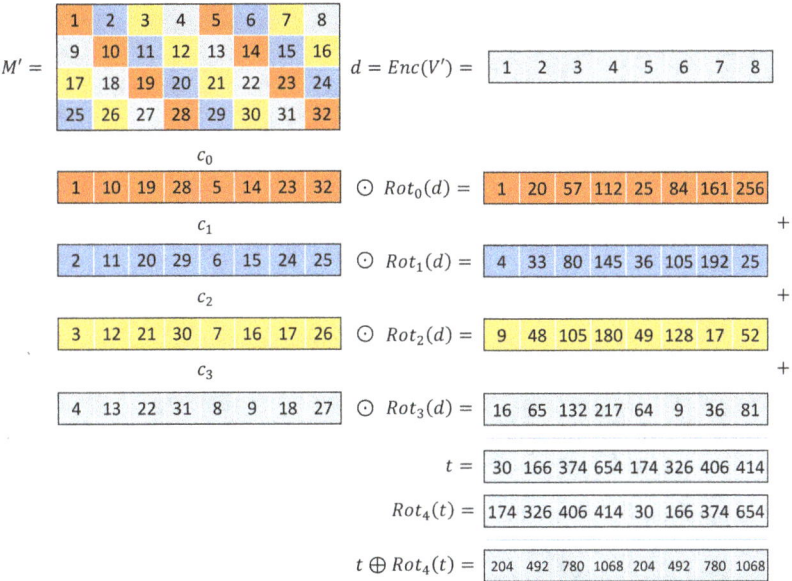

Fig. 7.12 A matrix-vector multiplication using the diagonal technique (see [4, 8]) over a rectangular matrix

element is $M[j \bmod n, (i+j) \bmod nk]$ and pack the vector V in a single ciphertext d in the ordinary manner.

Computing MV is done by first computing $t = \sum_i c_i \mathrm{Rot}_i(d)$. Then, interpreting t as a $[k, n]$ matrix M' flattened in a row-major order, and computing $\mathrm{Sum}^0(M')$ using Algorithm 10. The result is a single ciphertext containing MV duplicated k times.

Lemma 7.28 *Let M be a matrix of shape $[n, kn]$ and V a vector of shape $[kn]$. The hybrid diagonal method produces a single ciphertext with the result MV duplicated k times.*

Proof Let c_i be the encryption of M diagonals, d an encryption of V, and set $t = \sum_i c_i \mathrm{Rot}_i(d)$, where $0 \leq i < n$. Thus,

$$t[j] = \sum_{i=0}^{n-1} M[j \bmod n, (i+j) \bmod kn] \cdot V[(i+j) \bmod kn] \tag{7.7}$$

Interpreting t as a matrix M' of shape $[k, n]$ flattened in row-major order, we get

$$M'[x, y] = t[xn + y] \tag{7.8}$$

$$= \sum_{i=0}^{n-1} M[y, (i + xn + y) \bmod kn] \cdot V[(i + xn + y) \bmod kn]$$

Now, set $M'' = \text{Sum}^0(M)$, i.e., M'' is of shape $[1, n]$ and

$$M''[0, y] = \sum_{x=0}^{k-1}\sum_{i=0}^{n-1} M[y, (i+xn+y) \bmod kn] \cdot V[(i+xn+y) \bmod kn] \quad (7.9)$$

$$= \sum_{x=0}^{nk-1} M[y, x] \cdot V[x],$$

where the last equation holds because summing over x and i causes the expression $(i + xn + y) \bmod kn$ to consider all values in the range $0, \ldots, nk-1$.

M'' is packed flattened in t. To sum over the first dimension, we use Algorithm 10, except that according to Lemma 7.18, we can use real rotations and not pseudo-rotations, which produce duplicated results. □

The expression $\sum_{i=0}^{n-1} c_i \text{Rot}_i(d)$ obviously requires n multiplications and $n-1$ rotations. The number of rotations can be reduced to $2\sqrt{n}$ based on the baby-step/giant-step method [5, 7]. The idea is to decompose n to $n = mp$, where m and p are some factors of n, and compute

$$\sum_{i=0}^{m-1}\sum_{j=0}^{p-1} c_{ip+j}\text{Rot}_{ip+j}(d). \quad (7.10)$$

This is equivalent to the above summation, since $ip + j$ runs through all values $0, 1, \ldots, n-1$. Extracting a common rotation factor ip results in

$$\sum_{i=0}^{m-1} \text{Rot}_{ip}\left(\sum_{j=0}^{p-1} \text{Rot}_{-ip}(c_{ip+j})\text{Rot}_j(d)\right), \quad (7.11)$$

where the rotated diagonals $c'_{ip+j} = \text{Rot}_{-ip}(c_{ip+j})$ can be prepared in advance. Thus, the final computation reduces to

$$\sum_{i=0}^{m-1} \text{Rot}_{ip}\left(\sum_{j=0}^{p-1} c'_{ip+j}\text{Rot}_j(d)\right). \quad (7.12)$$

This costs $p-1$ rotations for $\text{Rot}_j(d)$) (the baby-steps) and additional $m-1$ rotations for the Rot_{ip} operator (the giant-steps). Choosing $p = \sqrt{n}$, we get $2\sqrt{n}-2$ rotations.

Remark 7.29 For computing the matrix vector multiplication MV using the diagonal technique, we assumed that the ciphertext size is large enough to contain V, whereas in Sect. 7.9, we assumed it is large enough to contain M. In the next chapter, we will extend both techniques to accommodate matrices and vectors of arbitrary size, independent from ciphertext size. This will allow us to compare the two techniques with ciphertexts of equal sizes. For that reason, we do not include a comparison of the different methods here and defer it to the next chapter.

7.11 Lab Exercise: Programming Encrypted Tensors and the Studied Primitives

The following lab exercise requires a programming environment such as Python. This environment should include libraries that can handle cleartext tensors such as NumPy [6] and also FHE libraries such as *pyhelayers* [2]. The exercises assume the FHE library allows for toy configuration with a small number of slots $s = 16$ or $s = 4$. Such configurations are usually not secure, but their use here will simplify initial experimentation with packing of small tensors. The **pyhelayers** library supports a mockup option that only emulates the FHE scheme without any encryption, allowing for simple configuration with any number of slots, and fast runtime. Such an option is ideal for the exercises below.

When using pyhelayers, we recommend using a mockup context with 16 slots of the CKKS scheme. In pyhelayers ciphertexts are named CTiles.

1. Create a vector V of length 16 with random elements from the cleartext space of the selected FHE scheme. Encrypt it to a ciphertext c. Implement Algorithm 7 to compute V's sum using RotateAndSum operations on c. Decrypt the obtained result, and compare it with a direct summation over V.
2. Create a random tensor A of shape [4, 4]. Flatten it in row-major order to a vector, and encrypt this vector to a ciphertext c. Implement Algorithm 10 to compute A's sum of rows and sum of columns. Decrypt the results and compare with the plaintext computation.
3. Create a random vector V of shape [4, 1]. Compute the vector AV using ciphertext operations on c and the methods of Sect. 7.9. Compare with the plaintext computation.
4. Repeat the last exercise, but with both inputs transposed. That is, compute $V^T A^T$, which should produce basically the same result, only transposed. Pack in c the matrix A transposed, multiply it with a matrix V' such that every column has a copy of V, and sum over the first dimension. What is the difference in the results? Why? Which method is better?
   [Hint:] See Sect. 7.7.
5. Compute again the vector AV, but this time use ciphertexts with only *four* slots, and use the diagonal technique of Sect. 7.10.
6. Using ciphertexts of 16 slots, create a random vector V of size 32, and use the complex-packing technique to pack it in a single ciphertext. Implement Function 11.

7. Create a second random vector U of size 32, and pack it in a second ciphertext. Compute their inner product as in Algorithm 12. Verify the correctness of the result by comparing with a plain computation.

References

1. Aharoni, E., Drucker, N., Ezov, G., Shaul, H., Soceanu, O.: Complex encoded tile tensors: accelerating encrypted analytics. IEEE Secur. Privacy **20**(5), 35–43 (2022)
2. Aharoni, E., Adir, A., Baruch, M., Drucker, N., Ezov, G., Farkash, A., Greenberg, L., Masalha, R., Moshkowich, G., Murik, D., Shaul, H., Soceanu, O.: HeLayers: a tile tensors framework for large neural networks on encrypted data. Privacy Enhancing Technology Symposium (PETs) 2023 (2023). https://petsymposium.org/popets/2023/popets-2023-0020.php
3. Cormen, T.H., Leiserson, C.E., Rivest, R.L., Stein, C.: Introduction to Algorithms, 3 edn. PHI Learning Pvt. Ltd. (Originally MIT Press), Cambridge (2010)
4. Halevi, S., Shoup, V.: Algorithms in HElib. In: Garay, J.A., Gennaro, R. (eds.) Advances in Cryptology – CRYPTO 2014, pp. 554–571. Springer, Berlin (2014)
5. Halevi, S., Shoup, V.: Faster homomorphic linear transformations in HElib. In: Annual International Cryptology Conference, pp. 93–120. Springer, Berlin (2018)
6. Harris, C.R., Millman, K.J., van der Walt, S.J., Gommers, R., Virtanen, P., Cournapeau, D., Wieser, E., Taylor, J., Berg, S., Smith, N.J., Kern, R., Picus, M., Hoyer, S., van Kerkwijk, M.H., Brett, M., Haldane, A., del Río, J.F., Wiebe, M., Peterson, P., Gérard-Marchant, P., Sheppard, K., Reddy, T., Weckesser, W., Abbasi, H., Gohlke, C., Oliphant, T.E.: Array programming with NumPy. Nature **585**(7825), 357–362 (2020). https://doi.org/10.1038/s41586-020-2649-2
7. Jiang, X., Kim, M., Lauter, K., Song, Y.: Secure outsourced matrix computation and application to neural networks. In: Proceedings of the 2018 ACM SIGSAC Conference on Computer and Communications Security, pp. 1209–1222 (2018)
8. Juvekar, C., Vaikuntanathan, V., Chandrakasan, A.: GAZELLE: a low latency framework for secure neural network inference. In: 27th USENIX Security Symposium (USENIX Security 18), pp. 1651–1669 (2018)
9. Knuth, D.E.: The Art of Computer Programming, vol. 2, Seminumerical Algorithms, vol. 2, 2 edn. Addison-Wesley Pub (Sd), Boston (1981)

Chapter 8
SIMD Packing Part II—Tile Tensor Basics

Abstract In this chapter we describe tile tensors, a versatile data structure used for working with tensors under encryption. It offers a layered approach that separates packing details from a high-level view of the computation. Its tile tensor shape notation allows to easily choose from among multiple packing options.

8.1 What Are Tile Tensors?

Tile tensors are data structures designed for computations involving tensors (see Definition 7.9) under encryption. A single tile tensor object represents a single tensor. For example, say A is some tensor, and T_A is a tile tensor that represents it. Then T_A contains all the data of A encrypted and stored within ciphertexts. It might be that more than one ciphertext is needed, either because A is too big to fit within one ciphertext, or there might be other reasons, as we will see later on.

A tile tensor T_A can pack the data of A using a wide variety of options. These options can be viewed and controlled by the developers or by automated optimization tools, using a special language and notation called *tile tensor shapes*. The tile tensor shape notation is a simple and accurate packing description language. It covers and generalizes upon many widely used packing techniques, e.g., for matrix multiplication [4, 6, 8, 13], for convolution [3, 8–11], and others.

A tile tensor T_A logically represents some tensor A. Performing operators on T_A is equivalent to performing the same operators on the tensor it represents. For example, computing the product of T_A and T_B will produce a tile tensor T_R, which logically represents the tensor $R = AB$. Here too, using tile tensor operators can succinctly describe and generalize upon a multitude of algorithms for dealing with packed data known in the literature.

Since tile tensors are for storing and operating on tensors, we open this chapter with some more details on tensor operators.

8.2 Elementwise Operators on Tensors

Many AI applications effectively translate to tensor-based computations. Consider, for example, a convolutional neural network (CNN) used for image processing. The input to such a network is often a rank-4 tensor of shape $[I_w, I_h, I_c, I_b]$, where I_w, I_h are the width and height of the image, I_c is the number of channels (e.g., three channels for red, green, and blue), and I_b is the number of samples batched together. Running this input through the network means applying a series of tensor operators on this input tensor: convolution, matrix-multiplication, pooling, summation, concatenation, rotation, flattening, etc.

We have already defined tensor rotation and tensor summation in Definitions 7.15 and 7.26, respectively. In this section, we extend our toolbox for handling tensors by defining elementwise operators and in particular binary elementwise operators. We focus here on plain (unencrypted) tensors, following conventions common in AI, and in the remaining sections we discuss encrypted tensors.

To perform elementwise operators on two tensors A and B their shapes must satisfy some constraints. Clearly, if they have identical shapes then we can operate on them elementwise. But a less strict requirement termed *compatibility* is usually sufficient.

Definition 8.1 (Compatible Shapes) A tensor shape $[n_0, \ldots, n_{k-1}]$ and a tensor shape $[m_0, \ldots, m_{k-1}]$ are compatible if they have the same number of dimensions and $m_i = n_i$ or either $n_i = 1$ or $m_i = 1$, for $0 \leq i < k$.

Example 8.2 (Compatible Shapes) Let A, B, and C be tensors of shape $[4, 5]$, $[1, 5]$, and $[4, 1]$, respectively. According to Definition 8.1 all three are pairwise compatible. Let D be a tensor of shape $[4, 6]$, then D is compatible with C, but it is not compatible with A and B because they mismatch on the size of the second dimension. See Fig. 8.1.

Next we define the broadcasting operation, which expands a given tensor to the dimensions of another compatible tensor, so that it is possible to perform elementwise operators on them.

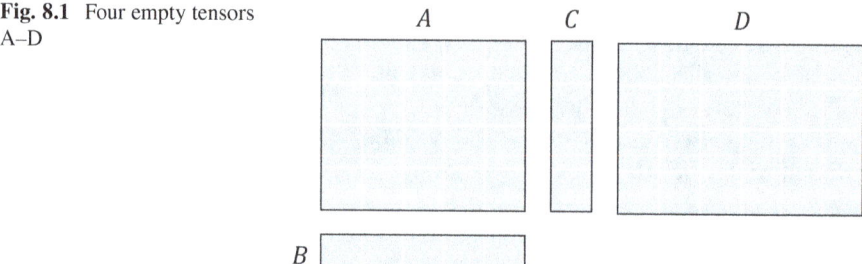

Fig. 8.1 Four empty tensors A–D

8.2 Elementwise Operators on Tensors

Definition 8.3 (Broadcasting) For a tensor A of shape $[n_0, \ldots, n_{k-1}]$ and another compatible shape $[m_0, \ldots, m_{k-1}]$, the operation

$$A' = \text{Broadcast}(A, [m_0, \ldots, m_{k-1}]) \tag{8.1}$$

replicates the content of A, m_i times along every dimension i where $n_i = 1$. The output tensor A' is of shape

$$[\max(n_0, m_0), \max(n_1, m_1), \ldots, \max(n_{k-1}, m_{k-1})]. \tag{8.2}$$

Because Broadcasting basically enables elementwise operators, it is usually implemented as an automatic operation that is performed before applying an elementwise operators on two compatible tensors. For example, let us compute $A + B$ with A and B from the last example. Since they are compatible this operation is allowed, but first they'll be automatically broadcasted to fit each other's shapes. In this case B will be broadcasted to the shape $[4, 5]$ (Fig. 8.2).

We can also compute $B + C$, since they are compatible. Here both B and C will be broadcasted to the shape $[4, 5]$ (Fig. 8.3).

Remark 8.4 We note that some definitions of compatibility and broadcasting allow for different ranks as well. For example, degenerated dimensions can be added to the tensor with the lower rank, and subsequently, these dimensions are broadcasted if needed. For our purposes of focusing on packing methods we will not consider

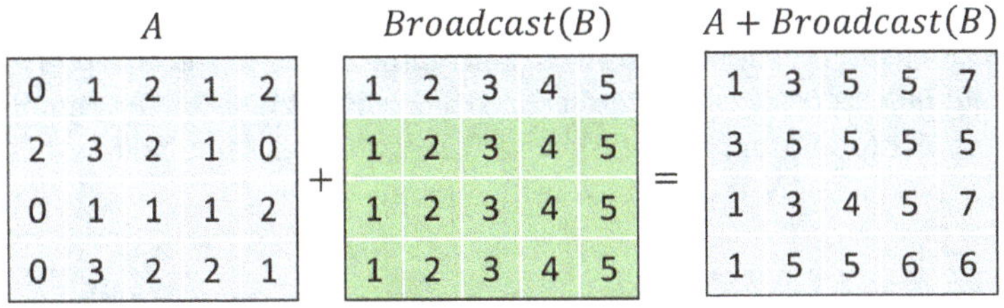

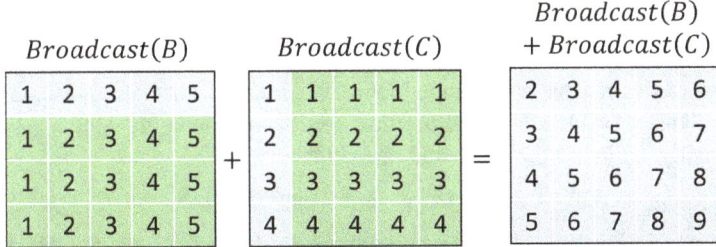

Fig. 8.2 Computing $A + B$ by broadcasting B along the second dimension

Fig. 8.3 Computing $B + C$ by broadcasting B and C along the first and second dimensions, respectively

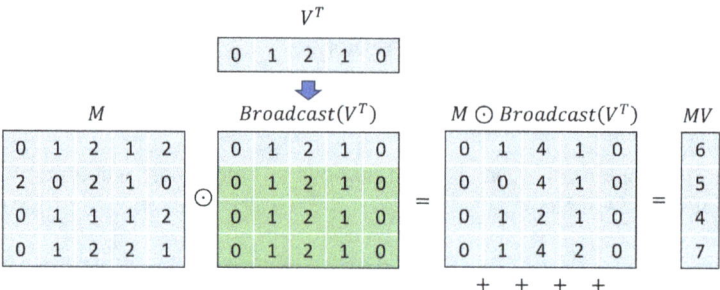

Fig. 8.4 Computing MV using broadcasting

tensors of different ranks compatible. Instead, we will explicitly state when we add degenerate dimensions, specifying their position in the shape.

Using compatibility and broadcasting we can define matrix-vector multiplication using more basic operators. Given a matrix M of shape $[a, b]$ and a vector V of shape $[b]$, we will first reshape V to $[1, b]$, then compute $\text{Sum}^1(M \odot V)$. This computation first broadcasts V to have the shape $[a, b]$, meaning a matrix where every row has a copy of V. Then we multiply it elementwise with M, and sum over the rows to obtain a vector of shape $[a, 1]$ as expected (Fig. 8.4).

We can similarly compute matrix-matrix multiplication. Given M of shape $[a, b]$ and N of shape $[b, c]$, we first reshape both of them to $[a, b, 1]$ and $[1, b, c]$, respectively, then compute $\text{Sum}^1(M \odot N)$. This results in the product MN shaped as $[a, 1, c]$. This is because $M \odot N$ first broadcasts both inputs to the shape $[a, b, c]$, and the result has the same shape as well, containing all the products $M[i, j] \cdot N[j, k]$ for all valid indices i, j, k. Then the second dimension is reduced by summing over it. This is akin to the function einsum of the Python package NumPy [7]. Using einsum the product MN is computed using numpy.einsum($'ij','jk'$, A, B).

8.3 Tile Tensor Overview

As mentioned in the introduction to this chapter, a single tile tensor object represents a single tensor. Now let us examine an example demonstrating how this is done.

Say we have a matrix M of shape $[5, 6]$. A tile tensor T_M that represents M will have the elements of M stored within *tiles*; see illustration in Fig. 8.5.

Informally, a tile has the following properties:

1. It has the same rank as the tensor, part of which it contains.
2. It has the same total number of elements as the number of ciphertext slots. Because of this equality, each tile can be stored flattened in a ciphertext.
3. All tiles of a single tile tensor have the same shape.

8.3 Tile Tensor Overview

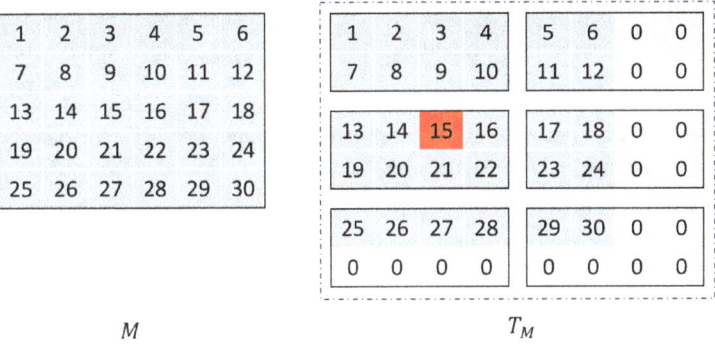

Fig. 8.5 A matrix M of shape [5,6] and a tile tensor T_M of shape $\left[\frac{5}{2}, \frac{6}{4}\right]$ with tiles of shape [2,4]

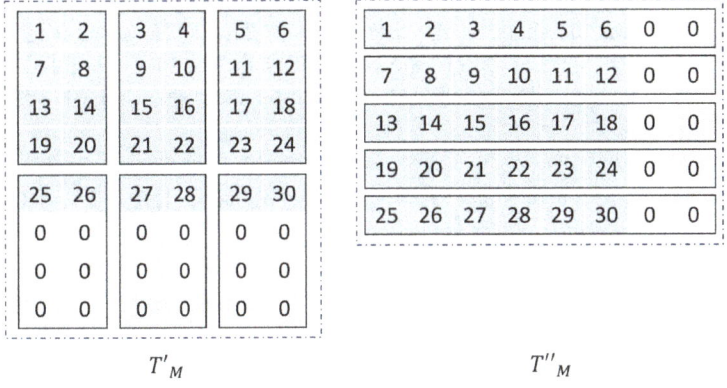

Fig. 8.6 A matrix M of shape [5,6] and tile tensors T'_M and T''_M with tiles of shape [4,2], and [1,8], respectively

In Fig. 8.5 every tile is a rank-2 tensor, or a matrix, containing a sub-block of matrix M. Some tiles are only partially filled: those at the bottom row and at the right column. The remaining unused slots are filled with zeroes (and colored in light gray). In this example, each tile has a total of *eight* elements; thus, also the underlying ciphertext has *eight* slots. Essentially, T_M stores the data of M divided between six ciphertexts, each containing the encryption of one flattened tile of shape [2, 4].

Of course, we can choose other tile shapes, but whatever we choose has to be consistent for all tiles for a given tile tensor. Figure 8.6 provides two more examples, with tile shapes [1, 8], and [4, 2].

The matrix M, like any other tensor, has a shape: [5, 6]. Also T_M, as any other tile tensor, has a shape. In Fig. 8.5 this shape is $\left[\frac{5}{2}, \frac{6}{4}\right]$. The numbers above the fraction line indicate the size of the tensor which T_M represents, which is called the *logical tensor*. In this case it is M, whose shape is [5, 6]. The numbers below the fraction line indicate the shape of the tiles inside the tile tensor. In the case of T_M it is [2, 4]. Similarly, T'_M has shape $\left[\frac{5}{4}, \frac{6}{2}\right]$, and T''_M has shape $\left[\frac{5}{1}, \frac{6}{8}\right]$, which can be abbreviated $\left[5, \frac{6}{8}\right]$.

Tile tensors are created from tensors using an operator TTEnc that receives a tensor, and a desired tile tensor shape. For example, T_M was created as follows: $T_M = \text{TTEnc}(M, \left[\frac{5}{2}, \frac{6}{4}\right])$. The operator in this case divides M into blocks of shape [2, 4], flattens each one, pads with zeroes as needed, and subsequently encrypts it in a ciphertext. The T_M data structure contains the resulting *six* ciphertexts together with the tile tensor shape.

The inverse operator TTDec takes a tile tensor and returns the logical tensor it represents. It does so using the following steps: decrypt each ciphertext, unflatten each resulting vector to a tile according to the tile shape stored in the tile tensor shape, and finally combine all tiles together to form one large tensor. The tile tensor shape also contains the shape of the logical tensor, and TTDec uses it to trim out the zero padding if there is any. In the case of T_M the *six* tiles are glued together to form a [6, 8] matrix, which is then trimmed to have the shape [5, 6]. Thus, $M = \text{TTDec}(T_M)$.

Now, say another matrix N of the same shape [5, 6] is packed to a tile tensor using the same tile tensor shape $T_N = \text{TTEnc}(N, \left[\frac{5}{2}, \frac{6}{4}\right])$. Tile tensors support operators similar to the tensor operators we discussed in Sect. 8.2, so, for example, we can now add the two tile tensors $T_S = T_M + T_N$. The result of this operation is a tile tensor that represents a tensor $S = M + N$. Alternatively, we can also write $\text{TTDec}(T_M + T_N) = \text{TTDec}(T_M) + \text{TTDec}(T_N)$.

So we can think of the tile tensors T_M and T_N as logical representations of M and N. Internally they are stored, packed, and encrypted with various metadata, but they allow us to operate on the tensors M and N as if we were performing the operators directly on them. This way we can perform a whole circuit of tensor operations by applying the same operators on tile tensors instead of on the tensors they represent.

This abstraction becomes handy also when writing programming code. The code should consider only the tensors that the tile tensors logically represent, and it can be agnostic to some parameters such as the specific packing details. At any time we can change the initial tile tensor shape to another, as long as we satisfy some compatibility constraints. Our circuit will run with the tensors stored using the new packing arrangements.

In this brief overview we have shown a few packing options, but tile tensor shapes can be used to describe many more. They describe the packing arrangement using a compact and precise language. This language covers most known packing solutions used in practice. It can not only serve to ease communication among researchers, but also to help with programming, log printouts, and debugging.

To summarize, here are the main advantages of using tile tensors:

1. **Writing packing-oblivious code.** Packing arrangements can be easily changed by modifying a tile tensor shape, without the need to modify the entire FHE-based program.
2. **Facilitating a search in the packing options space.** The ability to enumerate packing schemes allows designers to scan many different packing options in search for the most efficient ones. This can be done offline or online, manually or automatically by using an optimization tool.
3. **Simplicity.** Tile tensors offer a compact and precise language to describe packing arrangements.

8.4 Tile Tensor Basic Concepts

With the intuition obtained above, it will now be easier to understand some formal definitions.

Definition 8.5 (Tile Tensor) A *tile tensor* is a data structure containing data arranged in *tiles* and a *tile tensor shape*.

Definition 8.6 (Tile) A *tile* is a tensor stored flattened and encrypted in a ciphertext with s slots. The total number of elements in a tile is also s.

The tiles of a tile tensor data structure all have the same rank and shape. They are themselves arranged in a multi-dimensional array called the *external tensor*.

Definition 8.7 (External Tensor) A multi-dimensional array of tiles, with the same rank as the tiles' rank. All the tiles in this array have the same shape.

For example, in Fig. 8.5 each tile is of shape [2, 4], and there are *six* of them. They are themselves arranged in an array of shape [3, 2]. This is the external tensor.

Definition 8.8 (Logical Tensor) A tile tensor T_M logically represents a tensor M called the *logical tensor*, which is the output of applying TTDec on T_M.

A tile tensor data structure contains the main data which is the external tensor. It also contains the *tile tensor shape*. The tile tensor shape contains metadata defining how to interpret the information stored within the external tensor.

Definition 8.9 (Tile Tensor Shape) Tile tensor shape is part of a tile tensor object. It defines the shape of the tiles, the shape of the logical tensor, the shape of the external tensor, and how data is mapped from the logical tensor to the slots of the tiles.

For example, the shape $\left[\frac{5}{2}, \frac{6}{4}\right]$ of Fig. 8.5 defines that the logical tensor of T_M has shape [5, 6] and that the tile shape is [2, 4]. The external tensor's shape is implicitly defined to have enough tiles to accommodate the logical tensor, which is $\lceil \frac{5}{2} \rceil = 3$ rows of tiles and $\lceil \frac{6}{4} \rceil = 2$ columns; thus, the external tensor shape is [3, 2].

Definition 8.10 (Tile Tensor Rank) The rank of a tile tensor is the rank of its external tensor, which also equals the rank of each of its tiles and the rank of its logical tensor.

8.5 Dimension Independence and Basic Tiling

Consider a vector V of shape [6], packed into a tile tensor T_V of shape $\left[\frac{6}{4}\right]$. This means we separated it to *two* tiles of $s = 4$ slots each.

Note that each of the *six* rows in Fig. 8.5 is packed in the same way as in Fig. 8.7, except the last row, which is completely empty. This is no surprise, since in Fig. 8.5 the shape was $\left[\frac{5}{2}, \frac{6}{4}\right]$, where the second dimension is $\frac{6}{4}$, same as in Fig. 8.7.

This phenomenon is generally true. Whenever there is a tile tensor shape $[\ldots, \frac{6}{4}, \ldots]$ with $\frac{6}{4}$ describing dimension i, a slice across dimension i will be packed the same as in Fig. 8.7. This brings us to the following more refined definition of the structure of a tile tensor shape.

Definition 8.11 (Tile Tensor Shape Element) A rank-k tile tensor shape is a vector with k elements. The i'th element independently defines the size of the i'th dimension of the logical tensor, the external tensor, and each tile. It also specifies a mapping from each slot and an index j_i to the i'th dimension of the logical tensor.

Each tile tensor shape element can be one of several types, each specifying the packing along its respective dimension in a different way. We will start with the simplest type, *basic tiling*.

Definition 8.12 (Basic Tiling) A tile tensor shape element $\frac{n_i}{t_i}$ specifies that the size of the i'th dimension of the logical tensor is n_i, the tile size along this dimension is t_i, and the size of the external tensor is $\lceil \frac{n_i}{t_i} \rceil$. In the external tensor, in the tile with index l_i, the slot at index m_i is mapped to the logical tensor index $j_i = l_i t_i + m_i$.

In Definition 8.12, the indices j_i, l_i, m_i refer to indices along the i'th dimension of the corresponding tensor. For some slots it might happen that the j_i index comes out with an illegal value that exceeds the size of the i'th dimension of the logical tensor. In this case the slot is unused. We will define this more formally in Sect. 8.6.

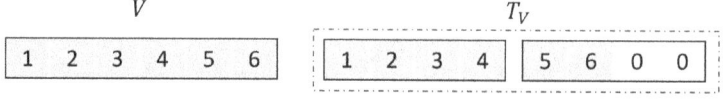

Fig. 8.7 A row vector V of shape [6] and a tile tensor T_V of shape $\left[\frac{6}{4}\right]$

8.6 Used, Unused, and Unknown Slots

Example 8.13 (Basic Tiling) Figure 8.5 illustrates T_M, a tile tensor of shape $\left[\frac{5}{2}, \frac{6}{4}\right]$, in which we highlighted one slot with the red color. Our marked slot lies in a tile whose external tensor indices are $(l_0, l_1) = (1, 0)$ (i.e., second row, first column), and its indices within its tile are $(m_0, m_1) = (0, 2)$ (i.e., first row, third column). Thus, the highlighted slot is mapped to $j_0 = 2l_0 + m_0 = 2$. Similarly the second dimension is mapped to $j_1 = 4l_1 + m_1 = 2$. As expected, this slot contains the value of the logical tensor at indices $(2, 2)$.

This may seem as an overly elaborate way of saying that we divide the logical tensor into contiguous blocks of equal size. However, writing the definitions independently for each dimension will generalize more easily, as we will see next.

8.6 Used, Unused, and Unknown Slots

In previous examples we saw that the logical tensor does not always fill all slots of all tiles. Some slots remain unused since the dimension sizes are not generally divisible by the tile sizes.

Definition 8.14 (Used Slots) *Used slots* are tile tensor slots that are mapped by the tile tensor shape to some element at indices $(j_0, j_1, \ldots, j_{k-1})$ within the valid range of the logical tensor.

Definition 8.15 (Unused Slots) *Unused slots* are slots of a tile tensor that are not defined as used slots.

For simplicity, we define the TTEnc operator to fill unused slots with zeroes. But we point out that they do not always remain zero. Due to various side effects of operations, they sometimes get filled with arbitrary values. This is harmless when decrypting a tile tensor, since the TTDec operator ignores unused slots. However, some other operators either fail to work or work differently if the unused slots are not zeroes.

For this purpose, the tile tensor shape includes an indication whether the unused slots are filled with zeroes or not. As usual, this is done per dimension.

Definition 8.16 (Unknown Slots Mark) When the i'th element of a tile tensor shape is marked with the suffix "?" then slots with invalid indices for dimension i are marked as unknown slots.

Definition 8.17 (Content of Unused Slots) An unused slot is guaranteed to hold 0, unless it is marked as unknown by any dimension, in which case there is no guarantee for its value.

Example 8.18 (Unknown Slots) Consider M as in our previous examples but this time the tile tensor shape of T_M is $\left[\frac{5}{2}, \frac{6?}{4}\right]$; see Fig. 8.8.

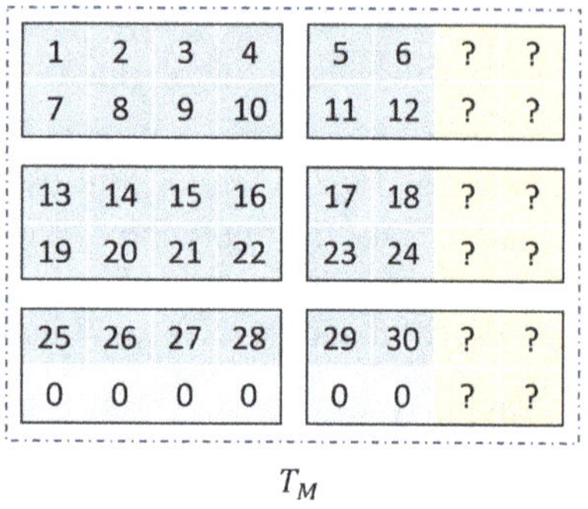

Fig. 8.8 A tile tensor T_M of shape $\left[\frac{5}{2}, \frac{6?}{4}\right]$

| 1 | 2 | 3 | 4 | 5 | 6 | ? | ? |
| 7 | 8 | 9 | 10 | 11 | 12 | ? | ? |

| 13 | 14 | 15 | 16 | 17 | 18 | ? | ? |
| 19 | 20 | 21 | 22 | 23 | 24 | ? | ? |

| 25 | 26 | 27 | 28 | 29 | 30 | ? | ? |
| 0 | 0 | 0 | 0 | 0 | 0 | ? | ? |

T_M

This means that the last two columns are filled with arbitrary values instead of zeroes. In contrast, the last row is filled with zeroes as before, except the last two slots at the bottom right which exceed the valid range both on the first and the second dimensions. These unused slots contain arbitrary values since the second dimension marks them as unknown.

In some cases, to allow further computations we need to clean the unknown slots and fill them with zeroes. The following operator is used for this task,

Definition 8.19 (TTClearUnknowns) Given a tile tensor T_A, the operator $T'_A =$ TTClearUnknowns(T_A) creates a copy of T_A except all the unused slots are filled with zeroes, and in the tile tensor shape, all the unknown slot marks ("?") are removed.

This operation is easily implemented by multiplying with masks. Since we can use the tile tensor shape to determine which slots are unused, and which of them (if at all) are not already set to zero, we can customize a 0–1 mask for each tile such that elementwise multiplication with these masks will ensure that all unused slots are zero. Each tile may need a different mask, as the pattern of used and unused slots varies between tiles.

Remark 8.20 Clearing unknowns can sometimes be piggybacked with some other operation, or at other times it can be a side effect of another operation. A simple example is multiplying a tensor by a scalar. There are two ways to implement such an operation for a tile tensor. The simplest is to multiply every tile with the required scalar, where multiplication by scalar is usually a primitive provided by FHE libraries. This will not clear the unknowns. However, we can also multiply each tile with a mask containing the scalar in all used slots, and 0 in all unused slots. It will be somewhat more costly to encode these masks, but they will clear the unknown slots as a side effect.

Remark 8.21 Unknown values may be perceived as harmless because they appear in slots that are trimmed during a TTDec$(\cdot)$ operation, i.e., these slots are ignored during decryption. However, keeping them uncleared for a long period of computation may have some devastating side effects. In many cases, ciphertexts are in fact

8.7 Duplications

polynomials, and when an unknown value crosses some bound it has an effect on the other polynomial coefficients. Consider, for example, packing a tensor M of shape [4, 5] with values in [0, 1] in a tile tensor T_M of shape $\left[\frac{4}{4}, \frac{5?}{2}\right]$. Raising the elements of T_M to some large power p will keep the values associated with M in the range [0, 1]. In contrast, the unknown values may go to infinity when p grows. When that happens there is no guarantee whether T_M truly represents M anymore. To avoid these cases, it is important to either keep track on the bound of the unknown values during the computation or clear them when they may cause problems.

8.7 Duplications

Another packing option for tile tensors allows us to store data duplicated across a dimension of a tile. This relates to the broadcasting operator for tensors we saw in Definition 8.3, which is supported also by tile tensors as follows.

Definition 8.22 (Duplicated Dimension) A tile tensor shape element $\frac{*}{t_i}$ specifies that the size of the i'th dimension of the logical tensor is 1, the tile size along this dimension is t_i, and the size of the external tensor is 1. Each slot is mapped to logical tensor index $j_i = 0$.

Example 8.23 (Duplicated Dimension) Figure 8.9 illustrates a vector V shaped as a matrix row [1, 6] and packed in a tile tensor T'_V with tile tensor shape $\left[\frac{*}{2}, \frac{6}{4}\right]$.

T'_V shape's first dimension, $\frac{*}{2}$, indicates that the tile size is 2, but each of its two slots are mapped to the same logical index $j_0 = 0$, meaning the first (and only) row of V. Consider, for example, the left tile of T'_V, and in it, the slot at the bottom right. The indices of the tile (0, 0) and the slot indices inside the tile are (1, 3). The first duplicated dimension maps it to the logical index $j_0 = 0$, and the second dimension which uses ordinary basic tiling (Definition 8.12) maps it to the logical index $j_1 = 3$. Hence, it contains the value $V(0, 3) = 4$. Thus, $\frac{*}{2}$ essentially means that V is duplicated twice inside T'_V.

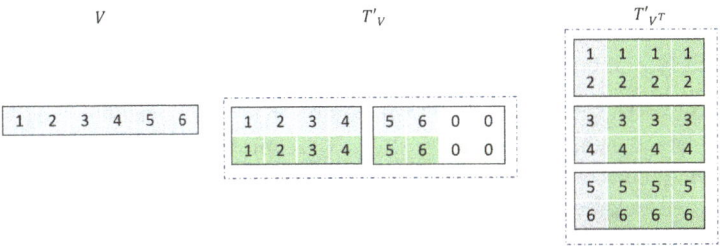

Fig. 8.9 Tile tensors T'_V and T'_{V^T} of shapes $\left[\frac{*}{2}, \frac{6}{4}\right]$ and $\left[\frac{6}{2}, \frac{*}{4}\right]$, respectively

Here is an alternative. We will reshape V to V^T of shape $[6, 1]$ and pack it into T'_{V^T} of shape $\left[\frac{6}{2}, \frac{*}{4}\right]$. Here, the second dimension $\frac{*}{4}$ specifies that each of the four columns of a tile is mapped to $j_1 = 0$, the only column of V^T, meaning that it is duplicated *four* times.

A tile tensor shape dimension $\frac{*}{t_i}$ essentially means that this dimension is a degenerate dimension of size 1. The single element it has along this dimension is duplicated t_i times. This is useful for certain operators as we will see later. It may also imply an excess use of memory by a factor of t_i, since we store each element duplicated t_i times. For example, if V is a vector of shape $[1, 16]$ and we pack it as $\left[\frac{*}{2}, \frac{16}{4}\right]$ it will require *four* tiles. However, if we pack it as $\left[\frac{16}{8}\right]$ it will require only *two* tiles.

The operator TTDec returns the logical tensor and ignores duplications. Therefore, for T'_V of Example 8.23, we will obtain $V = \text{TTDec}(T_V)$. The TTDec operator examines the metadata of the tile tensor, i.e., its tile tensor shape, understanding that the tiles contain duplications, and discards them to reconstruct the original shape of the logical tensor.

Remark 8.24 One way for TTDec to ignore duplications is to simply pick one value and discard the duplicates. However, note that in FHE, sometimes the encrypted values contain noise. Therefore, instead of discarding duplicated values it is more useful to take their average, which reduces the noise.

If a dimension is not duplicated and we want it to be, we can use the TTDup operator defined as follows.

Definition 8.25 (TTDup) Given a tile tensor T_A such that the i'th element of its shape is $\frac{1}{n_i}$, and there are no unknown slots along the other dimensions, then $T'_A = \text{TTDup}^i(T_A)$ creates a tile tensor T'_A with the same logical tensor as T_A and the same tile tensor shape except the i'th element changes to $\frac{*}{n_i}$.

Example 8.26 (Applying TTDup**)** Let V be a tensor of shape $[1, 6]$, packed to a tile tensor T_V of shape $\left[\frac{1}{2}, \frac{6}{4}\right]$. Then $T'_V = \text{TTDup}^0(T_V)$ is a tile tensor of shape $\left[\frac{*}{2}, \frac{6}{4}\right]$ as in Fig. 8.9.

Since TTDup does not change the logical tensor, only the way it is packed, then in Example 8.26 it holds that $\text{TTDec}(T_V) = \text{TTDec}(T'_V) = V$.

Remark 8.27 The definition of TTDup requires that no dimension in the tile tensor shape is marked as having unknown slots. We require this to simplify the methods that implement this operator and they are shown next. An alternative definition could have handled unknown values by first cleaning them up and then duplicating the values. We chose to define the operator as above since cleaning unknowns requires masking multiplication and consumes multiplication depth. Hence, we preferred for it to appear explicitly when it is needed.

8.7 Duplications

The operator TTDup can be implemented using a variant of RotateAndSum, applied on every tile. For duplicating dimension i, we sum the values along dimension i of each tile of the tile tensor. But, instead of using Algorithm 10, we use a variant that rotates the tile to the right, as in Algorithm 13. The end result is, of course, the same: each element along dimension i contains the sum of all elements along this dimension, which is simply the single non-zero element with which we started. An algorithm demonstration is given in Fig. 8.10

Algorithm 13: TensorRotateAndSumReverse(T,i,m)—sum the last 2^m elements of T along dim i

---
Input: A tensor A of n dimensions, and $0 \leq i < n$
Output: Sum the last 2^m elements of T along dimension i.
1 **Function** TensorRotateAndSumReverse(T, i, m):
2   for $j \leftarrow 0$ to $m - 1$ do
3          $T = T \oplus \text{Rot}^i_{-2^j}(T)$
4   end
5   return T
6 end

The reason that we use right rotations instead of left rotations is related to the limited ability to rotate along arbitrary dimensions. Recall from Chap. 7 that we can only pseudo-rotate a tensor's dimension (Lemma 7.17) for non-rotatable dimensions. However, since the dimension we want to duplicate on has trailing zeroes, we can rotate it to the right. The reason relies on Lemma 7.24. The lemma considers the zero padding on this dimension as constant values. Thus, it guarantees that these values will correctly rotate and cycle back from right to left. To make sure all the unused slots are zeroes, we must have no unknown values in any dimension, not just the dimension on which we duplicate. This is to prevent cases as in Example 8.18.

Algorithm 13 has the property that whenever we perform a right rotation by an offset of -2^j we have at least that amount of trailing zeroes; hence, we can do the

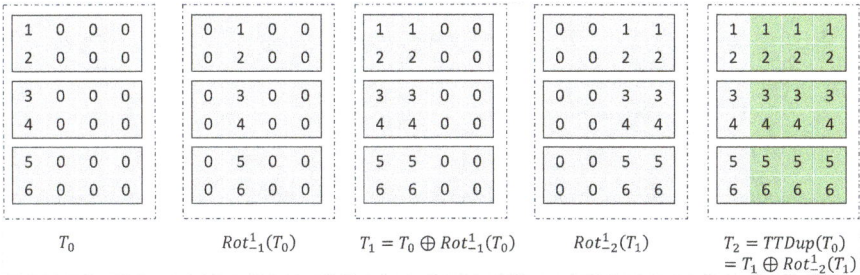

Fig. 8.10 A tensor T_0 of shape $\begin{bmatrix} 6 \\ 2 \end{bmatrix}, \frac{1}{4} \end{bmatrix}$ duplicated along dimension 1 to T_2 of shape $\begin{bmatrix} 6 \\ 2 \end{bmatrix}, \frac{*}{4} \end{bmatrix}$

rotation. If the tile's size along the i'th dimension t_i is a power of 2, then we keep halving the number of trailing zeroes while doubling the rotation offset, and in the last iteration both equal $t_i/2$. If t_i is not a power of 2, we need to use the more general summation techniques as in algorithm 8 or 9.

Remark 8.28 We showed how to take a tile tensor dimension $[\ldots, \frac{1}{t_i}, \ldots]$ and duplicate it to $[\ldots, \frac{*}{t_i}, \ldots]$. Sometimes a more general operation is needed, where a dimension needs to be first split, then duplicated. For example, a vector of shape [16] packed as $\left[\frac{16}{16}\right]$ needs to be converted to $\left[\frac{16}{4}, \frac{*}{4}\right]$. This requires multiplying by masks for the splitting, then duplicating. For the extreme case of splitting the vector to $\left[16, \frac{*}{16}\right]$ (every element moves to a separate tile), there are methods to reduce the number of rotations at the expense of increasing the multiplication depth (see [6]).

8.8 Elementwise Operators

In Sect. 8.2 we saw that we can apply elementwise operators on tensors if they have compatible shapes. The same is true for tile tensors as well. Our basic approach will be very simple: apply the operator tile-wise on the tile tensors' external tensors, including broadcasting if needed, then update the resulting tile tensor shape. Let's start with an example.

Example 8.29 (Multiplying Two Tile Tensors) Consider a matrix M of shape [5, 6] and a vector V of shape [1, 6]. The result of their elementwise product $R = M \odot V$ is a tensor of shape [5, 6]. It is computed by first broadcasting V to have *five* rows (Definition 8.3), then multiplying elementwise. Consider that M is packed into a tile tensor T_M with shape $\left[\frac{5}{2}, \frac{6}{4}\right]$, and V is packed in T_V with shape $\left[\frac{*}{2}, \frac{6}{4}\right]$.

As shown in Fig. 8.11 T_M has an external tensor of shape [3, 2] and T_V has an external tensor of shape [1, 2]. To multiply them we first broadcast the smaller one, the external tensor of T_V, to have the shape [3, 2], then do the operation tile-wise. The result is a tile tensor T_R with shape $\left[\frac{5}{2}, \frac{6}{4}\right]$. Once we decrypt and unpack it, we obtain the matrix R, which is T_R's logical tensor.

We also compute $R' = M + V$ using the same technique. Here too, we get a tile tensor T'_R containing R' as its logical tensor. However, this time the tile tensor shape is $\left[\frac{5?}{2}, \frac{6}{4}\right]$. The last row which is unused is now filled with non-zero values; hence, the first dimension is marked with "?".

The scenario in Example 8.29 is well defined because the following conditions hold:

- Both T_M and T_V have tiles with identical shapes [2, 4], allowing us to operate tile-wise.

8.8 Elementwise Operators

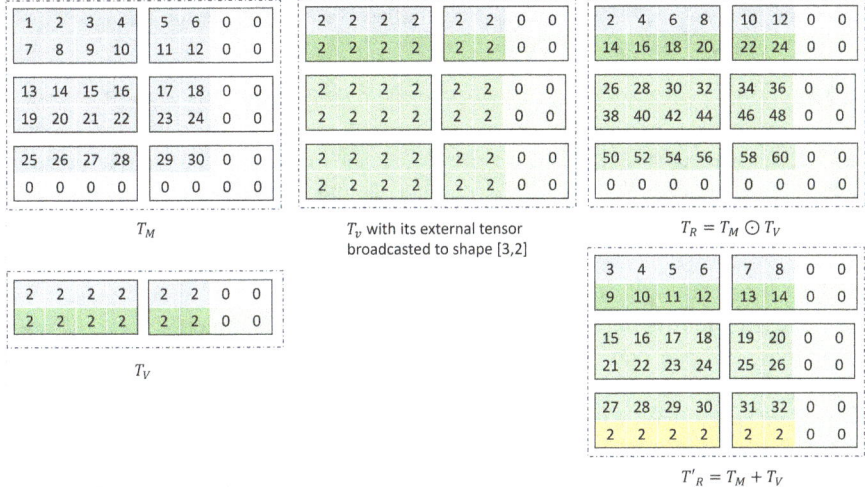

Fig. 8.11 Multiplying and adding a tile tensor T_M of shape $\left[\frac{5}{2}, \frac{6}{4}\right]$ to a tile tensor T_V of shape $\left[\frac{*}{2}, \frac{6}{4}\right]$, resulting with the tile tensors T_R and T'_R, of shapes $\left[\frac{5}{2}, \frac{6}{4}\right], \left[\frac{5?}{2}, \frac{6}{4}\right]$ respectively

- T_V was already broadcasted along the first dimension (the $\frac{*}{2}$ element of its shape), which allowed us to further broadcast its external tensor as much as we want by simply duplicating entire tiles.
- The second dimension of both T_M and T_V is identical.

We can generalize this list to the following condition for compatibility of two tile tensor shapes.

Definition 8.30 (Compatible Tile Tensor Shapes) Two tile tensor shapes are compatible if their logical tensor shapes are compatible, and for all i their i'th elements are compatible. Specifically, the three elements $\frac{n_i}{t_i}$, $\frac{n_i?}{t_i}$, and $\frac{*}{t_i}$ are mutually compatible with each other.

We can now formally define how to perform elementwise operators.

Definition 8.31 (Elementwise Operators on Tile Tensors) Given two tile tensors T_A and T_B with compatible tile tensor shapes, the result of applying an elementwise operator $T_A \diamond T_B$ is a tile tensor T_C such that:

1. Its external tensor is the result of applying $\diamond$ on T_A and T_B's external tensors. This means broadcasting the two external tensors to have the same shape, then applying $\diamond$ tile-wise.
2. The i'th element of its tile tensor shape is $\frac{*}{t_i}$ when both inputs have $\frac{*}{t_i}$ in their i'th element of their tile tensor shape.
3. Otherwise, at least one of the inputs has a non-duplicated dimension in its i'th element, either $\frac{n_i}{t_i}$ or $\frac{n_i?}{t_i}$. The i'th element of the resulting tile tensor shape will

also be either $\frac{n_i}{t_i}$ or $\frac{n_i?}{t_i}$, depending on whether we can deduce the unused slots to be all zeroes or not.

Definition 8.31 does not specify whether the resulting tile tensor shape elements are marked with ? or not. We can leave the ? out in cases where we can be sure that the unused slots are filled with zeroes. One such case is when the output shape element is $\frac{n_i}{t_i}$ with n_i that is a multiple of t_i. In this case there cannot be unused slots at all. A second case is when both inputs are $\frac{n_i}{t_i}$ (with no unknown slots), and a third is when one input has unknowns and the other does not, but the operator is multiplication. Multiplication will further prevent unknowns from being created in a fourth case, multiplying $\frac{n_i}{t_i}$ and $\frac{*}{t_i}$, as demonstrated in Example 8.29.

The elementwise operators on tile tensors as defined here have an important property that will be shared with other tile tensor operators: applying them to the tile tensors is equivalent to applying the same operators to the logical tensors that they represent. The following lemma formalizes this.

Lemma 8.32 *If T_A and T_B are two tile tensors with compatible tile tensor shapes, then for any elementwise operator $\diamond$ it holds that* $\mathrm{TTDec}(T_A \diamond T_B) = \mathrm{TTDec}(T_A) \diamond \mathrm{TTDec}(T_B)$.

8.9 Sum Over Dimension

Section 8.2 showed how for regular tensors we can perform matrix multiplications and other linear algebraic operations by using elementwise multiplication and summation. Section 8.8 showed how to perform elementwise operators using tile tensors. In this section, we add the sum operator for tile tensors.

Figure 8.12 demonstrates how to sum along dimension 1 of the tile tensor T_M of shape $\begin{bmatrix} \frac{5}{2} & \frac{6}{4} \end{bmatrix}$. The first step is basically summing over T_M's external tensor, which

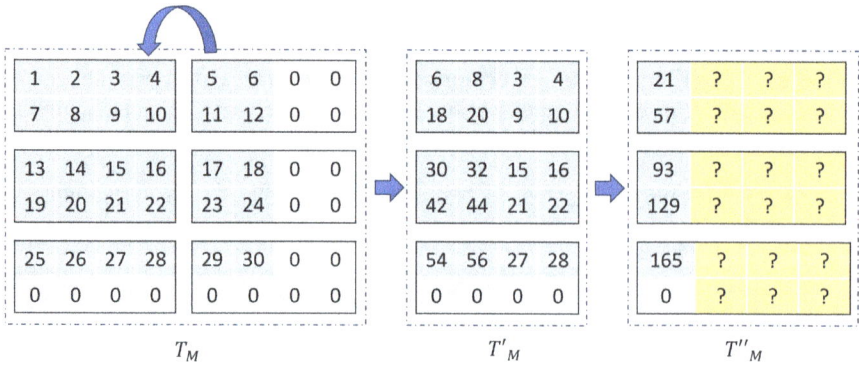

Fig. 8.12 Accumulating the elements of T_M along dimension 1

8.9 Sum Over Dimension

is summing over the [3, 2] matrix of tiles and summing the tiles of each column together.

The second step is summing inside each tile, same as in Sect. 7.7. Recall that this step is basically a RotateAndSum algorithm, which uses either rotations or pseudo-rotations. The above example uses pseudo-rotations; hence, the result comes out in the first column, and the other cells are filled with unknown values. The resulting shape is $\left[\frac{5}{2}, \frac{1?}{4}\right]$.

Definition 8.33 (Tile Tensor Summation) Let T_A be a tile tensor with a tile tensor shape $[\ldots, \frac{n_i}{t_i}, \ldots]$. The operation $T_B = \text{Sum}^i(T_A)$ sums the external tensor of T_A tile-wise as in Definition 7.26, then sums each remaining tile using Algorithm 10, either using rotations if i is a rotatable dimension or pseudo-rotations otherwise. If pseudo-rotations are used then the i'th element of the tile tensor shape is reduced to $\frac{1?}{t_i}$, and if real rotations are used then it is reduced to $\frac{*}{t_i}$.

Remark 8.34 Definition 8.33 uses Algorithm 10 to sum inside a tile. This function assumes the tile size is a power of 2. If it is not we can use other variants of the rotate-and-sum algorithm, as discussed in Sect. 7.7. Similarly, the definition above assumes the dimension being summed over is free of unknown values (no "?" in its shape at the i'th place). This constraint can be relaxed as well using the same variants of the rotate-and-sum algorithm which can sum only over the used slots (see Algorithms 8 and 9).

The following lemma states that summing over a tile tensor is logically equivalent to summing over the logical tensor.

Lemma 8.35 *For any tile tensor T_A and dimension i, it holds that*

$$\text{TTDec}(\text{Sum}^i(T_A)) = \text{Sum}^i(\text{TTDec}(T_A)), \tag{8.3}$$

and the tile tensor shape of $\text{Sum}^i(T_A)$ correctly reflects its content with respect to unknown values and duplication.

Proof Our proof proceeds by examining the possible cases, as shown in Fig. 8.13.

Figure 8.13a shows a tile tensor with shape $\left[\frac{9}{4}, \frac{3}{4}\right]$, as a representative of a general case $\left[\ldots, \frac{n_i}{t_i}, \ldots, \frac{n_j}{t_j}, \ldots\right]$, where i is the dimension being summed over is i, and j is a second dimension we examine in the proof. Our goal is to show dimension i is correctly being summed over. We further show that dimension j retains its shape, i.e., that after the summation the shape $\left[\frac{n_j}{t_j}\right]$ remains the correct description of this dimension.

The summation operator sums tiles along dimension i, then sums inside the resulting tile. Thus, the slot marked by (x) now contains the sum of all the values in the second column of T_A: $(x) = 1+4+7+10+13+16+19+22+25+0+0+0$. If the summation inside the tile uses real rotations, then the slots marked by (y) will be duplicates of (x); otherwise, if pseudo-rotations are used, then they will contain

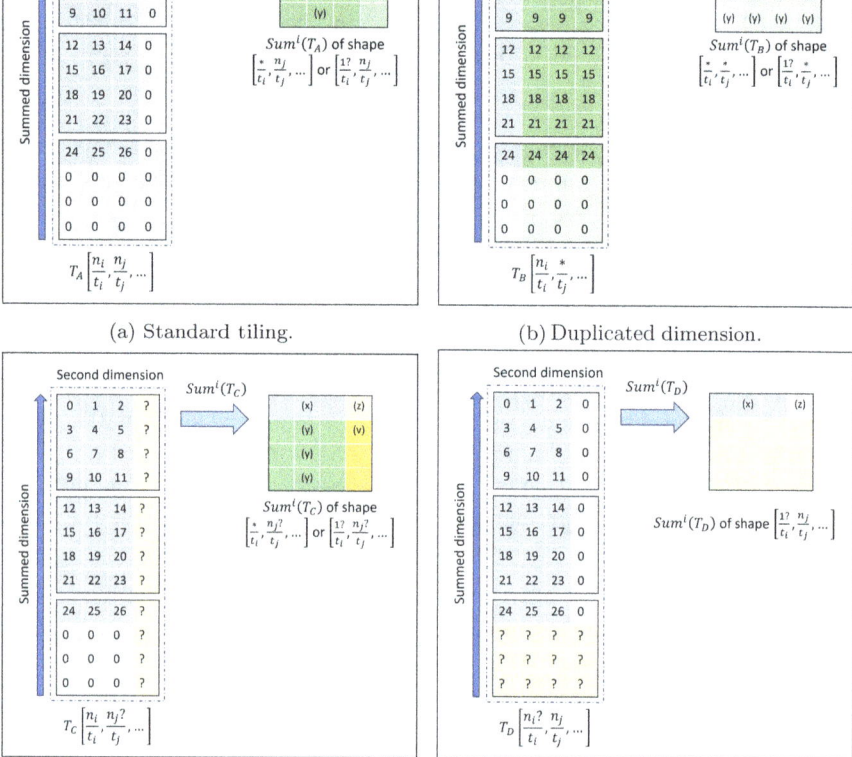

Fig. 8.13 Tile tensors with different tile shapes for the proof of Lemma 8.35. For readability, we use [dim i, dim j, ...] instead of [..., dim i, ..., dim j, ...], so that i and j here do not necessarily refer to dimensions 0, 1, respectively. (**a**) Standard tiling. (**b**) Broadcast dimension. (**c**) Unknowns in the jth dimension. (**d**) Unknowns in the ith dimension

unknown values. Hence, the i dimension is now correctly described by either $\left[\frac{*}{t_i}\right]$ or $\left[\frac{1?}{t_i}\right]$, and it contains the correct sum. The slot marked by (z) is the sum of all the zeroes in the right-most column of T_A; hence, it equals zero. Therefore, the j dimension retains its shape $\left[\frac{n_j}{t_j}\right]$: it has the same number of used slots and same number of unused slots, which are filled with zeroes. The slot marked by (v) is again either duplicate of (z) or unknown.

Figure 8.13b shows a similar case, except the second dimension now has shape $\begin{bmatrix} * \\ t_j \end{bmatrix}$. By the same considerations, the i dimension is correctly being summed over, and the j dimension retains its duplicated structure.

Figure 8.13c shows the case that the j dimension has unknown values. These unknown values do not enter the range of values being summed into the slot marked by (x); hence, the summation is still computed correctly (with or without duplication). The slot marked by (z) is now the sum of unknown values; hence, it too contains an unknown value. Therefore, dimension j retains its shape in this case as well.

Figure 8.13d shows the case that the dimension being summed over contains unknown values. The simple summation algorithm will include these unknown values in the summation, causing an incorrect value in the slot marked by (x). However, as mentioned in Remark 8.34, more sophisticated algorithms can sum over just the used slots. Using these, (x) will contain the correct sum, and (z) will contain zero, and all slots in rows below will contain unknown values. The exact details of this computation are left as an exercise (see Exercise 7). □

8.10 Linear Algebra Using Tile Tensors

We can now demonstrate how to perform various linear algebraic operations using elementwise multiplication and summation.

A simple example is an inner product computation. Assume that we have vectors V and U of size n each, packed into tile tensors T_V and T_U, both with the tile tensor shape $\begin{bmatrix} \frac{n}{s} \end{bmatrix}$, where s is the number of slots in a ciphertext. The inner product is computed using $\text{Sum}^0(T_V \odot T_U)$ and the result is a tile tensor T_R with a tile tensor shape $\begin{bmatrix} \frac{*}{s} \end{bmatrix}$. The result is duplicated due to Observation 7.8.

Let us write this equation where we attach the associated shape of a tile tensor to its name.

$$\text{Sum}^0 \left(T_U \begin{bmatrix} \frac{n}{s} \end{bmatrix} \odot T_V \begin{bmatrix} \frac{n}{s} \end{bmatrix} \right) = T_R \begin{bmatrix} \frac{*}{s} \end{bmatrix}. \tag{8.4}$$

Similarly, we can perform matrix-vector multiplication of a matrix M of shape $[a, b]$ and a vector V of shape $[b]$. Here are two possible equations for doing this computation:

$$\text{Sum}^1 \left(T_M \begin{bmatrix} \frac{a}{t_0}, \frac{b}{t_1} \end{bmatrix} \odot T_V \begin{bmatrix} \frac{*}{t_0}, \frac{b}{t_1} \end{bmatrix} \right) = T_R \begin{bmatrix} \frac{a}{t_0}, \frac{1?}{t_1} \end{bmatrix} \tag{8.5}$$

$$\text{Sum}^0 \left(T_M \begin{bmatrix} \frac{b}{t_0}, \frac{a}{t_1} \end{bmatrix} \odot T_V \begin{bmatrix} \frac{b}{t_0}, \frac{*}{t_1} \end{bmatrix} \right) = T_R \begin{bmatrix} \frac{*}{t_0}, \frac{a}{t_1} \end{bmatrix}. \tag{8.6}$$

In Eq. 8.5, the tile shapes are $[t_0, t_1]$, such that $t_0 t_1 = s$, the number of slots. Both matrix and vector use the same tile shapes, to ensure compatibility. Both tensors have a dimension of size b, and we place it at the same position: dimension 1. This means each element of V will be multiplied with the corresponding element along M's second dimension. We add a duplicated dimension to T_V for broadcasting purposes. The $\frac{*}{t_0}$ in T_V aligned with $\frac{a}{t_0}$ of T_M essentially means we want the operation to be applied to each of the a rows of the matrix. Finally, the summation reduces the second dimension to 1 and leaves unknown values in the unused slots, hence the "?" mark (Definition 8.33).

In Eq. 8.6 we packed M transposed, so now we sum over the first dimension and not the second. According to Definition 8.33 we obtain duplication in the result. This is useful: the tile tensor shape of T_R has its first dimension duplicated, which matches the requirement from the shape of the input vector in Eq. 8.5, and hence we can do a further matrix multiplication on T_R without any additional preprocessing. We will show a concrete demonstration near the end of this section.

To continue with another matrix multiplication with the result of Eq. 8.5, we first need to duplicate the second dimension using the TTDup operator, which requires us to first clear it from unknowns. Hence,

$$T'_R = \text{TTDup}^1(\text{TTClearUnknowns}(T_R)). \tag{8.7}$$

This does not change its logical tensor, but only changes its tile tensor shape to $\left[\frac{a}{t_0}, \frac{*}{t_1}\right]$, making it ready as the input vector of Eq. 8.6.

Equations 8.5 and 8.6 are well defined for any combination of the input sizes a, b, and any choice of tile shapes t_0, t_1. We discuss the performance implications of different choices in Sect. 8.11. The above method extends to many other linear algebraic operators according to these simple rules:

1. Identify the common dimension. This is the dimension that has the same size in both tensors and the one over which values will be summed.
2. Align the common dimension. By aligning we mean placing it in the same dimension index of both tile tensor shapes.
3. Prefer placing the common dimension as the first dimension, so the summation will produce duplicated results.
4. All other dimensions of each tensor should be aligned with duplicated dimensions.

Example 8.36 (Sequence of Matrices) Consider, for example, the *four* matrices X, Y, Z, W of shapes $[a, b], [b, c], [c, d], [d, e]$. Our goal is to compute the product $XYZW$. Here is what we can do. We first pack these matrices as:

$$T_X\left[\frac{b}{t_0}, \frac{a}{t_1}, \frac{*}{t_2}\right] \qquad T_Y\left[\frac{b}{t_0}, \frac{*}{t_1}, \frac{c}{t_2}\right]$$

$$T_Z\left[\frac{*}{t_0}, \frac{d}{t_1}, \frac{c}{t_2}\right] \qquad T_W\left[\frac{e}{t_0}, \frac{d}{t_1}, \frac{*}{t_2}\right].$$

Computing two intermediate products, $T_T = \text{Sum}^0(T_X \odot T_Y)$ and $T'_T = \text{Sum}^1(T_Z \odot T_W)$, results with tile tensors of the following shapes:

$$T_T \left[\frac{*}{t_0}, \frac{a}{t_1}, \frac{c}{t_2} \right] \qquad T'_T \left[\frac{e}{t_0}, \frac{1?}{t_1}, \frac{c}{t_2} \right].$$

We can then clear and duplicate the second dimension of T'_T by performing

$$T''_T = \text{TTDup}^1(\text{TTClearUnknowns}(T'_T)) \qquad (8.8)$$

changing its tile tensor shape to $\left[\frac{e}{t_0}, \frac{*}{t_1}, \frac{c}{t_2}\right]$. Finally, $T_R = \text{Sum}^2(T_T \odot T''_T)$ obtains the desired result with a tile tensor shape of $\left[\frac{e}{t_0}, \frac{a}{t_1}, \frac{1?}{t_2}\right]$. Applying TTDec and transposing returns R of shape $[a, e]$, which is the expected final product.

Remark 8.37 This matrix-vector multiplication method, or special cases of it, was described in diverse ways in the literature [4, 6, 8, 13]. The tile tensor framework and shape notation generalizes and unifies these descriptions.

8.11 Performance Considerations

We already saw different ways to perform linear algebraic operations. In particular, we could choose different ways to order the dimensions and use alternative sizes for the tiles. Next, we will present some performance considerations for using the different configurations.

First, let us consider the simple case of packing a matrix M of shape $[a, b]$ into tile tensor T_M with tile tensor shape $[\frac{a}{t_0}, \frac{b}{t_1}]$. The number of tiles used in this case is $\left\lceil \frac{a}{t_0} \right\rceil \left\lceil \frac{b}{t_1} \right\rceil$, where we can choose any value for t_0, t_1 as long as $t_0 t_1 = s$, the number of slots in a ciphertext.

For example, say the matrix shape is $[7, 11]$, and $s = 16$. Table 8.1 shows the number of tiles we obtain for various choices of the tile sizes. The differences are of course due to the the tile sizes not cleanly dividing our matrix dimension sizes, leaving a varying number of unused slots in each case. The resulting differences are substantial, as can be seen in the table. Obviously, the $[4, 4]$ and $[8, 2]$ tile size

Table 8.1 Number of tiles when packing matrix M of shape $[7, 11]$ for different choices of tile sizes $[t_0, t_1]$, where $t_0 t_1 = s = 16$

$[t_0, t_1]$	Number of tiles
[1, 16]	7
[2, 8]	8
[4, 4]	6
[8, 2]	6
[16, 1]	11

options are the most efficient memory-wise and will also be the most efficient time-wise for any operation that is performed tile-wise on this matrix, e.g., scalar product or bootstrapping. These operations will be more efficient since they will be applied on a smaller number of tiles. However, when doing more complicated operations, additional considerations affect performance and we will get to this later on.

Now let's consider the case of multiplying M with a vector V of shape $[b]$, by first packing them into tile tensors and following Eq. 8.5. The number of tiles in T_V is $\lceil \frac{b}{t_1} \rceil$, and the computation follows the following steps:

1. The elementwise multiplication step requires $\lceil \frac{a}{t_0} \rceil \lceil \frac{b}{t_1} \rceil$ multiplications between every matrix tile and the corresponding vector tile.
2. The summation over the second dimension starts by accumulating tiles together, i.e., reducing their number to $\lceil \frac{a}{t_0} \rceil$ by performing $\lceil \frac{a}{t_0} \rceil (\lceil \frac{b}{t_1} \rceil - 1)$ additions.
3. Summing inside the tiles using RotateAndSum requires $\lceil \frac{a}{t_0} \rceil \log_2(t_1)$ rotations and additions.
4. The number of tiles in the resulting tile tensor is $\lceil \frac{a}{t_0} \rceil$.

Table 8.2 shows the performance profile for different tiling options. Interestingly, there's no single optimal choice that is best for all situations. The [4, 4] option excels in memory: it requires a total of 11 tiles, counting 3 for T_V, 2 for T_R, and 6 for T_M as shown in Table 8.1. This is the best option in this respect. We may choose this option if we run on a machine with limited memory. The option [16, 1] is likely to excel in time performance: it performs 0 rotations, which are usually the slowest operation. The option [8, 2] may be a little better or worse in time performance, depending on the ratio between the cost of rotations and the other operators. If an application requires to transmit T_V over the network, then transmission time might be the bottleneck, and in this case option [1, 16] is the optimum. Alternatively, if the result is expected to be sent back, then considering the total number of tiles of both T_V and the result then again [4, 4] is the optimum.

This demonstrates that choosing the optimal tile sizes depends on multiple factors: the sizes of the tensors involved, the operations being performed, and the optimization goals and constraints. In more complicated circuits, we should also consider time spent on various intermediate operations required between the circuit

Table 8.2 Number of operations for multiplying M of shape [7, 11] with V of shape [11] for different choices of tile sizes $[t_0, t_1]$, where $t_0 t_1 = s = 16$

$[t_0, t_1]$	# of tiles in T_V	Multiplications	Rotations	Additions	# of tiles in the result
[1, 16]	1	7	28	28	7
[2, 8]	2	8	12	16	4
[4, 4]	3	6	4	8	2
[8, 2]	6	6	1	6	1
[16, 1]	11	11	0	10	1

nodes: for example, TTClearUnknowns and TTDup may be required in a chain of matrix multiplications. Optimization goals may include latency, throughput, memory, network usage, or some combination thereof. We may also choose to optimize our chosen measure under constraints on memory usage or other resources.

A useful approach for achieving optimal or near-optimal results in complicated scenarios is to simulate runs using various tile shapes and count the number of tiles and the number of operations being performed (without actually executing them, to speed up simulations). These stats can then be used to estimate time and memory performance. The space of possible tile shapes can be scanned by brute force if possible, or using some local search heuristic.

Section 7.10 showed a diagonalization method for matrix-vector multiplication different from the approach shown in this chapter. In the next chapter we will extend tile tensors to encompass it as well, but here let us present a preliminary performance comparison for our test matrix of shape [7, 11]. Padding it to form a square [16, 16] matrix, each row filling all the $s = 16$ slots, and then diagonalizing it, the multiplication will require 15 rotations, 16 multiplications, and 15 additions. Using the baby-step/giant-step method the number of rotations reduces to 6, and generally, the diagonalization method requires $O(\sqrt{b})$ rotations for an input vector of size b. Therefore, the performance is pretty far from the optimal achievable with the approach shown in this chapter, which can reach even zero rotations. On the other hand, the diagonalization method does excel in memory utilization of both input and output vectors; both take just one tile. This is important in cases where the input and output vectors are sent over a slow network, making the transmission time the bottleneck of the computation.

We will show a more extensive comparison between matrix multiplication techniques in Chap. 9.

8.12 Example: A Simple Neural Network

Say we have a NN with k fully connected (FC) layers. The i'th layer is defined by a matrix M_i of shape $[x_i, y_i]$ and a bias B_i of shape $[x_i]$. Each layer receives as input the vector V_i and computes $M_i V_i + B_i$. For an activation function, we use simple squaring. The output serves as the input to the next layer, or as the final output of the network for the k'th layer. We can pack M_i as T_{M_i} of shape $\left[\frac{x_i}{t_0}, \frac{y_i}{t_1}, \frac{*}{t_2}\right]$ for odd layers and transposed for even layers: $\left[\frac{y_i}{t_0}, \frac{x_i}{t_1}, \frac{*}{t_2}\right]$ (including the first layer which is at index 0). The third dimension will be used as the batch dimension.

The input is a vector of shape $[y_0]$. We assume we run inference for a batch of b samples. Thus, a batch of b inputs is a tensor A_0 of shape $[y_0, b]$, which we can pack as $T_{A_0} = \left[\frac{y_0}{t_0}, \frac{*}{t_1}, \frac{b}{t_2}\right]$.

Computing the matrix multiplication of even layers is done by $\text{Sum}^0(T_{A_i} \odot T_{M_i})$ (note the compatibility of the batch dimensions), where T_{A_i} is the input to the i'th layer, which results with tile tensor shape $\left[\frac{*}{t_0}, \frac{x_i}{t_1}, \frac{b}{t_2}\right]$. To this we add the bias that can be packed as $\left[\frac{*}{t_0}, \frac{x_i}{t_1}, \frac{*}{t_2}\right]$ (note that here too the batch dimensions are compatible). Similarly, odd layers are computed using $\text{Sum}^1(T_{A_i}(*)T_{M_i})$ and adding an appropriately packed bias. The square activation can be done by multiplying the tile tensor with itself elementwise or by calling a native square operator tile-wise.

The choice of tile sizes t_0, t_1, t_2 now depends on the sizes of the matrices and the batch size b. For example if $b = 8$, then it is natural to set $t_2 = 8$. This will allow the system to process the *eight* samples together.

If the total number of slots is s, the extreme case is setting $t_0 = 1, t_1 = 1, t_2 = s$. This results in what is sometimes called "packing across the batch dimension" (or other similar terms) [2, 5, 12]: each element of an input sample is stored in a separate tile, and the slots are used for different samples in the batch. Performing the computation this way requires no rotation operations, since the different slots never interact with each other. This brings throughput to a maximum, but also requires high memory consumption.

The other extreme is setting $t_2 = 1$, and tuning t_0, t_1 to reduce the latency as much as possible. This tuning can be done manually or automatically by scanning all the options. This extreme usually brings latency and memory usage to a minimum, as the system is tuned to process one sample at a time.

There are also options that fall between these alternatives. For example, setting $t_2 = 8$ will be optimal for processing *eight* samples at a time. As before t_0, t_1 will need to be tuned to achieve optimal results.

8.13 Encrypted vs. Encoded Tile Tensors

Throughout this chapter we assumed each tile is stored encrypted in a ciphertext. However, this is not necessarily the case. For example, in the NN scenario of the Sect. 8.12, it might be that the input to the network is encrypted, but the network weights are not.

Unencrypted tensors can be handled by tile tensors in much the same way: each tile is stored encoded in a plaintext object. Such an object has the same number of slots as a ciphertext, and hence all the tools and terminology we develop here apply equally to such encoded tile tensors.

Two differences however are noteworthy:

1. Packing a tensor into a tile tensor can increase the memory consumption significantly. If some dimensions are duplicated, this directly duplicates the memory size, and the FHE scheme adds its own overhead, even if we are only encoding the data and not encrypting it. If a tensor is given to us in the clear, we may choose to defer packing it into an encoded tile tensor until the point that it is needed. Alternatively, a more efficient way is to encode only specific tiles when they are needed and immediately discard them after use.

2. The memory size of a ciphertext and the time performance of the operations running on it do not depend on its content. Otherwise, we could have used the differences to obtain information about the data it is encrypting. On the other hand, this rule does not always apply to encoded objects. A plaintext object that stores a single real scalar duplicated all over its slots is usually much cheaper to store in memory and faster to encode. Many FHE schemes have a native multiply-by-real-scalar operator on ciphertexts, which is usually more efficient than multiplying with a whole plaintext object. This allows us to avoid encoding the scalar in the first place. An integer scalar can be even more efficient. For example, in CKKS, multiplying with an integer does not require a subsequent *rescaling* operation. Our choice of tile tensor shapes can be influenced by such considerations. For example, for encoded tile tensors, we might prefer duplicating the same value along all slots, whereas in encrypted tile tensors this would be highly inefficient.

8.14 Lab Exercise: Program with Tile Tensors

The following lab exercise requires a programming environment such as Python. This environment should include libraries that can handle cleartext tensors such as NumPy [7] and also libraries that can handle tile tensors such as *pyhelayers* [1]. If the latter environment supports a mockup option that only emulates the ciphertext operations but does not truly run them under encryption, then consider using it as it can speed up the computation without losing the ability to practice and get the correct results.

When using pyhelayers we recommend using a mockup context with 16 slots of the CKKS scheme. In pyhelayers encoded tile tensors are called PTileTensors and encrypted tile tensors are called CTileTensors.

1. Create a random matrix A of shape $[11, 7]$. Encrypt it in a tile tensor object T_A with tile tensor shape $\left[\frac{11}{4}, \frac{7}{4}\right]$. Print the content of T_A, and see how the elements of A are divided among the tiles. (Specifically for pyhelayers, use $pyhelayers.get\_print\_options().tt\_demo\_tiles = True$.)
2. Create a random tensor V of shape $[7]$. Encrypt it in a tile tensor such that it is possible to perform matrix-vector multiplication with T_A, and perform the required multiplication. (In pyhelayers, use the method $multiply\_and\_sum$.) Print the context of the result, decrypt it and compare with the plaintext computation.
3. Repeat the last two exercises, but this time pack A transposed, and use Eq. 8.6.
4. Create random matrices X, Y, Z, W of shapes $[70, 40], [40, 100], [100, 20], [20, 80]$. Encrypt them in tile tensors, and compute the product $XYZW$ as explained in Example 8.36. Pack them using tile sizes $[2, 4, 4]$.
5. Repeat the previous exercise, but this time, change the placement of the brackets. For example, instead of computing $(XY)(ZW)$, compute $((XY)Z)W$.

Change the tile tensor shapes to accommodate the alternative order. In terms of multiplication depth, which order is better?
6. Implement the simple NN example of Sect. 8.12. Use $k = 3$ layers and matrices of shape [30, 30]. Find the best performing tile sizes for batch sizes $b = 1, 4, 16$. Try other values for b. Which value provides the best throughput? For this example use a secure context with 8,192 slots and depth 7. Draw random values in the range [0, 0.1] for the tensors to prevent numeric overflow.
7. Tile tensor summation (Definition 8.33) requires that the dimension being summed over will not have unknown values. If it does have unknown values, then we can first clear them using TTClearUnknowns. This requires multiplication by masks and increases the multiplication depth. Remark 8.34 suggests an alternative approach that does not require multiplication. Design a summation algorithm following this suggestion (see Fig. 8.13d for some further guidance). What are the pros and cons of each alternative?

References

1. Aharoni, E., Adir, A., Baruch, M., Drucker, N., Ezov, G., Farkash, A., Greenberg, L., Masalha, R., Moshkowich, G., Murik, D., Shaul, H., Soceanu, O.: HeLayers: a tile tensors framework for large neural networks on encrypted data. Privacy Enhancing Technology Symposium (PETs) 2023 (2023). https://petsymposium.org/popets/2023/popets-2023-0020.php
2. Boemer, F., Lao, Y., Cammarota, R., Wierzynski, C.: nGraph-HE: a graph compiler for deep learning on homomorphically encrypted data. In: Proceedings of the 16th ACM International Conference on Computing Frontiers, pp. 3–13 (2019)
3. Cheon, J.H., Kang, M., Kim, T., Jung, J., Yeo, Y.: High-throughput deep convolutional neural networks on fully homomorphic encryption using channel-by-channel packing. Cryptology ePrint Archive (2023)
4. Crockett, E.: A low-depth homomorphic circuit for logistic regression model training. Cryptology ePrint Archive (2020). https://eprint.iacr.org/2020/1483
5. Gilad-Bachrach, R., Dowlin, N., Laine, K., Lauter, K., Naehrig, M., Wernsing, J.: Cryptonets: applying neural networks to encrypted data with high throughput and accuracy. In: International Conference on Machine Learning, pp. 201–210. PMLR (2016)
6. Halevi, S., Shoup, V.: Algorithms in HElib. In: Garay, J.A., Gennaro, R. (eds.) Advances in Cryptology – CRYPTO 2014, pp. 554–571. Springer, Berlin (2014)
7. Harris, C.R., Millman, K.J., van der Walt, S.J., Gommers, R., Virtanen, P., Cournapeau, D., Wieser, E., Taylor, J., Berg, S., Smith, N.J., Kern, R., Picus, M., Hoyer, S., van Kerkwijk, M.H., Brett, M., Haldane, A., del Río, J.F., Wiebe, M., Peterson, P., Gérard-Marchant, P., Sheppard, K., Reddy, T., Weckesser, W., Abbasi, H., Gohlke, C., Oliphant, T.E.: Array programming with NumPy. Nature **585**(7825), 357–362 (2020). https://doi.org/10.1038/s41586-020-2649-2
8. Juvekar, C., Vaikuntanathan, V., Chandrakasan, A.: GAZELLE: a low latency framework for secure neural network inference. In: 27th USENIX Security Symposium (USENIX Security 18), pp. 1651–1669 (2018)
9. Kim, M., Jiang, X., Lauter, K., Ismayilzada, E., Shams, S.: HEAR: human action recognition via neural networks on homomorphically encrypted data. e-prints. arXiv–2104 (2021)
10. Kim, D., Park, J., Kim, J., Kim, S., Ahn, J.H.: HyPHEN: A hybrid packing method and its optimizations for homomorphic encryption-based neural networks. IEEE Access **12**, 3024–3038 (2024). https://doi.org/10.1109/ACCESS.2023.3348170

11. Lee, E., Lee, J.W., Lee, J., Kim, Y.S., Kim, Y., No, J.S., Choi, W.: Low-complexity deep convolutional neural networks on fully homomorphic encryption using multiplexed parallel convolutions. In: International Conference on Machine Learning, pp. 12403–12422. PMLR (2022)
12. Nandakumar, K., Ratha, N., Pankanti, S., Halevi, S.: Towards deep neural network training on encrypted data. In: Proceedings of the IEEE/CVF Conference on Computer Vision and Pattern Recognition Workshops, pp. 0–0 (2019)
13. Rizomiliotis, P., Triakosia, A.: On matrix multiplication with homomorphic encryption. In: Proceedings of the 2022 on Cloud Computing Security Workshop, CCSW'22, pp. 53–61. Association for Computing Machinery, New York (2022). https://doi.org/10.1145/3560810.3564267

Chapter 9
SIMD Packing Part III: Advanced Tile Tensors

Abstract Chapter 8 introduced the tile tensor data structure, basic packing options, and simple operators it supports. This chapter covers additional packing options that incorporate more techniques from Chap. 7: complex packing and diagonal packing. Another important packing technique not previously discussed is interleaved packing, which extends the operator toolbox to include rotations and pseudo-rotations. This in turn enables the convolution operator.

9.1 Interleaved Packing and Rotations

Our discussion about tensor rotations was initiated in Sect. 7.4, e.g., Definition 7.15, where we assumed that a tensor is flattened into a single ciphertext (or in tile tensor terminology, stored within one tile). Specifically, we distinguished between rotations and pseudo-rotations (Definition 7.16), where Theorem 7.21 established that we can always rotate the tensor's first dimension and pseudo-rotate all of its other dimensions. This section deals with rotating (or pseudo-rotating) a tensor that is packed into a tile tensor and spans multiple tiles. For this we will use tile tensors again and extend the packing options they support.

As always, let us start with a toy example. Let V be a vector of shape $[12]$, packed into T_V with tile tensor shape $\left[\frac{12}{4}\right]$ as in Fig. 9.1. We would like to compute $V' = \text{Rot}_1^0(V)$, that is, rotate V along its first (and only) dimension by an offset of 1. We can perform this in two steps. First, rotate each tile to get T'_V. This correctly places most of the elements, except the last element of each of the three tiles. Subsequently, we correct these entries by multiplying each tile with masks to separate the last element from the rest, and then using additions to place each element in its correct tile to get T''_V.

This costs three rotations, three additions, and six mask multiplications (two for each tile), which means also an increase of the multiplication depth by 1. Consequently, one rotation of a tensor can become a costly operation when several tiles are required to accommodate the logical tensor. Here is a much better way

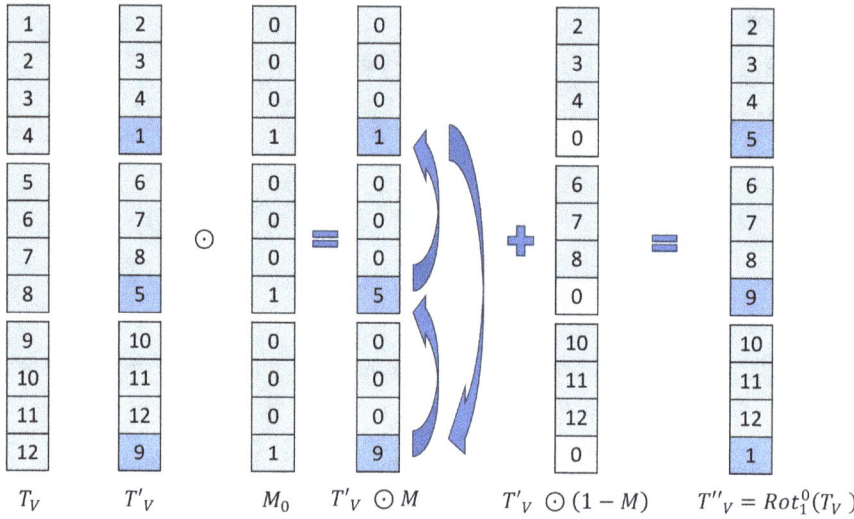

Fig. 9.1 Rotating T_V of shape $\left[\frac{12}{4}\right]$ by 1

to do so. Instead of the ordinary tiling of a tensor dimension, we consider a more sophisticated way called *interleaved tiling* [1].

Definition 9.1 (Interleaved Tiling) A tile tensor shape element $\frac{n_i\sim}{t_i}$ specifies that the size of the i'th dimension of the logical tensor is n_i, the tile size along this dimension is t_i, and the size of the external tensor $\lceil\frac{n_i}{t_i}\rceil$. In the external tensor, in the tile with index l_i, the slot with index m_i is mapped to the logical tensor index $j_i = l_i + m_i \lceil\frac{n_i}{t_i}\rceil$.

Example 9.2 (Interleaved Tiling) Given a vector V of shape $[12]$, packed into T_V using interleaved packing $\left[\frac{12\sim}{4}\right]$, we get $\left[\frac{12}{4}\right]$ the the tile with *four* slots each. In tile with index 1 (the second tile), the slot with index 2 is mapped to logical index $1 + 2\left\lceil\frac{12}{4}\right\rceil = 7$.

Definition 9.1 is similar to Definition 8.12 of basic tiling, with only two differences: a) the $\sim$ symbol is added to the tile tensor shape notation, which indicates we are using interleaved tiling, and b) the formula for computing j_i is different.

Example 9.3 (Interleaved Rotation) Figure 9.2 illustrates an example of packing V into T_V using interleaved packing $\left[\frac{12\sim}{4}\right]$. As before, we use *three* tiles; only now, we fill the first slot in each tile, so they contain the values 1, 2, 3 of V. Then we continue to the second slot of each tile, containing the values 4, 5, 6, and so on. Rotating T_V by 1 is done as follows. First, rotate the external tensor, i.e., move the first tile to be the last tile. Subsequently, rotate the last tile. Unpacking the result in the same interleaved order returns the vector $Rot_1^0(V)$.

9.1 Interleaved Packing and Rotations

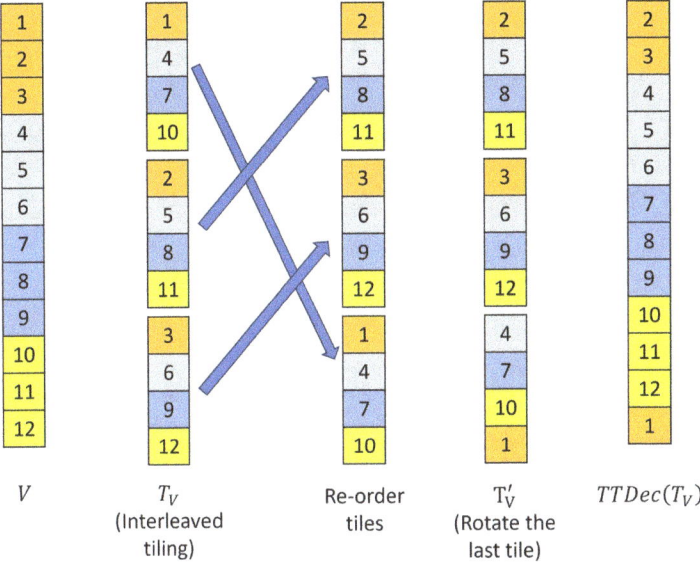

Fig. 9.2 Rotating T_V of shape $\left[\frac{12\sim}{4}\right]$ by 1

The next definition generalizes this example to any number of dimensions, by defining the Rot operator for tile tensors. This operator is applicable if the rotated dimension is rotatable, packed using interleaved tiling, and has no unused slots.

Definition 9.4 (Tile Tensor Rotation) Let A be a rank-k tensor, packed into a tile tensor T_A. Let the i'th dimension of the tile tensor be a rotatable interleaved dimension $\frac{n_i\sim}{t_i}$, and let n_i be divisible by t_i (i.e., no unused slots). Then $T'_A = \text{Rot}^i_m(T_A)$ is computed by:

1. Rotating T_A's external tensor along dimension i by an offset of m.
2. Performing rotations on each tile along dimension i as follows: for a tile whose indices in the external tensor of T_A are $(l_0, \ldots, l_{k-1})$, the offset is $\left\lceil \frac{m - l_i}{e_i} \right\rceil$, where $e_i = \frac{n_i}{t_i}$ (recall we assume here n_i is divisible by t_i).

The following lemma shows that rotating a tile tensor according to Definition 9.4 is equivalent to rotating the logical tensor when setting $m = 1$.

Lemma 9.5 *Let* $T'_A = \text{Rot}^i_1(T_A)$, $A = \text{TTDec}(T_A)$, *and* $A' = \text{TTDec}(T'_A)$. *Then* $A' = \text{Rot}^i_1(A)$.

Proof We need to show that $A[j_0, \ldots, j_{k-1}] = A'[j_0, \ldots, (j_i - 1) \bmod n_i, \ldots, j_{k-1}]$. Let E be the external tensor of T_A. W.l.o.g., fix an element $A[j_0, \ldots, j_{k-1}]$, and select a tile $t = E[l_0, \ldots, l_{k-1}]$ and a slot $t[m_0, \ldots, m_{k-1}]$ that is mapped by T_A's shape to the element $A[j_0, \ldots, j_{k-1}]$. Note that there might be more than one such slot, if some of the dimensions are duplicated (dimension

i must be interleaved according to Definition 9.4, but other dimensions may be duplicated). We consider two cases $l_i \geq 1$ and $l_i = 0$.

If $l_i \geq 1$ then by Definition 9.1, $j_i \geq 1$, and we have

$$Rot_1^i(E)[l_0, \ldots, \mathbf{l_i-1}, \ldots, l_{k-1}] = t.$$

Subsequently, because $0 < l_i < e_i$, t is rotated by an offset of $\left\lceil \frac{1-l_i}{e_i} \right\rceil = 0$; hence, t does not change. The change of a tile index from l_i to $l_i - 1$ changes its mapped index for j_i to $j_i - 1$, as needed.

When $l_i = 0$, we have that

$$Rot_1^i(E)[l_0, \ldots, \mathbf{e_i-1}, \ldots, l_{k-1}] = t$$

and t is rotated by an offset of 1, changing the slot index in the tile to $(m_i - 1)$ mod n_i. Also here there are two cases: $m_i > 0$ and $m_i = 0$.

If $m_i > 0$, then $j_i > 0$, and the total effect on j_i is increasing by $e_i - 1$ because of the change in l_i, and decreasing by e_i due to the change in m_i, totaling in reducing by 1 as required.

If $m_i = 0$, j_i increases by $e_i - 1$ and further increases by $(t_i - 1)e_i$, totaling in $t_i e_i - 1 = t_i \frac{n_i}{t_i} - 1 = n_i - 1$ as required. Note that we rely here on the assumption in Definition 9.4 that t_i divides n_i.

□

Corollary 9.6 *Similar considerations extend Lemma 9.5 to an offset of* -1.

To see the correctness of this method for arbitrary rotation offsets, note that rotating a tile tensor by an offset of a positive m is equivalent to performing m rotations by 1. The external tensor in this case is rotated m times, and each tile is rotated once every time its i'th index is 0 before the rotation. If its original tile index is l_i, this occurs $\left\lceil \frac{m-l_i}{e_i} \right\rceil$ times as defined in Definition 9.4. Similar considerations extend this result for negative m.

Corollary 9.7 *Extending Lemma 9.5 to every positive or negative m value is done by repeatedly rotating by 1 or -1, respectively.*

Similarly we define a pseudo-rotation operator. But here, unlike in Definition 9.4, we have no constraints: any tile tensor can be pseudo-rotated along any interleaved dimension.

Definition 9.8 (Tile Tensor Pseudo-Rotation) Let A be a rank-k tensor, packed into a tile tensor T_A with external tensor E. Let the i'th dimension of the tile tensor be an interleaved dimension $\frac{n_i \sim}{t_i}$, or possibly with unknown values, $\frac{n_i \sim ?}{t_i}$. Let e_i be the number of tiles $\left\lceil \frac{n_i}{t_i} \right\rceil$ in E along dimension i, then $T_A' = \text{PseudoRot}_m^i(T_A)$ is computed by:

1. Rotating E along dimension i by an offset of m

9.1 Interleaved Packing and Rotations

2. For every tile along dimension i of E with indices $(l_0, \ldots, l_{k-1})$ performing a pseudo-rotation with an offset of $\lceil \frac{m-l_i}{e_i} \rceil$

The resulting tile tensor shape has "?" marking added to all dimensions where there are unused slots.

The proof that $\text{TTDec}(\text{PseudoRot}_m^i(T_A)) = \text{PseudoRot}_m^i(\text{TTDec}(T_A))$ is similar to the proof of Lemma 9.5 and its corollaries. In Lemma 9.5 the case $l_i = m_i = 0$ can be ignored, since in this case $j_i = 0$ and in pseudo-rotations we do not require the first element to roll back to the end of the tensor. In the lemma, this is the case that requires the constraint that t_i divides n_i. Since we can discard this case, the constraint is not required anymore. These elements cycling back in a permuted way may end up in unused slots of any dimension; hence, the "?" marking is turned on everywhere.

Another interesting property of our pseudo-rotations method is that if the logical tensor ends with an all-zero padding along some dimension and fills the tiles entirely, then pseudo-rotating to the right along this dimension is equivalent to rotating.

Lemma 9.9 *Let A be a rank-k tensor, packed into a tile tensor T_A. Let the j'th dimension of T_A be an interleaved dimension $\frac{n_j}{t_j}\tilde{\,}$, and let n_j be divisible by t_j (i.e., no unused slots). Also, assume no other dimension has the unknown slot mark ("?"). If A has zeroes in all elements whose j'th index i_j satisfies $i_j \geq n_j - l$ then $\text{PseudoRot}_{-x}^i(T_A) = \text{Rot}_{-x}^i(T_A)$ for $0 \leq x \leq l$.*

Proof We can present $\text{PseudoRot}_{-x}^i(T_A)$ as a sequence of at most l applications of $\text{PseudoRot}_{-1}^i(T_A)$. In this sequence of rotate right operations, all the rotated tiles have at least a depth of one zero padding at the end of the j'th dimension. By Lemma 7.24, pseudo-rotating a tile with zero padding is equivalent to rotating it. □

Example 9.10 (Interleaved Rotation—Cont') Using T_V from Example 9.3. Let us compute its rotations with multiple offsets: $\text{Rot}_1^0(T_V)$, $\text{Rot}_2^0(T_V)$, and $\text{Rot}_3^0(T_V)$ as in Fig. 9.3. The first rotation costs *one* tile rotation, the second *two* tile rotations, and the third *three*. But each operation can reuse some tile rotations of the previous operations. In fact, each operation can reuse all previous tile rotations, and add just one more. So the three tile tensor rotations cost together only *three* tile rotations.

This is a time optimization available whenever we perform multiple rotation operations along the same dimension of the same tile tensor. To gain this optimization benefit, we need to set up a rotation cache, storing the tiles we already rotated, and reuse them when needed. This cache can be discarded once the rotations are no longer required.

Next, we extend Definition 8.30 of tile tensor shape compatibility, used for elementwise operators:

Definition 9.11 (Compatible Tile Tensor Shapes Including Interleaved Tiling)
Two tile tensor shapes are compatible if their logical tensor shapes are compatible,

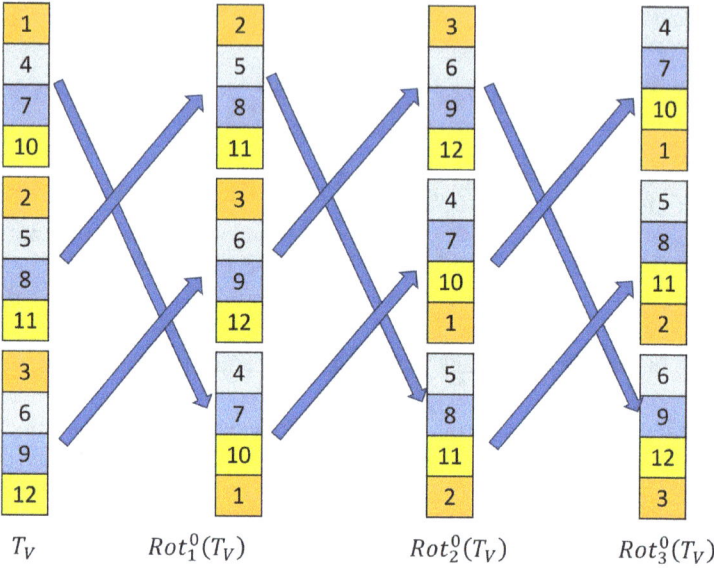

Fig. 9.3 Rotating T_V of shape $\left[\frac{12\sim}{4}\right]$ by 1, three times

and for all i their i'th elements are compatible respectively. The three elements $\frac{n_i}{t_i}$, $\frac{n_i?}{t_i}$, and $\frac{*}{t_i}$ are mutually compatible with each other. Also $\frac{n_i\sim}{t_i}$, $\frac{n_i\sim?}{t_i}$, and $\frac{*}{t_i}$ are mutually compatible.

This extended definition of compatibility allows us to apply the elementwise operator of Definition 8.31 on tile tensors with interleaved packing. The output tile tensor shape has interleaved tiling in dimensions where any of the inputs has interleaved tiling. Also, the summation operator (Definition 8.33) can be applied on interleaved dimensions as both its parts, summing tiles and summing inside a tile, do not assume any particular order of elements.

Observation 9.12 $\frac{n_i\sim}{t_i}$ where $n_i \leq t_i$ is equivalent to $\frac{n_i}{t_i}$. In particular, $\frac{1\sim}{t_i}$ is equivalent to $\frac{1}{t_i}$.

9.2 Simulating Large Ciphertexts

Consider the case where some clever packing scheme is designed for some specific task, e.g., computing some kind of image processing algorithm. Assume also that the packing scheme requires that the input fits within a single ciphertext. For example, if the input has 65,536 pixels then the FHE scheme should be configured for ciphertexts with at least $s = 65{,}536$ slots.

9.2 Simulating Large Ciphertexts

Such a constraint on a minimum size for the number of slots has several disadvantages. Firstly, some FHE implementations cannot support arbitrarily large numbers of slots. This is may be due to some constraint on buffer sizes, for example, or due to security evaluation not defined for large ciphertexts in the relevant FHE standard. Consequently, it might be that $s = 65{,}536$ is not feasible for some implementations.

Secondly, some homomorphic operations on ciphertexts have a time complexity that is nonlinear in the number of slots. For example, rotate operations usually have a time complexity of at least $o(s \log s)$. Therefore, it is usually more efficient if we can work with small ciphertexts, assuming, of course, that our packing scheme allows it.

This section shows a general approach to simulate large ciphertexts using smaller ones. That is, we show a wrapper layer that presents to its users virtual ciphertexts that can be configured to be arbitrarily large. Internally, the data of each such virtual ciphertext is divided among smaller ciphertexts. This removes the constraints on the ciphertext size. We can design a packing scheme based on any ciphertext size that we want, and configure the FHE scheme to ciphertexts of a smaller size, if the security constraints allow it. This concept was first introduced in [2].

Our approach is based on the interleaved packing method presented in the previous section. Say the FHE scheme is configured for ciphertexts with s slots, and we want to simulate ciphertexts of size ms slots, for some positive integer m. Therefore, the content of each large ciphertext is a vector of length ms. We will pack it into a tile tensor of tile tensor shape $\left[\frac{ms\sim}{s}\right]$. This means each simulated ciphertext is stored as a tile tensor with m tiles.

We now need to show how to simulate the basic operations: addition, multiplication, and rotation. However, we have already shown them: addition and multiplication are elementwise operators, explained in Sect. 8.8. In this simple scenario, each simulated operation costs m operations on actual ciphertexts. Since the time complexity of these operators is at least linear, we can only benefit from dividing the task into m tasks each smaller by a factor of m.

Regarding rotations, we already explained how to perform them in the previous section. A single rotation on the simulated ciphertext costs between 1 and m rotations on the actual ciphertexts. The worst-case scenario of m rotations is still an improvement, since rotations usually have $O(s \log s)$ time complexity. Therefore, m rotations of ciphertexts of size s are more efficient than 1 rotation of a ciphertext of size ms. In the average case scenario a single simulated rotation will cost less than m actual rotations, and further efficiency can be gained by using the rotation cache mentioned in Sect. 9.1.

To summarize, this technique shows that it is beneficial to reduce the ciphertext size as much as possible due to the following reasons:

- It improves the ability to parallelize by breaking each operator into multiple smaller ones.
- It reduces the cost of rotation and possibly other operators due to the above-linear time complexity.
- It reduces the number of ciphertext rotations performed.

Table 9.1 Performance benchmark for simulating a 320,000 size ciphertext using smaller ciphertexts. The benchmark includes a square operation, then rotate by 8. The benchmark uses HEaaN [5] and runs in single-thread CPU

Ciphertext size (s)	Time (sec)
32,768	2.8
16,384	2.43
8,192	2.3
4,096	2.2

However, some non-packing-related constraints prevent us from reducing ciphertext size arbitrarily. Security considerations usually require ciphertexts to be of at least a certain size, as is the availability and accuracy of the bootstrap operator.

Table 9.1 shows our benchmark results, where we simulated a ciphertext of size 320,000 using smaller ciphertexts. Specifically, we executed a square operation, followed by a rotate left by an offset of 8 on a single-thread CPU and the CKKS FHE scheme.

9.3 Slicing

In this section we will explore another common tensor operator called *slicing*, which extracts elements along a dimension.

Definition 9.13 (Slice) Let A be a tensor of shape $[n_0, \ldots, n_{k-1}]$. Then $A' = \text{Slice}^i_{a,b}(A)$ is a tensor of shape $[n_0, \ldots, n_{i-1}, b, \ldots, n_{k-1}]$, such that

$$A'[j_0, \ldots, j_i, \ldots, j_{k-1}] = A[j_0, \ldots, j_i + a, \ldots, j_{k-1}]. \tag{9.1}$$

Remark 9.14 In our notation, $\text{Slice}^i_{a,b}(A)$ means slice along dimension i, starting at index a, and the size of the slice is b.

For example, given a matrix M of shape $[5, 6]$, then $\text{Slice}^0_{1,3}(M)$ returns a matrix of shape $[3, 6]$, taking *three* rows out of M starting from row 1. Applying the slice operator on a tile tensor has the usual effect of being equivalent to applying it on the logical tensor, so

$$\text{TTDec}(\text{Slice}^i_{a,b}(T_A)) = \text{Slice}^i_{a,b}(\text{TTDec}(T_A)). \tag{9.2}$$

We consider two cases for applying $\text{Slice}^i_{a,b}(T_A)$ on a dimension whose tile tensor shape element is $\frac{n_i}{t_i}$. If $a = 0$, simply adjust the tile tensor shape to $\frac{b}{t_i}$ and discard unused tiles if such exist. If the last tile contains unused slots that are filled with trimmed data, then set the tile tensor shape element to $\frac{b?}{t_i}$.

9.3 Slicing

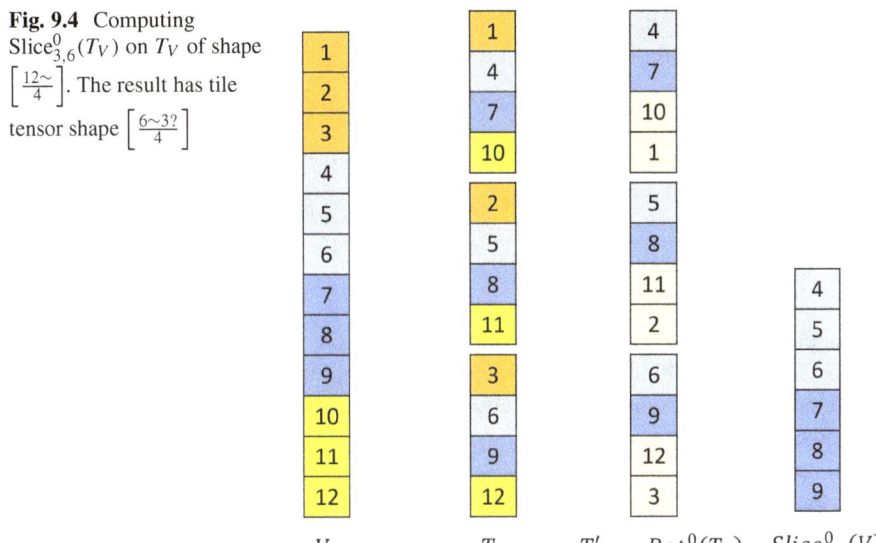

Fig. 9.4 Computing $\text{Slice}^0_{3,6}(T_V)$ on T_V of shape $\left[\frac{12\sim}{4}\right]$. The result has tile tensor shape $\left[\frac{6\sim 3?}{4}\right]$

When $a > 0$, we first perform a $\text{PseudoRot}^i_a(T_A)$, which is easier to perform on interleaved dimensions. But, when a is also a multiple of t_i we can implement it simply by rotating the order of the tiles.

Applying $\text{Slice}^i_{a,b}(T_A)$ on a dimension whose tile tensor shape element is interleaved, $\frac{n_i\sim}{t_i}$ is simple, as mentioned above, but requires introducing a new notation that will be defined in Definition 9.16. To get an intuition for this definition, we first consider an example. The main idea is that we first need to perform a pseudo-rotation: $T'_A = \text{PseudoRot}^i_a(T_A)$, and then slice out the last $n_i - b$ values from the end, which can be done by simply adjusting the tile tensor shape.

Example 9.15 (Slicing an Interleaved Tile Tensor) Applying $\text{Slice}^0_{3,6}(T_V)$ to T_V from Example 9.3, i.e., extract 6 elements starting from element index 3. We start by performing a (real) rotation $T'_V = \text{Rot}^0_3(T_V)$; see Fig. 9.4. Subsequently, we need to discard the six elements $\{10, 11, 12, 0, 1, 2\}$ at the end, but because of the interleaved packing, they are distributed among all tiles and we cannot just discard whole tiles. Also, if we adjust the tile tensor shape to $\left[\frac{6\sim?}{4}\right]$, it would seem as though there are only 2 tiles, but the *six* remaining elements are distributed among *three* tiles. To this end, we introduce the following notation $\left[\frac{6\sim 3?}{4}\right]$, indicating the external size along this dimension has a custom value of 3, and not the minimal possible value of 2.

Definition 9.16 (Interleaved Tiling with Custom External Size) A tile tensor shape element $\frac{n_i\sim e_i}{t_i}$ specifies that the size of the i'th dimension of the logical tensor is n_i, the tile size along this dimension is t_i, and the size of the external tensor is e_i, such that $e_i > \lceil \frac{n_i}{t_i} \rceil$. In the external tensor, a tile with index l_i and slot with index m_i is mapped to the logical tensor index $j_i = l_i + m_i e_i$.

Observation 9.17 *For purposes of compatibility, tile tensor shape elements $\frac{n_i \sim e_i}{t_i}$, $\frac{n_i \sim e_i?}{t_i}$, and $\frac{*}{t_i}$ are mutually compatible.*

9.4 Convolution

A popular building block in NNs is 2D convolution, which is the main topic of this section. We study it by first considering its simplest form: a plain single input, single output (SISO) convolution. Subsequently, we will extend the study with some additional common features: padding, strides, and then multiple inputs (channels) and outputs.

9.4.1 SISO

In the SISO setting, the input to the 2D convolution is a matrix I of shape $[h_I, w_I]$ representing an image, and another matrix called filter or kernel, F of shape $[h_F, w_F]$. The convolution between them is defined as follows.

Definition 9.18 (Convolution (SISO)) *Let I and F be two input tensors of shapes $[h_I, w_I]$ and $[h_F, w_F]$, representing an image and a filter, respectively. The result of the convolution operation $O = \text{Conv}_{\text{siso}}(I, F)$ is the tensor O of shape $[h_O, w_O]$, where $h_O = h_I - h_F + 1$, $w_O = w_I - w_F + 1$, and*

$$O[i, j] = \sum_{k=0}^{h_F-1} \sum_{l=0}^{w_F-1} I[i+k, j+l] F[k, l]. \tag{9.3}$$

Alternatively, we can define the SISO convolution as a sequence of two steps:

Lemma 9.19 *Let I and F be two input tensors of shapes $[h_I, w_I]$ and $[h_F, w_F]$, then the following computation:*

$$O_1 = \sum_{k=0}^{h_F-1} \sum_{l=0}^{w_F-1} \text{PseudoRot}_l^1 (\text{PseudoRot}_k^0(I)) F[k, l]. \tag{9.4}$$

$$O_2 = \text{Slice}_{0, w_O}^1 (\text{Slice}_{0, h_O}^0 (O_1)), \tag{9.5}$$

results with $O_2 = \text{Conv}_{\text{siso}}(I, F)$.

9.4 Convolution

Proof W.l.o.g, fix some valid indices i, j of O_1. The pseudo-rotations act as ordinary rotations when $i + h_F - 1 < h_I$ and $j + w_F - 1 < w_I$. Hence, in this case, by Eq. 9.4

$$O_1[i, j] = \sum_{k=0}^{h_F-1} \sum_{l=0}^{w_F-1} I[i+k, j+l] F[k, l] = \text{Conv}_{\text{siso}}(I, F)[i, j].$$

Otherwise, because either $i \geq h_I - h_F + 1 = h_O$ or $j \geq w_I - w_F + 1 = w_O$ then the element $O_1[i, j]$ is trimmed out by the slice operators in Eq. 9.5. □

From Lemma 9.19 we learn that it is possible to use both rotations or pseudo-rotations for performing the $\text{Conv}_{\text{siso}}$ operator, where for simplicity, we prefer using pseudo-rotations, since these are easier to implement in HE.

Given the $\text{Conv}_{\text{siso}}$ alternative definition of Lemma 9.19, implementing convolution is now straightforward. Let us pack I in a tile tensor T_I with shape $\left[\frac{h_I\sim}{t_0}, \frac{w_I\sim}{t_1} \right]$. Tile tensors support all the required operators: PseudoRot, Slice, and elementwise addition. Finally, assuming the filter F is given in plaintext, we can implement the scalar multiplication by multiplying every tile by a scalar. This can be written in mathematical notation as follows.

$$T_{O_1} = \sum_{k=0}^{h_F-1} \sum_{l=0}^{w_F-1} \text{PseudoRot}_l^1 (\text{PseudoRot}_k^0 (T_I)) F[k, l]. \tag{9.6}$$

$$T_{O_2} = \text{Slice}_{0, w_O}^1 (\text{Slice}_{0, h_O}^0 (T_{O_1})) \tag{9.7}$$

where the shape of T_{O_2} is $\left[\frac{h_O\sim?}{t_0}, \frac{w_O\sim?}{t_1} \right]$.

Remark 9.20 It is usually the case that the slicing operator adds the unknown marking, because it reduces the size of the tile tensor by changing metadata only. The pseudo-rotations of the first step are likely to produce some junk values outside the valid output range. Removing these unknown marks by using masks will be considered later on.

Remark 9.21 In the general case, the slicing operator keeps the external size of the input and uses the additional notation of Definition 9.16 as the output shape. This additional notation is omitted for the sake of brevity.

Equation 9.6 performs multiple pseudo-rotations of the same tile tensor. Therefore, a rotation cache as mentioned in Sect. 9.1 is crucial for avoiding repeating the same tile rotations.

Lemma 9.22 *Computing convolution under encryption using Eq. 9.6 requires r ciphertext rotations, where*

$$r = \left\lceil \frac{w_I}{t_1} \right\rceil (h_F - 1) + \left\lceil \frac{h_I}{t_0} \right\rceil (w_F - 1) + (h_F - 1)(w_F - 1). \tag{9.8}$$

Proof The inner pseudo-rotation operator is used with $h_F - 1$ different offsets (excluding 0). By similar considerations as in Example 9.10, every invocation of the pseudo-rotation operator rotates only a single row of tiles, therefore requiring $\lceil \frac{w_I}{t_1} \rceil$ tile rotations. The total number of rotations is thus $\lceil \frac{w_I}{t_1} \rceil (h_F - 1)$.

Similarly, the outer pseudo-rotation operator performs $\lceil \frac{h_I}{t_0} \rceil (w_F - 1)$ tile rotations. Finally, for every row rotated by the inner operator, we need to recompute all $w_F - 1$ rotations of the outer ones, each now adding a single tile rotation. Hence, we get the last term of $(h_F - 1)(w_F - 1)$ rotations. □

If the input image fits within a single tile, that is, $h_I \leq t_0$ and $w_I \leq t_1$, then the above method reduces to the more common and simpler algorithm where the pseudo-rotate tile tensor operators are simply pseudo-rotations of the image tensor as, e.g., in [4, 8, 9, 11, 12]. The generalization to tile tensor is beneficial for the reasons explained in Sect. 9.2: the image may be too big for the largest ciphertext supported by the underlying FHE implementation, and even when it is not, it is usually more efficient to break it down to smaller ciphertexts. In some scenarios, even if the image fits within the smallest ciphertext allowed by the security constraints, it might still be useful to add a batch dimension and handle multiple images in parallel.

Until now we considered an unencrypted filter, but how can we handle an encrypted filter? One idea is to pack it in a tile tensor T_F with tile tensor shape $[h_F, w_F, \frac{*}{t_0}, \frac{*}{t_1}]$ as in Fig. 9.5. Recall that the h_F and w_F elements in this shape are abbreviations for $\frac{h_F}{1}$ and $\frac{w_F}{1}$, indicating a tile size of 1, or that each filter element is in a separate tile. The $\frac{*}{t_0}, \frac{*}{t_1}$ elements indicate that this element is duplicated all over the tile.

We use T_F in the convolution by replacing $F[k, l]$ in Eq. 9.6 with $T_F[k, l, :, :]$, which denotes slicing tiles out of the tile tensor T_F. In our case, it simply means taking the single tile at coordinates k, l. Subsequently, we perform an elementwise multiplication between this tile and every tile of the rotated input tile tensor, achieving the same result as multiplying each slot by the plaintext filter value.

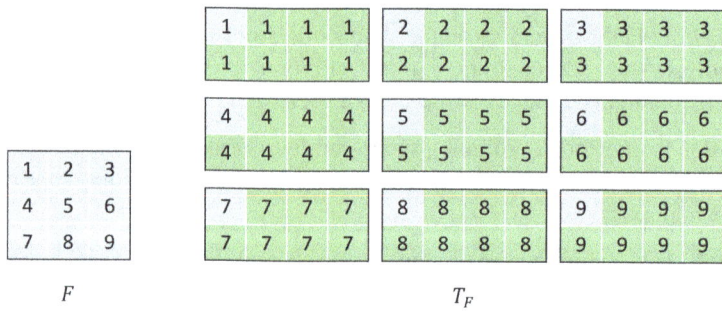

Fig. 9.5 Packing a filter F of shape [3, 3] in a tile tensor T_F of shape $[3, 3, \frac{*}{2}, \frac{*}{4}]$

9.4 Convolution

Equation 9.9 shows the updated computation of O_1 for this case (which should still be followed by the slicing step, as before).

$$T_{O_1} = \sum_{k=0}^{h_F-1} \sum_{l=0}^{w_F-1} \text{PseudoRot}_l^1(\text{PseudoRot}_k^0(T_I)) \odot T_F[k, l, :, :]. \tag{9.9}$$

9.4.2 Padding

A 2D convolution can support different padding modes. Until now, we only considered the *valid padding* mode, where the filter is placed at all valid positions of the image, i.e., where all the filter elements are placed on valid image elements. Because of that, the size of the output was smaller than the input: $h_O = h_I - h_F + 1$ and $w_O = w_I - w_F + 1$. For example, when the filter size is [3, 3], and the image size is [10, 10], the output size is [8, 8].

A second padding mode is the *same padding* mode. Here, we zero pad the input image from all of its sides. For example, if we add to our image one row of zeroes from above, one row from below, one column to the left, and one to the right, then a filter of size [3, 3] produces an image with the same size as the input.

More generally, the padding can be specified by a quadruple $p = (p_u, p_d, p_l, p_r)$, which indicates how much zero padding should be added to a 2D image from above, below, left, and right directions, respectively. Let I_p be the image I padded according to some padding configuration p, then a simple way to pad an encrypted image is to encrypt I_p instead of I, and compute the convolution on it.

Under the reasonable assumption that the padding is not larger than the filter, a small improvement can be gained by realizing that it is enough to pad on two edges only, i.e., given $p = (p_u, p_d, p_l, p_r)$, set

$$p' = (0, \max(p_u, p_d), 0, \max(p_l, p_r)). \tag{9.10}$$

Working with an encryption of $I_{p'}$ is equivalent to working with an encryption of I_p when using the slightly modified convolution equation, defined as follows.

$$T_{O_1} = \sum_{k=0}^{h_F-1} \sum_{l=0}^{w_F-1} \text{PseudoRot}_{l-p_l}^1(\text{PseudoRot}_{k-p_u}^0(T_{I_{p'}})) \odot F[k, l], \tag{9.11}$$

$$T_{O_2} = \text{Slice}_{0, w_O}^1(\text{Slice}_{0, h_O}^0(T_{O_1})). \tag{9.12}$$

To see why this formula works, let's focus on $k = l = 0$. For these indices the input is pseudo-rotated by the negation of p_u and p_l. By Lemma 9.9, because of the zero padding, this is equivalent to real rotations. Hence, it moves the last p_u rows of padding to be the first rows, and similarly the last p_l columns of padding.

Thus, the first element of the filter "sees" an image properly padded in its first rows and columns. Recall that we assume the padding is not larger than the filter. This guarantees the first element of the filter is never placed against zero-padded elements in the last rows and columns, and therefore, it does not matter if any padding remained there.

An alternative to adding padding is to use masks as follows:

$$O_1 = \sum_{k=0}^{h_F-1} \sum_{l=0}^{w_F-1} \text{PseudoRot}^1_{l-p_l}(\text{PseudoRot}^0_{k-p_u}(I))M_{k,l}F(k,l), \qquad (9.13)$$

where each $M_{k,l}$ is a tensor of shape $[h_I, w_I]$, determining for each element (i, j) whether it is a valid element after the (pseudo-)rotation or if it has cycled from the other side, in which case we should treat it as padding. Formally, $M_{k,l}[i, j] = 1$ if $[i + k - p_u, j + l - p_l]$ is a valid index to I, and 0 otherwise. Each $M_{k,l}$ can be tiled according to the same tile tensor shape as T_I, in order to have a tile-by-tile correspondence with it. However, the tiles need not be encrypted; they may be kept as plaintext for faster computation.

Remark 9.23 Using masks guarantees that O_1 will have zeroes outside the valid output range. Therefore, the slicing we perform to compute O_2 will slice out only zeroes. This means that the unknown marking mentioned in Remark 9.20 can be omitted.

9.4.3 Strides

Another common feature of convolution is called strides. Instead of moving the filter from one pixel to the next, we move it in strides of h_s pixels along the height dimension, and w_s along the width dimension. For example, if $h_s = w_s = 3$ then we move it three pixels at a time along each dimension as illustrated in Fig. 9.6.

To compute a 2D convolution with strides we can first compute the convolution according to Eqs. 9.4 and 9.5, assuming $h_s = w_s = 1$. Then, add a third

Fig. 9.6 A 2D convolution with a [2, 2] filter applied with strides of $h_s = w_s = 3$ to an image of size [7, 6]

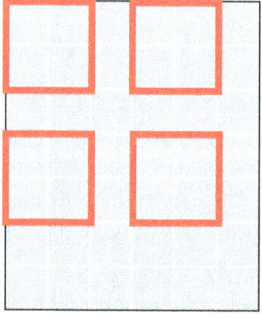

9.4 Convolution

step to compute O_3 by taking from O_2 only rows with strides of h_s (i.e., rows $0, h_s, 2h_s, 3h_s, \ldots$), and only columns with strides of w_s. Under encryption, this can easily be accomplished in some special cases, which we explore next. In the next paragraphs let T_{O_2} be the tile tensor containing O_2.

First, assume that h_s divides the external size of T_{O_2} along the first dimension. In that case, we can discard rows by discarding entire tiles.

Example 9.24 (When h_s Divides the External Size of T_{O_2} Along the First Dimension) Let $h_s = 4$ and let the first dimension of T_{O_2} be defined by the tile tensor shape element $\frac{125\sim}{8}$. The external size along this dimension is $16 = \left\lceil \frac{125}{8} \right\rceil$, that is, there are 16 tiles of size 8, together containing enough space to hold the 125 rows. Because of using interleaved packing, the first tile contains rows with indices $(0, 16, 32, \ldots, 112)$. All of these rows are included in the final output O_3. The next tile contains rows with indices $(1, 17, 33, \ldots, 113)$, all of which should be discarded. The next tile that should be kept is the 4th one, containing $(4, 20, 36, \ldots, 116)$.

As the example shows, if we want to keep every h_s-th row of O_2 and discard the rest, we should keep every h_s-th row of the external tensor of T_{O_2}, while adjusting its tile tensor shape accordingly. However, it is more efficient to not compute these tiles in the first place, i.e., skip them during the computation of Eq. 9.6.

Another relatively simple case is when the external tensor size is 1, and the stride divides the tile size.

Example 9.25 Let $h_s = 4$, and let T_{O_2} have the tile tensor shape $\left[\frac{15}{16} \right]$. For simplicity we use only one dimension. This technique extends to multiple dimensions by handling each dimension separately. The external size along this dimension is 1 (hence the interleaved packing has no effect). To discard all but every 4th row, we can simply adjust the tile tensor shape, splitting this single dimension to two: $\left[\frac{4}{4}, \frac{1?}{4} \right]$. Recall that we use the row-major order convention; hence, in the new shape the second dimension indicates each element is followed consecutively by three unknown ones; see Fig. 9.7.

In general, if T_{O_2} has a tile tensor shape $\left[\frac{h_{O_2}}{t_0}, \frac{w_{O_2}}{t_1} \right]$, it changes to $\left[\frac{h_{O_3}}{t_0/h_s}, \frac{1?}{h_s}, \frac{w_{O_3}}{t_1/w_s}, \frac{1?}{w_s} \right]$. However, we can do a bit better: instead of having the unknown indication in the newly created dimensions, we can zero out the discarded elements in advance, by replacing Eq. 9.4 with:

$$O_1 = \sum_{k=0}^{h_F-1} \sum_{l=0}^{w_F-1} \text{PseudoRot}_l^1(\text{PseudoRot}_k^0(I)) M_{h_s,w_s} F(k,l), \qquad (9.14)$$

where M_{h_s,w_s} is a mask of shape $[h_I, w_I]$, zeroing out all elements that will be discarded once we get to O_3.

Fig. 9.7 A tile tensor T_{O_2} with shape $\left[\frac{15}{16}\right]$ reshaped to T_{O_3} with shape $\left[\frac{4}{4}, \frac{1?}{4}\right]$

Remark 9.26 The resulting shape $\left[\frac{h_{O_3}}{t_0/h_s}, \frac{1}{h_s}, \frac{w_{O_3}}{t_1/w_s}, \frac{1}{w_s}\right]$ does not utilize all the slots. The dimensions $\frac{1}{h_s}$ and $\frac{1}{w_s}$ indicate only one out of $h_s w_s$ slots is used. This inefficiency can grow large in a sequence of convolutions with strides. A possible solution is to utilize these unused dimensions for multiple channels (we discuss channels in the next section). Such an approach was taken by [9, 11]. Another is to use them for multiple images [4], thus increasing the throughput of the computation by batching several inputs together. In both approaches the unused slots are created after each convolution with a stride step, and then they need to be filled with either channels or images. This is done by a step called gather or merge, in which the multitude of outputs from different channels/images are rotated along the unused dimensions $\frac{1}{h_s}, \frac{1}{w_s}$, and added together, creating dimensions $\frac{h_s}{h_s}, \frac{w_s}{w_s}$ filled with $h_s w_s$ channels/images.

9.4.4 Multiple Channels and Filters

In this section we extend the $O = \text{Conv}_{\text{siso}}(I, F)$ operator to support multiple channels and filters. We do that for tensors as well as for tile tensors. In this case,

9.4 Convolution

the input image I has three dimensions $[c, h_I, w_I]$, such that c is the number of channels. A simple example for channels is a color image having red, green, and blue channels, and each channel is specified as a matrix of scalars specifying the intensity of each pixel. More generally an image can have any number of channels.

The filter F is now four dimensional, with shape $[f, c, h_F, w_F]$, where f is the number of filters, and c is again the number of channels, the same as the number of channels of the input image. A multiple inputs multiple outputs (MIMO) convolution is defined as follows.

Definition 9.27 (Convolution (MIMO)) Let I and F be two input tensors for the convolution operator representing images and filters, with shapes $[c, h_I, w_I]$ and $[f, c, h_F, w_F]$, respectively. The result of the operation $O = \text{Conv}_{\text{mimo}}(I, F)$ is the tensor O of shape $[f, h_O, w_O]$, where $h_O = h_I - h_F + 1$, $w_O = w_I - w_F + 1$, and

$$O[i, :, :] = \sum_{j=0}^{c-1} \text{Conv}_{\text{siso}}(I[j, :, :], F[i, j, :, :]).$$

The output O is an image with f channels; hence, the number of input filters determines the number of output channels. The i'th output channel is the sum of SISO convolutions between the i'th filter and the image, where we sum over all input channels.

Observation 9.28 *A MIMO convolution is a case of matrix-vector multiplication. If we think of I as a vector of shape $[c]$ such that each of its elements is an image instead of a scalar, and F as a matrix of shape $[f, c]$ such that each of its element is an image, then we can compute the convolution using $O = FC$, using the usual definition of matrix-vector multiplication, except we replace multiplying scalar elements with $\text{Conv}_{\text{siso}}$ operators.*

This observation suggests a method for computing $\text{Conv}_{\text{mimo}}$ under encryption, by combining $\text{Conv}_{\text{siso}}$ with matrix-vector multiplication methods. For example, consider the method presented in Sect. 8.10. We pack I as a tile tensor T_I with shape $\left[\frac{c}{t_0}, \frac{*}{t_1}, \frac{h_I\sim}{t_2}, \frac{w_I\sim}{t_3}\right]$, and the filter F as a tile tensor T_F with shape $\left[\frac{c}{t_0}, \frac{f}{t_1}, h_F, w_F, \frac{*}{t_2}, \frac{*}{t_3}\right]$. Adapting Definition 9.27 to work with tile tensors, we get the equation,

$$T_{O_1}[i, j, :, :] = \text{Conv}_{\text{siso}}(T_I[i, :, :, :], T_F[i, j, :, :, :, :])), \tag{9.15}$$

$$T_{O_2} = \text{Sum}^0(T_{O_1}). \tag{9.16}$$

Equation 9.15 computes a tile tensor T_{O_1} of shape $\left[\frac{c}{t_0}, \frac{f}{t_1}, \frac{h_O\sim?}{t_2}, \frac{w_O\sim?}{t_3}\right]$ using tile-wise slicing: $T_I[i, :, :, :]$ is a slice of T_I with all tiles whose first index is i. Such a slice contains t_0 channels. The slice $T_F[i, j, :, :, :]$ contains t_1 filters and t_0 channels. We compute all their SISO convolutions, obtaining $t_0 t_1$ convolution

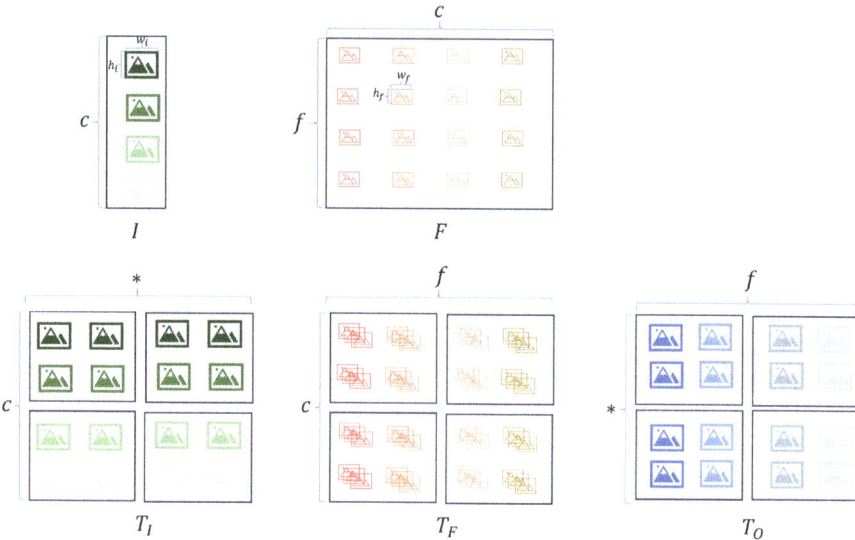

Fig. 9.8 An input I of shape $[c, h_i, w_i]$ and a filter F of shape $[c, f, h_f, w_f]$, accommodated in tile tensors T_I and T_F of shapes $\left[\frac{c}{t_0}, \frac{*}{t_1}, \frac{h_I\sim}{t_2}, \frac{w_I\sim}{t_3}\right]$, and $\left[\frac{c}{t_0}, \frac{f}{t_1}, h_F, w_F, \frac{*}{t_2}, \frac{*}{t_3}\right]$, respectively. After the convolution the output T_O has a shape of $\left[\frac{*}{t_0}, \frac{f}{t_1}, \frac{h_O\sim?}{t_2}, \frac{w_O\sim?}{t_3}\right]$

results in parallel (using SIMD), filling the slice $T_{O_1}[i, j, :, :]$. Repeating this for all $0 \le i < \lceil\frac{c}{t_0}\rceil$ and $0 \le j < \lceil\frac{f}{t_1}\rceil$ we fill all of T_{O_1}. Subsequently, Eq. 9.16 sums over the channel dimension, reducing this dimension's logical size to 1. Since it is the first dimension, we obtain a duplicated result, and the resulting tile tensor shape of T_{O_2} is $\left[\frac{*}{t_0}, \frac{f}{t_1}, \frac{h_O\sim?}{t_2}, \frac{w_O\sim?}{t_3}\right]$. See Fig. 9.8.

The unknown marking "?" in the output tile tensor shape in the image dimensions can be omitted; see Remarks 9.20 and 9.23. Also, similar to previous sections, the resulting tile tensor shape will generally require the additional notation for defining external size; see Remark 9.21.

Each application of the $\text{Conv}_{\text{siso}}$ operator performs some number x of (pseudo-)rotations on I (see Lemma 9.22 for an exact analysis). Let's further denote by y the number of tiles a single image channel takes (i.e., $y = \lceil\frac{h_I}{t_2}\rceil\lceil\frac{w_I}{t_3}\rceil$). We repeat the $\text{Conv}_{\text{siso}}$ operator $\lceil\frac{c}{t_0}\rceil\lceil\frac{f}{t_1}\rceil$ times. However, the $\text{Conv}_{\text{siso}}$ operator applies these rotations directly on the input, before multiplying it with the filter values. Hence, these rotations can be reused for all filters, and the total number of rotations required for the application of the SISO convolution is $\lceil\frac{c}{t_0}\rceil x$. The summation across the channel dimension adds $\log_2(t_0)\lceil\frac{f}{t_1}\rceil y$ rotations, reaching a total of $\lceil\frac{c}{t_0}\rceil x + \log_2(t_0)\lceil\frac{f}{t_1}\rceil y$.

Different ordering of the operations can lead to different rotation counts. The $\text{Conv}_{\text{siso}}$ operator first performs the x rotations then multiplies by the filter weights.

9.4 Convolution

We sum over the channels only at the end. Instead, we can multiply with the filter weights first, creating $h_F w_F$ tiles from each input tile. Then, we can reduce the channel dimension by summation, now requiring $\log_2(t_0) \lceil \frac{f}{t_1} \rceil y h_F w_F$ rotations. Finally, we complete the $\text{Conv}_{\text{siso}}$ operator computation by performing the pseudo-rotations, costing a total of x rotation for each of the $\lceil \frac{f}{t_1} \rceil$ tiles along the filters' dimension. Depending on the constants involved, c, f, h_F, w_F, one method can outperform the other.

Remark 9.29 The method above can be extended to handle the case where there is not a single channel dimension, but multiple dimensions holding channel data. This may arise when filling gaps after convolution with strides, as explained in Remark 9.26. Handling these multiple dimensions can be done using the guidelines detailed in Sect. 8.10.

A special case of the method described here for handling multiple channels is described in [10]. An alternative method is to perform the matrix-vector multiplication part using diagonal packing, as will be discussed in Sect. 9.7.1.

9.4.5 Image to Columns

A different approach to convolution is to reduce it to matrix-matrix multiplication as, e.g., in [3].

Given an input image with channels I of shape $[c, h_I, w_I]$, and filters F of shape $[f, c, h_F, w_F]$, we first compute an intermediate representation of the image, I' of shape $[p, h_O, w_O]$. It has the width and height of the output of the convolution, and a new dimension of size $p = c \cdot h_F \cdot w_F$, the total number of elements in a single filter. For each output pixel, I' contains the p pixels of I involved in computing this output. In addition, F is reshaped to F' of shape $[f, p]$. This is a simple reshape operation that maintains the number of elements and their order in a flattened representation.

Now, the convolution can be computed by viewing it as a matrix-matrix multiplication. Further reshaping I' as a two-dimensional matrix I'' of shape $[p, h_O \cdot w_O]$, the convolution is $O = F'I''$, resulting with a shape $[f, h_O w_O]$, which can be unflattened to $[f, h_O, w_O]$.

Remark 9.30 We can skip the flattening of I' to obtain I'', as well as the unflattening of the result. While using I'' makes it easier to see this computation as a matrix-matrix multiplication, it is not really necessary. We can directly use the product $F'I'$, interpreting it as the matrix-vector multiplication $F'I'(:, i, j)$ for each output pixel (i, j), which is equivalent to the above computation.

Consequently, the convolution can be computed using any method to compute matrix-matrix multiplication under encryption. For example, using the methods described in Sect. 8.10, we can pack I' as $T_{I'}$ with tile tensor shape $\left[\frac{*}{t_0}, \frac{p}{t_1}, \frac{h_O}{t_2}, \frac{w_O}{t_3} \right]$, and F' as $T_{F'}$ with tile tensor shape $\left[\frac{f}{t_0}, \frac{p}{t_1}, \frac{*}{t_2}, \frac{*}{t_3}\right]$. The result is computed using $T_O = \text{Sum}^1(T_{F'} \odot T_{I'})$, having a tile tensor shape $\left[\frac{f}{t_0}, \frac{1?}{t_1}, \frac{h_O}{t_2}, \frac{w_O}{t_3}\right]$.

This method can be highly efficient when the input is already packed and encrypted as in I'; however, it is inefficient otherwise. The reason is that processing I to I' can be costly under encryption. As a result, this method is mostly useful for the first convolutional layer, where preprocessing I' is done in the clear. In this case, using the method can boost time performance, at the cost of expanding the input size, which would increase the network usage time if the input is sent over a network.

9.5 Complex-Packed Tile Tensors

Complex packing was introduced in 7.8. This technique can also be combined with tile tensors and the tile tensor shape notation. The goal is to leverage the capability of an FHE scheme that natively works with complex numbers, in cases where the computation uses only real numbers. In this case, we can view each slot as if it has two slots: the real number is one and the imaginary number is another one.

Figure 9.9 presents the matrix M of shape $[5, 6]$ packed with its second dimension complex-packed. The notation in the tile tensor shape is adding (c) in the complex-packed dimension. In this example the tile tensor shape is $\left[\frac{5}{2}, \frac{6(c)}{4}\right]$.

The following definition formalizes this notion.

Definition 9.31 (Complex Packing) A tile tensor shape element $\frac{n_i(c)}{t_i}$ specifies that the size of the i'th dimension of the logical tensor is n_i, the tile size along this dimension is t_i, and the size of the external tensor $e_i = \lceil \frac{n_i}{2t_i} \rceil$. In the external tensor, in the tile with index l_i, the slot with index m_i contains a complex number whose real value is mapped to the logical tensor index $j_i = 2(l_i t_i + m_i)$, and imaginary value to $j_i = 2(l_i t_i + m_i) + 1$

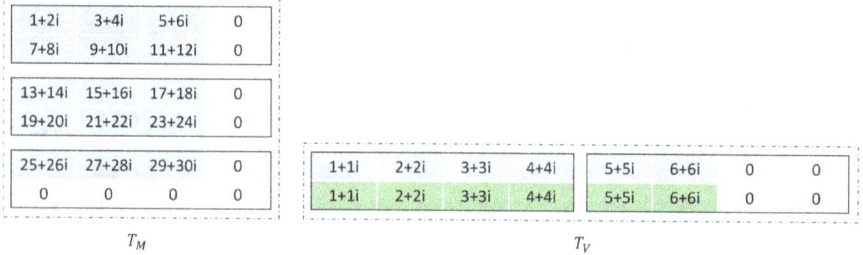

Fig. 9.9 Two tile tensors T_M, T_V complex-packed with shapes $\left[\frac{5}{2}, \frac{6(c)}{4}\right]$ and $\left[\frac{*(c)}{2}, \frac{6}{4}\right]$, respectively

9.5 Complex-Packed Tile Tensors

Similarly, we can complex-pack a duplicated dimension. For example, given a row vector V of shape $[1, 6]$, we can pack it to a tile tensor with tile tensor shape $\left[\frac{*(c)}{2}, \frac{6}{4}\right]$. This results with the vector duplicated *four* times, twice in each of the two slots along the first dimension; see Fig. 9.9.

Definition 9.32 (Duplicated Complex-Packed Dimension) A tile tensor shape element $\frac{*(c)}{t_i}$ specifies that the size of the i'th dimension of the logical tensor is 1, the tile size along this dimension is t_i, and the size of the external tensor is 1. Each slot contains a complex number where both the real value and imaginary value are mapped to logical tensor index $j_i = 0$.

Similarly to Definition 8.22, the definition above states that along dimension i the logical index remains constant, $j_i = 0$; hence, data is duplicated along this dimension. Here, however, the data is duplicated in each slot twice: once in the real part and once more in the imaginary part.

Remark 9.33 At most one dimension can be complex-packed. The reason is that we effectively split each slot to act as *two* slots. We should decide which dimension gets this ×2 boost. If we assign complex packing to two dimensions at once, then we need each slot to act as *four* slots, which cannot be done natively with complex numbers.

Based on the properties of complex packing detailed in Sect. 7.8, we can deduce the following compatibility rules:

1. $\frac{n_i(c)}{t_i}$ is compatible for additive operators with both $\frac{n_i(c)}{t_i}$ and $\frac{*(c)}{t_i}$. This is due to property (1) of Sect. 7.8.
2. $\frac{n_i(c)}{t_i}$ is compatible for multiplicative operators with $\frac{*}{t_i}$ with the additional condition that no dimension of the second tile tensor is complex-packed. This works correctly due to property (2) of Sect. 7.8. The constraint that the second tile tensor has no complex-packed dimensions is important, since otherwise we will be multiplying two complex numbers, which will not give us the elementwise semantics we require.
3. $\frac{n_i(c)}{t_i}$ is compatible with $\frac{n_i(c)}{t_i}$ for the operation of inner product over this dimension. This is done by extending Algorithm 12 to operate over a tile tensor dimension, which is simply changing Step 3 to use the $Sum^i()$ operator over the tile tensor's i'th dimension.

We can use the real and imaginary extraction functions of Sect. 7.8 to split a complex-packed tile tensor $[\ldots, \frac{n_i(c)}{t_i}, \ldots]$ into two tile tensors each having the tile tensor shape $[\ldots, \frac{n_i/2}{t_i}, \ldots]$. One contains every even index along the complex-packed dimension, and the other every odd index. It is also possible to combine complex packing and interleaved packing.

Definition 9.34 (Interleaved Complex Packing) A tile tensor shape element $\frac{n_i \sim (c)}{t_i}$ specifies that the size of the i'th dimension of the packed tensor is n_i, the tile size along this dimension is t_i, and the size of the external tensor $e_i = \lceil \frac{n_i}{2t_i} \rceil$. In the external tensor, in the tile with index l_i, the slot with index m_i contains a complex number whose real value is mapped to the logical tensor index $j_i = l_i + 2m_i e_i$, and imaginary value to $j_i = l_i + (2m_i + 1)e_i$.

Interleaved complex packing has the benefit that the complex packing can be removed more cleanly, resulting in a single tile tensor, as the following lemma shows.

Lemma 9.35 *Given a tile tensor T of shape $[\ldots, \frac{n_i \sim (c)}{t_i}, \ldots]$, the following method removes the complex-packed dimension.*

1. *Extract the real and imaginary parts of T (as in Sect. 7.8) to T_0 and T_1, respectively.*
2. *Concatenate T_0's and T_1's external tensors along dimension i to create a single external tensor.*
3. *Update the tile tensor shape of the resulting tile tensor T_2 to $[\ldots, \frac{n_i \sim}{t_i}, \ldots]$.*

Proof W.l.o.g. consider a tile t at index l_i of the external tensor of T and consider its slot index m_i. Its real value is mapped to logical index $a = l_i + 2m_i e_i$, and its imaginary part to logical index $b = l_i + (2m_i + 1)e_i$.

After the concatenation, the real part value ends up in T_2 at tile index l_i and slot index m_i. Because the external size of T_2 along dimension i is $2e_i$, it is mapped to logical index $l_i + 2m_i e_i$, which equals a.

Its imaginary value ends up in T_2 at tile index $l_i + e_i$ and slot index m_i, so it is mapped to the logical index $l_i + e_i + 2m_i e_i = l_i + (2m_i + 1)e_i$, which equals b. □

Example 9.36 (Complex-Packed Batch Dimension in NNs) Let us revisit the example of Sect. 8.12, but instead of packing a batch of inputs of shape $[y_0, b]$ using tile tensor shape $\left[\frac{y_0}{t_0}, \frac{*}{t_1}, \frac{b}{t_2}\right]$, we will use complex packing for the batch dimension, $\left[\frac{y_0}{t_0}, \frac{*}{t_1}, \frac{b(c)}{t_2}\right]$.

Computing the matrix multiplication is still done in exactly the same way, because for multiplication, $\frac{b(c)}{t_2}$ is compatible with $\frac{*}{t_2}$, with which the matrix batch dimension is packed. So, for example, we compute the even layers by $\text{Sum}^0(T_{A_i} \odot T_{M_i})$. The result remains complex-packed: $\left[\frac{*}{t_0}, \frac{x_i}{t_1}, \frac{b(c)}{t_2}\right]$. The bias is then packed in a tile tensor of shape $\left[\frac{*}{t_0}, \frac{x_i}{t_1}, \frac{*(c)}{t_2}\right]$. The complex-packed duplicated dimension is compatible for additive operators with the complex-packed batch dimension of the input.

For the square activation, we first split the output to two non-complex-packed tile tensors Re and Im, square each one of them, then merge them back together using $Re + i \cdot Im$, which result in a complex-packed tile tensor.

9.6 Complex Packing for Matrix-Matrix Multiplication

Let M and N be matrices of shapes $[x, y]$ and $[y, z]$. Using the matrix multiplication technique of Sect. 8.10, we will pack them in tile tensors T_M and T_N with shapes $\left[\frac{x}{t_0}, \frac{y}{t_1}, \frac{*}{t_2}\right]$ and $\left[\frac{*}{t_0}, \frac{y}{t_1}, \frac{z}{t_2}\right]$. The result is computed using $T_R = \text{Sum}_1 T_M \odot T_N$.

The number of ciphertext-ciphertext multiplications this computation includes is $\left[\frac{x}{t_0}\right]\left[\frac{y}{t_1}\right]\left[\frac{z}{t_2}\right]$. This can be sliced in half using complex packing in one of three ways.

1. Pack T_M as $\left[\frac{x(c)}{t_0}, \frac{y}{t_1}, \frac{*}{t_2}\right]$ and perform the same computation. The first dimension of $\frac{x(c)}{t_0}$ is still compatible with the other tile tensor's first dimension $\frac{*}{t_0}$. This guarantees the elementwise $\odot$ operator will work correctly. Then we sum over a non-complex-packed dimension which is done by the usual summation operator. The result has tile tensor shape $\left[\frac{x(c)}{t_0}, \frac{1?}{t_1}, \frac{z}{t_2}\right]$, so it remains complex-packed.

2. Pack T_N as $\left[\frac{*}{t_0}, \frac{y}{t_1}, \frac{z(c)}{t_2}\right]$. This works similarly to the method above, only now the result will have its last dimension complex-packed.

3. Pack T_M as $\left[\frac{x}{t_0}, \frac{y(c)}{t_1}, \frac{*}{t_2}\right]$ and T_N as $\left[\frac{*}{t_0}, \frac{y(c)}{t_1}, \frac{z}{t_2}\right]$. This works because $\frac{y(c)}{t_1}$ and $\frac{y(c)}{t_1}$ are compatible for inner product computations, that is, performing $\odot$ and summing over this dimension, which is what we do here.

This complex-packing optimization can be utilized for other matrix multiplication techniques, as it can be viewed as acting directly on the logical tensors. For example, we can squeeze N to have half the rows by combining every two rows into one using complex packing. Let N_c be the complex-packed version of N. Then $N_c M$ contains the complex-packed version of the result NM.

9.7 Diagonal Packing and Tile Tensors

Similar to other concepts that were extended to tile tensors, this section revisits the diagonal packing techniques of Sect. 7.10. Integrating them with tile tensors increases their flexibility and allows us to combine them with the previous methods that were discussed. In addition, it can provide us with some additional insights into how and why these techniques work.

Example 9.37 (Diagonal Tile Tensors I) Consider a matrix M of shape $[6, 8]$ packed transposed in a tile tensor T_{M^T} using diagonal packing, as illustrated in Fig. 9.10. In tile tensor shape notation we denote this as $\left[\frac{8(d)}{1}, \frac{6}{8}\right]$. We formally define the notation below. Informally, every tile has the shape $[1, 8]$, where the "(d)" indicates that the tiles are placed diagonally, so that each tile takes a diagonal slice out of M, wrapping around the bottom edge of M. Consider also a vector V of

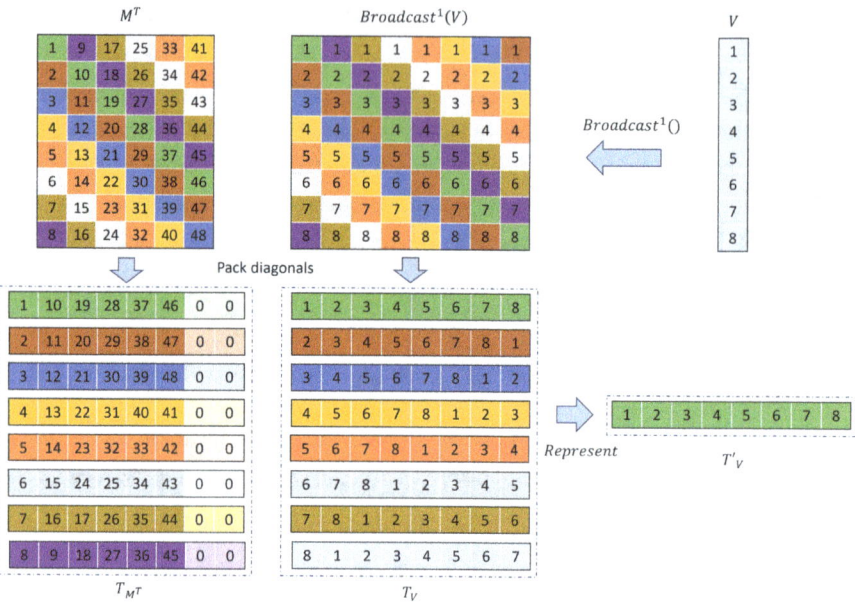

Fig. 9.10 Two tile tensors T_{M^T}, T_V with packed shapes $\left[\frac{8(d)}{1}, \frac{6}{8}\right]$ and $\left[\frac{8(d)}{1}, \frac{*}{8}\right]$, respectively

shape [8] packed in a tile tensor T_V of shape $\left[\frac{8(d)}{1}, \frac{*}{8}\right]$. This is the same packing as we used for M, except the second dimension is first duplicated and only then we consider the diagonals of V.

It is easy to see from Fig. 9.10 that $\text{Sum}^0(T_{M^T} \odot T_V)$ computes the matrix-vector multiplication: the elementwise multiplication multiplies V with every row of M. Summing the tiles together sums all the elements in each column, resulting with a single tile containing the result.

But a closer look on T_V reveals that all its tiles contain the same values, except rotated. That is, the first tile contains V, and the i'th tile is V rotated with an offset of i. So instead of packing V with diagonal packing, we can pack it simply as T'_V of shape $\left[\frac{8}{8}\right]$. Computing the product of M and V now requires a slightly modified algorithm. Multiply the i's tile of M with the single tile of $\text{Rot}_i(T'_V)$, then sum the results into T_R, a tile tensor of shape $\left[\frac{6}{8}\right]$. This is equivalent to the basic technique we have shown in Sect. 7.10, except we placed the common dimension of V and M as the first dimension which will have an advantage in the next example.

Example 9.38 (Diagonal Tile Tensors II) Consider now a matrix M of shape [4, 8] and V of shape [8]. The same diagonal tiling we used above still works, but is now more wasteful: all of T_{M^T}'s tiles are only half-full. For that reason, we select

9.7 Diagonal Packing and Tile Tensors

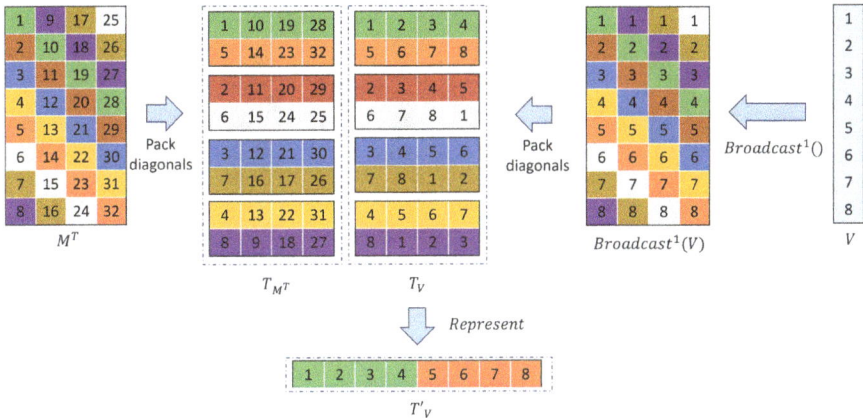

Fig. 9.11 Two tile tensors T_{M^T}, T_V with packed shapes $\left[\frac{8(d)}{2}, \frac{4}{4}\right]$ and $\left[\frac{8(d)}{2}, \frac{*}{4}\right]$, respectively

a different tile shape: [2, 4]. As before M is packed transposed in a tile tensor T_{M^T} with a tile tensor shape $\left[\frac{8(d)}{2}, \frac{4}{4}\right]$, and V in T_V of shape $\left[\frac{8(d)}{2}, \frac{*}{4}\right]$.

Diagonal tiles which have multiple rows, as in the current example, behave differently than ordinary multi-row tiles. The two rows are not placed adjacently, but one after the other. That is, one diagonal row starts after the other one ends. But as Fig. 9.11 shows, all the properties of the previous example are retained, and $\text{Sum}^0(T_M \odot T_V)$ computes the matrix-vector multiplication. The summation here involves also summing the two rows of the remaining tile; hence, the result has tile tensor shape $\left[\frac{*}{2}, \frac{4}{4}\right]$. Note also that the first tile of T_V, flattened (in row-major order), contains V, and all other tiles when flattened are rotated versions of V; hence, we can apply the same algorithm again. This is the hybrid method described in Sect. 7.10.

Note that a single tile of shape $[t_0, t_1]$ has t_0 diagonal rows, spanning together t_1 columns and $t_0 t_1$ rows, as shown in Fig. 9.11. With this observation, let us first formally define diagonalized tiling for tensors up to size $[t_0 t_1, t_1]$.

Definition 9.39 (Diagonalized Tiling) A tile tensor shape element pair $\frac{n_0(d)}{t_0}, \frac{n_1}{t_1}$ for $n_0 \leq t_0 t_1$ and $n_1 \leq t_1$ specifies that the size of dimensions 0, 1 of the logical tensor is n_0, n_1, the tile sizes along these dimensions are t_0, t_1, and the size of the external tensor $e_0 = t_1, e_1 = 1$, respectively. In the external tensor, in the tile with indices $l_0, 0$, the slot with indices m_0, m_1 is mapped to the logical tensor indices $j_0 = (l_i + m_0 t_1 + m_1) \mod (t_0 t_1), j_1 = m_1$.

To generalize this to a matrix M of arbitrary size $[a, b]$ we adopt the following strategy: we split the matrix into blocks of size $[t_0 t_1, t_1]$. Each of these blocks will be tiled with diagonal tiles according to Definition 9.39. See Fig. 9.12.

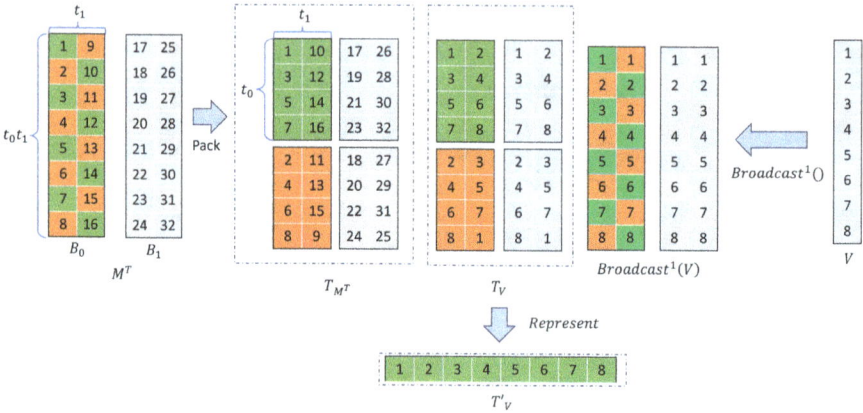

Fig. 9.12 A tile tensor T_{M^T} that packs M^T using diagonalization, by splitting M^T to two blocks of shape $[t_0 t_1, t_1]$. The tile tensor shape of T_{M^T} in this example is $\left[\frac{8(d)}{4}, \frac{4}{2}\right]$

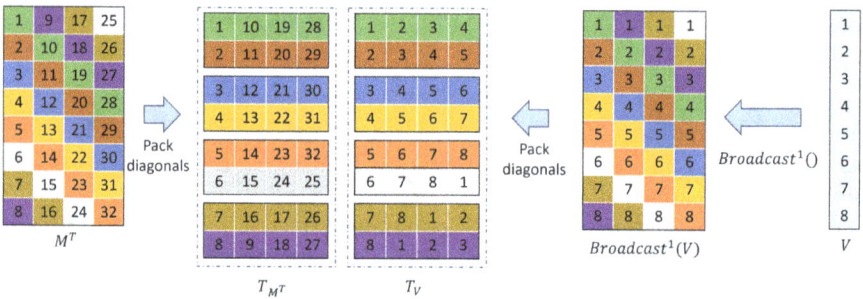

Fig. 9.13 Two tile tensors T_{M^T}, T_V but unlike in Fig. 9.11 the diagonals are packed in a consecutive order, which makes it harder to use one rotatable representative

Definition 9.40 (General Diagonalized Tiling) A tile tensor shape element pair $\frac{n_0(d)}{t_0}, \frac{n_1}{t_1}$ for arbitrary n_0, n_1 specifies that the size of dimensions 0, 1 of the logical tensor is n_0, n_1. The logical tensor is first split to contiguous blocks of size $[t_0 t_1, t_1]$ along the first two dimensions, adding zero padding where needed. Each of these blocks is tiled according to Definition 9.39. The total size of the external tensor along the first two dimensions is thus $e_0 = t_1 \left\lceil \frac{n_0}{t_0 t_1} \right\rceil, e_1 = \left\lceil \frac{n_1}{t_1} \right\rceil$.

Remark 9.41 Figure 9.13 shows another way to pack a diagonalized matrix. For brevity, and because we do not use this packing method below, we skip its formal description.

Let's pack M as a tile tensor T_M with tile tensor shape $\left[\frac{a(d)}{t_0}, \frac{b}{t_1}\right]$. Let V be a vector of shape $[a]$, which we pack as T_V with shape $\left[\frac{a}{t_0 t_1}\right]$. T_M is created by first

9.7 Diagonal Packing and Tile Tensors

dividing M to blocks $B_{i,j}$ of shape $[t_0 t_1, t_1]$, and then tiling each block diagonally. T_V is created as usual, by dividing V to tiles T_i of shape $[t_0 t_1]$.

To compute the matrix-vector multiplication we will compute it blockwise, as follows $R_j = \sum_i B_{i,j} T_i$, where $B_{i,j} T_i$ is a matrix-vector multiplication of a single diagonally tiled block with a single vector tile, computed with repeated rotations of T_i as described above. The results are tiles R_j that form the resulting tile tensor of shape $\left[\frac{*}{t_0}, \frac{b}{t_1}\right]$.

Lemma 9.42 *Computing matrix-vector multiplication (MV) with tile tensor shapes $\left[\frac{a(d)}{t_0}, \frac{b}{t_1}\right]$ and $\left[\frac{a}{t_0 t_1}\right]$, respectively, requires $\lceil \frac{a}{t_0 t_1} \rceil (t_1 - 1) + \lceil \frac{b}{t_1} \rceil \log_2 t_0$ rotations.*

Proof V is divided into $\lceil \frac{a}{t_0 t_1} \rceil$ blocks. Each is rotated $t_1 - 1$ times, where we exclude the rotation by 0. This accounts for the first part of the rotation count. The second part arises from the summation along the first dimension. After summing the resulted tiles together, which reduces their number to $\lceil \frac{b}{t_1} \rceil$, each one is summed inside using RotateAndSum, requiring $\log_2 t_0$ rotations. □

Compare this with the technique shown in Sect. 8.10 (see Eq. 8.5), where we used tile tensor shapes $\left[\frac{a}{t_0}, \frac{b}{t_1}\right]$ and $\left[\frac{a}{t_0}, \frac{*}{t_1}\right]$ for the matrix and vector. This technique required $\lceil \frac{b}{t_1} \rceil \log_2 t_0$ rotations for summing over each output tile using RotateAndSum. So the diagonal method adds an additional $\lceil \frac{a}{t_0 t_1} \rceil (t_1 - 1)$ rotations. However, it is not necessarily slower. While rotations usually dominate the runtime, other operators also contribute to the total latency, and different methods may differ also in the ability of parallelizing the derived circuit. We will see in Sect. 9.8 that sometimes the diagonal method ends up being faster and has the additional advantage of memory efficiency. Note also that here too the baby-step/giant-step method can be applied to further reduce the number of rotations.

9.7.1 Diagonal Packing for Convolution

Following Observation 9.28, we can utilize diagonal packing to compute a MIMO convolution [8]. For that, we pack the input I and filters F as the tile tensors T_I and T_F with tile tensor shapes $\left[\frac{c}{t_0 t_1}, \frac{h_I \sim}{t_2}, \frac{w_I \sim}{t_3}\right]$, and $\left[\frac{c(d)}{t_0}, \frac{f}{t_1}, h_F, w_F, \frac{*}{t_2}, \frac{*}{t_3}\right]$, respectively.

To compute the matrix-vector multiplication, we treat T_I as playing the role of the vector. We divide it into slices along its first dimension: $T_I[i, :, :]$, with $0 \le i < \lceil \frac{c}{t_0 t_1} \rceil$. Each such slice contains $t_0 t_1$ channels of the image I.

T_F plays the role of the matrix, with the matrix columns in the first dimension and rows in the second. In accordance with definition 9.40, we first divide the filters tensor F to blocks $F_{i,j}$ where each block contains $t_0 t_1$ channels and t_1 filters. Each such block is further divided into t_1 diagonals. Let $T_{F_{i,j}}[r]$ be the slice of T_F containing the r'th diagonal of block $F_{i,j}$. The matrix-vector multiplication over these slices is computed as:

$$O_{i,j} = \sum_r \text{Conv}_{\text{siso}}\left(\text{Rot}_r^0(T_I[i,:,:]), T_{F_{i,j}}[r]\right). \quad (9.17)$$

This is the usual formula for matrix-vector multiplication with diagonal packing, except we replaced the multiplication with SISO convolution. To complete the computations, we sum over the first dimension of the result:

$$O_j = \sum_i O_{i,j}, \quad (9.18)$$

and then sum over the first dimension of O_j, which results with a tile tensor of shape $\left[\frac{*}{t_0}, \frac{f}{t_1}, \frac{h_O \sim ?}{t_2}, \frac{w_O \sim ?}{t_3}\right]$.

We defer computing rotation counts to an exercise. As with the previous method, different ordering of the operations can obtain different rotation counts. (see a detailed example in [8]. The "input rotations" case is a special case of the above approach with $t_0 = 1$. The "output rotations" case describes how the rotations along the channel dimension can be deferred until after the SISO convolutions are performed).

9.8 Matrix Multiplication Method Comparison and Summary

This chapter and the previous one explored different methods to perform matrix multiplications. There may of course be more, but we believe that this is a good point to stop and compare the different methods that we have learned so far.

The first method we have seen is the duplication-based method of Sect. 8.10. This is a versatile method that encompasses inner products, outer products, matrix-vector, matrix-matrix, and other linear operators. It is also versatile in terms of performance. By appropriately choosing tile sizes the number of rotations can drop to 0, which saves memory on key sizes. Generally, the number of operations varies and can be adjusted for a particular environment. The multiplication depth is 1, and it can be easily used in a sequence of operators with little intermediate processing of cleaning up unknowns and duplication. Its main disadvantage is consuming more memory due to duplication. This is especially a problem when bootstrap operations are needed in the computation or when computing deep elementwise operators such as NN activation functions.

A second method is the diagonal method of Sect. 9.7. The diagonal method for matrix-vector multiplication can be easily extended for matrix-matrix multiplication. This is based on the observation that multiplying matrices A and B of shape $[x, y]$ and $[y, z]$ is the same as multiplying A with a batch of z vectors of shape $[y]$. Hence, A can be packed transposed as $\left[\frac{y(d)}{t_0}, \frac{x}{t_1}, \frac{*}{t_2}\right]$ and B as $\left[\frac{y}{t_0 t_1}, \frac{z}{t_2}\right]$. Now the same multiplication algorithm can be performed, treating the last dimension as a batch dimension.

This method of diagonalization is especially useful when the matrix in the matrix-vector or one of the matrices in the matrix-matrix multiplication is not encrypted and therefore can be packed appropriately using diagonalization. The method takes one multiplication depth and does not require any intermediate processing in a sequence of operators. However, it does require performing multiple rotations. This can be reduced using the baby-step/giant-step method. Also, in matrix-matrix multiplication duplication is added to the diagonalized matrix, causing increased encoding time and increased memory consumption.

A different diagonalization based method for matrix-matrix multiplication is shown in [7]. There, both matrices A and B are packed simply as $\left[\frac{x}{t_0}, \frac{y}{t_1}\right]$ and $\left[\frac{y}{t_0}, \frac{z}{t_1}\right]$. The multiplication process involves first diagonalizing both matrices, then repeatedly rotating one row-wise and the other column-wise, multiplying in pairs. This approach requires a multiplication depth of 3, but has the benefit of having both matrices packed tightly without duplications.

This method takes more multiplication depth than the previous methods, but has the benefit that both inputs and outputs are packed simply and tightly. It is useful when both matrices are encrypted, in scenarios where a complicated and deep sequence of operators is performed, including bootstrapping and deep activation functions. This is where the tight and stable packing comes in useful.

9.9 Summary

We presented an overview of tile tensors as a general packing solution. This section concludes Part III, by reiterating the benefits of using tile tensors, and summarizing the various options covered.

Tile tensors cover all the basic packing options described in Chap. 7, generalizing them to arbitrary sizes, while supporting additional options, such as interleaved packing. Many packing techniques described in the literature are equivalent to special cases of tile tensor use. The dimension independence idiom allows us to add more packing techniques, in a way that is interoperable with all existing ones.

Tile tensor shapes are a simple language to describe packing details precisely and concisely. They allow users to specify the desired packing details and track how they evolve as the computation progresses. Tile tensors also ease the planning and design of high-dimensional computation through the predictable changes that different operators have on them.

The concept of the logical tensor that a tile tensor represents allows separating the design of a computation task into the higher-level logical aspect and the lower-level implementation details. On the logical side, the computation is defined over ordinary tensors, with ordinary operators. Regarding the lower-level details, this computation is implemented using tile tensors, where specific tile tensor shapes are chosen for each tile tensor. These choices, done manually or automatically, can be optimized for specific goals: e.g., latency or throughput, memory constraints, network bandwidth, etc.

Even for additional, more sophisticated packing techniques, tile tensors can serve as a low-level underlying framework. Consider, for example, the image-to-column technique explained in Sect. 9.4.5. This technique takes an image M and transforms it to different matrix M'. This specific transformation is not natively supported by tile tensors. However, we can treat this transformation as preprocessing and pack the preprocessed result M' in a tile tensor. Thus, we enjoy the tiling and flattening capabilities of tile tensors, as well as the ability to use more advanced features such as interleaving, complex packing, and diagonalization.

Table 9.2 summarizes the various options allowed as elements of a tile tensor shape.

9.10 Lab Exercise: Program Advanced Circuits with Tile Tensors

The following lab exercise requires a programming environment such as Python. This environment should include libraries that can handle cleartext tensors such as NumPy [6] and also libraries that can handle tile tensors such as *pyhelayers* [1]. If

Table 9.2 Tile tensor shape: listing of all possible element options

Tile tensor shape element	Meaning	Compatible with
$\frac{n_i}{t_i}$	Basic tiling. Definition 8.12	$\frac{n_i}{t_i}, \frac{*}{t_i}$
n_i	Equivalent to $\frac{n_i}{1}$	
$\frac{*}{t_i}$	Duplication. Definition 8.22	For all operations: $\frac{n_i}{t_i}, \frac{*}{t_i}, \frac{n_i\sim}{t_i}, \frac{n_i\sim e_i}{t_i}$
		For multiplicative operations: $\frac{n_i(c)}{t_i}$
$\frac{n_i\sim}{t_i}$	Interleaved tiling. Definition 9.1	$\frac{n_i\sim}{t_i}, \frac{*}{t_i}$
$\frac{n_i\sim e_i}{t_i}$	Interleaved tiling with custom external size. Definition 9.16	$\frac{n_i\sim e_i}{t_i}, \frac{*}{t_i}$
$\frac{n_i(c)}{t_i}$	Complex-packed. Definition 9.31	For multiplicative operations: $\frac{*}{t_i}$ without complex packing. For additive operations: $\frac{n_i(c)}{t_i}, \frac{*(c)}{t_i}$
$\frac{*(c)}{t_i}$	Complex-packed with duplication. Definition 9.32	For additive operations: $\frac{n_i(c)}{t_i}, \frac{*(c)}{t_i}$
$\frac{n_i\sim(c)}{t_i}$	Complex-packed interleaved. Definition 9.34	For multiplicative operations: $\frac{*}{t_i}$ without complex packing. For additive operations: $\frac{n_i\sim(c)}{t_i}, \frac{*(c)}{t_i}$
$\left[\frac{n_0(d)}{t_0}, \frac{n_1}{t_1}, \ldots\right]$	Diagonal packing. Section 9.7	For matrix-vector multiplication: $\left[\frac{n_0}{t_0 t_1}, \ldots\right]$
?	Suffix indicating unknown slots. Definition 8.16	Does not affect compatibility

the latter environment supports a mockup option that only emulates the ciphertext operations but does not truly run them under encryption, then consider using it as it can speed up the computation without losing the ability to practice and get the correct results.

When using pyhelayers we recommend using a mockup context with 16 slots of the CKKS scheme. In pyhelayers encoded tile tensors are called PTileTensors and encrypted tile tensors are called CTileTensors.

1. Create a random matrix A of shape $[11, 7]$. Encrypt it in a CTileTensor object (HElayers' implementation of encrypted tile tensors) T_A with tile tensor shape $\left[\frac{11\sim}{4}, \frac{7\sim}{4}\right]$. Print the content of T_A, and see how the elements of A are divided among the tiles. For example, in pyhelayers, use

$$\text{pyhelayers.get\_print\_options().tt\_demo\_tiles} = \text{True}.$$

2. Use the PseudoRot method to perform pseudo-rotations. Verify that Theorem 7.21 correctly predicts whether a rotation or pseudo-rotation takes place.
3. Implement a SISO convolution operator using pseudo-rotations and the slice operator (use the slice method). Assume a plaintext filter of shape $[3, 3]$. Try it out on T_A.
4. Compute a matrix-vector multiplication using complex packing as explained in Sect. 9.6 (no need to use interleaved packing in this exercise). Try a matrix of shape $[12, 8]$, a vector of shape $[8]$, and pack them using tile shapes $[4, 4]$. Complex-pack the matrix rows.
5. Compute matrix-matrix multiplication using the diagonalized approach explained in Sects. 9.7 and 9.8. Pack a matrix A of shape $[12, 14]$ as $\left[14(d), \frac{12}{4}, \frac{*}{4}\right]$ and a matrix B of shape $[14, 10]$ as $\left[\frac{14}{4}, \frac{10}{4}\right]$. Try adding complex packing of B's columns
6. Compute the number of rotations required by the convolution approach of Sect. 9.7.1. Compare with that of Sect. 9.4.4.

References

1. Aharoni, E., Adir, A., Baruch, M., Drucker, N., Ezov, G., Farkash, A., Greenberg, L., Masalha, R., Moshkowich, G., Murik, D., Shaul, H., Soceanu, O.: HeLayers: a tile tensors framework for large neural networks on encrypted data. Privacy Enhancing Technology Symposium (PETs) 2023 (2023). https://petsymposium.org/popets/2023/popets-2023-0020.php
2. Aharoni, E., Drucker, N., Shaul, H.: Demo: rotating wide tensors with helayers. In: Proceedings of the 2023 Tutorial on Advanced HE Packing Methods with Applications to ML, Tutorial-HEPack4ML '23, pp. 3–5. Association for Computing Machinery, New York (2023). https://doi.org/10.1145/3605774.3625524
3. Benaissa, A., Retiat, B., Cebere, B., Belfedhal, A.E.: TenSEAL: a library for encrypted tensor operations using homomorphic encryption. Preprint. arXiv:2104.03152 (2021)
4. Cheon, J.H., Kang, M., Kim, T., Jung, J., Yeo, Y.: High-throughput deep convolutional neural networks on fully homomorphic encryption using channel-by-channel packing. Cryptology ePrint Archive (2023)

5. CryptoLab: HEaaN: homomorphic encryption for arithmetic of approximate numbers, version 3.1.4 (2022). https://www.cryptolab.co.kr/eng/product/heaan.php
6. Harris, C.R., Millman, K.J., van der Walt, S.J., Gommers, R., Virtanen, P., Cournapeau, D., Wieser, E., Taylor, J., Berg, S., Smith, N.J., Kern, R., Picus, M., Hoyer, S., van Kerkwijk, M.H., Brett, M., Haldane, A., del Río, J.F., Wiebe, M., Peterson, P., Gérard-Marchant, P., Sheppard, K., Reddy, T., Weckesser, W., Abbasi, H., Gohlke, C., Oliphant, T.E.: Array programming with NumPy. Nature 585(7825), 357–362 (2020). https://doi.org/10.1038/s41586-020-2649-2
7. Jiang, X., Kim, M., Lauter, K., Song, Y.: Secure outsourced matrix computation and application to neural networks. In: Proceedings of the 2018 ACM SIGSAC Conference on Computer and Communications Security, pp. 1209–1222 (2018)
8. Juvekar, C., Vaikuntanathan, V., Chandrakasan, A.: GAZELLE: a low latency framework for secure neural network inference. In: 27th USENIX Security Symposium (USENIX Security 18), pp. 1651–1669 (2018)
9. Kim, M., Jiang, X., Lauter, K., Ismayilzada, E., Shams, S.: HEAR: human action recognition via neural networks on homomorphically encrypted data. e-prints. arXiv–2104 (2021)
10. Kim, D., Park, J., Kim, J., Kim, S., Ahn, J.H.: HyPHEN: a hybrid packing method and its optimizations for homomorphic encryption-based neural networks. IEEE Access 12, 3024–3038 (2024). https://doi.org/10.1109/ACCESS.2023.3348170
11. Lee, E., Lee, J.W., Lee, J., Kim, Y.S., Kim, Y., No, J.S., Choi, W.: Low-complexity deep convolutional neural networks on fully homomorphic encryption using multiplexed parallel convolutions. In: International Conference on Machine Learning, pp. 12403–12422. PMLR (2022)
12. Zhang, Q., Xin, C., Wu, H.: GALA: greedy computation for linear algebra in privacy-preserved neural networks. Network and Distributed System Security (NDSS) Symposium (2021). https://www.ndss-symposium.org/ndss-paper/gala-greedy-computation-for-linear-algebra-in-privacy-preserved-neural-networks/

Part IV
Use Cases and Other Approaches

With the knowledge of all the algorithmic and cryptographic tools that were described in previous parts, the reader should now be able to approach the implementation of ML models over FHE with all their FHE-related challenges.

Various trade-offs and algorithmic options for scheme selection, bootstrapping, packing techniques, and function approximations are reviewed in the context of a whole system and are chosen in a way that optimizes performance and accuracy for the entire analytical model.

Chapter 10 describes how these techniques can be applied to several specific privacy-preserving ML examples. Chapter 11 discusses training considerations under FHE and then focuses on NNs and describes in more detail the challenges of implementing them in an FHE environment.

Chapter 10
Privacy-Preserving Machine Learning with HE

Abstract This chapter brings together the previously discussed techniques to the data science world. It includes a description of some specific ML models and use cases and explains how these can be handled privately with HE, what are the limitations of every model and how to overcome them under HE.

10.1 Training Under Encryption

The two tasks of machine learning: training (or fitting) and predicting (or inference), have a lot in common. Specifically, training of many models is performed by iteratively applying prediction on training data and then updating the model after comparing the predictions to the expected output. The details of how the model is updated are different from one model to another. The iterative updating process ends when the model performs accurately enough.

Under FHE the process of training is harder than training in the cleartext domain, and it is also much harder compared to prediction under FHE for three reasons:

1. **Stopping condition.** Unlike the cleartext case, when training under FHE one cannot test whether the model is accurate enough and use this info as a stopping condition. Instead, estimating the maximal number of iterations is needed in order to train an accurate enough model (see Sect. 4.6 for a discussion on stopping conditions of loops).
2. **Depth.** The training process involves many iterations, which result in a very deep computation. In contrast, the prediction process, which is sometimes deep by itself, is often just a single step of the training process.
3. **Ranges.** Training, like many other procedures, often involves non-polynomial functions. For example, NNs often use ReLU or Sigmoid as activation functions. Even simpler models like autoregressive integrated moving average (ARIMA) need to compute to compute non-polynomial functions such as the reciprocal inverse. As discussed in Chaps. 5 and 6, under FHE we replace these functions with polynomial approximations that are accurate (up to some small error) only

within a predefined range. If an input to these approximations is outside of the predefined range, the output may result with undefined behavior.

10.1.1 Resilience to Noise

Training ML models is done by searching the model parameters that minimize a cost function that measures how much the proposed model deviates from the training data. For simple models (e.g., linear regression (LR) models), the parameters that minimize the cost function can be computed analytically. For more complex models such as NNs the parameters that minimize the cost function are harder to compute. Instead, parameters that "approximately" minimize the cost function are found usually by applying some iterative method such as stochastic gradient descent (SGD). In their core, these methods start with an arbitrary set of parameters and in each iteration they try to improve the parameters to a set that yield a smaller cost function. The "self-correcting" nature of these algorithms where they converge to (approximately) the same results even if some noise is added between iterations makes these algorithms resilient to noise that is added in each iteration and therefore suitable to approximated encryption schemes such as CKKS.

An example for such an iterative method is the well-known SGD which we now briefly describe. This algorithm searches for a point near some local minimum of a predefined multivariate function. In the context of ML, this function is the $cost(x)$ function that associates an input $x \in \mathbb{R}^d$ with a value in $\mathbb{R}_{\geq 0}$. The input domain, i.e., the search space $\mathbb{R}^d$ involves different vectors x of d model parameters. The function $cost(x)$ estimates the model accuracy based on the input parameters, where $cost(x) = 0$ means the model exactly represents the data. The training algorithm starts with a random $cost(x) = 0$ means the model exactly represents the data. The training algorithm starts with a random model $x \in \mathbb{R}^d$ and then iteratively updates the model using a rule such as

$$x := x - \alpha \nabla cost(x), \qquad (10.1)$$

where $\alpha \in \mathbb{R}^+$ is called the *learning rate* (see Algorithm 14). The intuition behind SGD is that the gradient of a function points towards the minimum. A lot of research has been done on SGD optimization, addressing topics such as ways to avoid local minimum, or defining heuristics to compute the learning rate. These optimizations affect the learning process and are orthogonal to the optimization of FHE algorithms discussed in this book.

Since the SGD algorithm approaches a local minimum of $cost(x)$, any noise added in a single iteration only introduces a small perturbation. The next iteration will then start from a point close to where it would have been if there had been no noise but still continue to approach a local minimum. See a one-dimensional illustration in Fig. 10.1.

10.1 Training Under Encryption

Algorithm 14: Gradient descent (a simplification of [15])

Input: A cost function $cost : \mathbb{R}^d \mapsto \mathbb{R}_{\geq 0}$. A target upper bound cost ϵ.
Output: $x \in \mathbb{R}^d$ such that $\text{cost}(x) < \epsilon$.

1 $x \leftarrow \mathbb{R}^d$
2 **while** $\text{cost}(x) > \epsilon$ **do**
3 $\quad\quad x := x - \alpha \nabla \text{cost}(x)$
4 **end**
5 **return** x

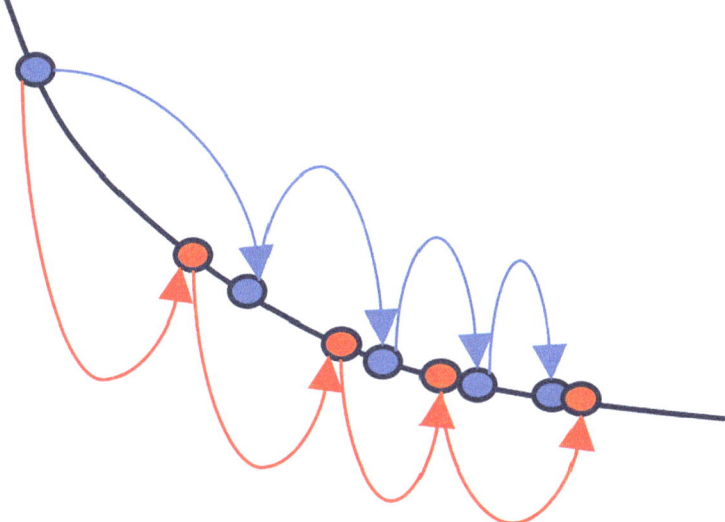

Fig. 10.1 An illustration of the SGD algorithm, starting from some point at the top. Blue points are points visited by the SGD process. Red points are points visited by the algorithm when noise is added to the computation, e.g., when using an approximated HE scheme such as CKKS

10.1.2 Training Approaches and Their Privacy Concerns

To expedite the training process, several more permissive approaches are sometimes considered. We discuss these together with their privacy concerns below.

10.1.2.1 Outsourcing to a Server

The vanilla case for privacy-preserving training is when a single party completely outsources the training processes to a single untrusted server. The server who cannot see the input and intermediate training data cannot check the stopping condition and thus needs to run some predefined (often large) number of iterations. One way to avoid this is to send the encrypted accuracy of the model to the client and have the client inform the server when the model is accurate enough and that no more computations are needed. This interactive process leaks some information about the training dataset to both the client and the server. Specifically, it leaks how easy it is to train on the data, which may be indicative to how homogeneous the training data is. Also here, dummy training iterations can be added to mask the real number of iterations needed, but these can only relax the leaked information, stating that the real number of iterations is within some bounded range.

To expedite the training, the server can send tasks that are hard to compute under FHE to the client. For example, in [5] the authors showed a system that (among other things) used the client to evaluate the activation functions in an execution of a neural network.

10.1.2.2 Federated Learning

A different type of training happens when data is partitioned between multiple parties. To this end, FL protocols allow training a shared model without uploading the data to untrusted environments. As mentioned in Sect. 3.1.5, data can be horizontally partitioned (i.e., the data records are partitioned between parties) or vertically partitioned (i.e., the features are partitioned between parties).

Algorithm 15: Vanilla FL

1 **the aggregator:**
2   initialize a model u_0
3   sends u_0 to all data owners
4 **end**
5 **for** *training iteration* $i = 1, 2, \ldots, \tau$ **do**
6   **every data owner** c:
7   computes a model update $u_i^c = Update(u_{i-1})$
8   sends u_i^c to the aggregator
9   **end**
10   **the aggregator:**
11   computes $u_i = \sum_c u_i^c$
12   sends u_i to all data owners
13   **end**
14 **end**
15 **return** u_n

The plain vanilla horizontal federated algorithm (see, e.g., [16]) involves n data owners c_i for $i < n$ and an agreed aggregator. The algorithm starts with an initial model u_0 sent by the aggregator to all data owners. In every learning iteration each model owner locally computes a model update and uploads it to the aggregator. After receiving all updates, the aggregator aggregates all the updates together (usually by summing or averaging them) and then returns the aggregated update to the data owners who use it to locally update their model; see Algorithm 15. Even though the private data of the data owners never leaves their premises, it has been shown (e.g., [12, 25]) that the model updates sent by every data owner leak significant information on their private data. Some works proposed to use DP to protect the data from leaking any individual data from any single record (e.g. [22]). Some works proposed to hide the model updates from the aggregator by using HE (e.g. [24]) or other MPC protocols (e.g. [6]). Although this hides the model updates from the aggregator, the intermediate models still leak to the different data owners, and when the number of data owners is small, this may lead to considerable data leakage.

When the data is vertically partitioned, a preliminary step is required to remove records for which not all features exist. Since the features are distributed between the parties, we need to remove records that do not appear in the databases of all the parties. This can be done, for example, by computing the intersection of sets of records the parties own. Doing this set intersection in the clear reveals to the parties which rows also appear in the datasets of the other parties. When this information is also private, the set intersection can done in a secure manner, e.g. under HE to protect this data leakage as well.

10.2 Machine Learning Models

We now review several machine learning models and discuss specific details in their FHE implementation.

10.2.1 Regressions (Linear and Logistic)

LR is a statistical model that shows a relation between multiple independent variables to a dependent variable. A logistic regression (LoR) model does the same, but instead of showing a relation, its goal is to identify the class for which a dependent variable belongs. In this section, we review both models in the context of FHE.

10.2.1.1 Linear Regression

LR finds a linear relation between d variables $x_1, \ldots, x_d$ and a dependent variable y. More formally, we are given n records $x^{(1)}, \ldots, x^{(n)}$, labels $y^{(1)}, \ldots, y^{(n)}$, and a distance function $dist(\cdot, \cdot)$, where $x^{(i)} = (x_1^{(i)}, \ldots, x_d^{(i)})$ and wish to find parameters $w_1, \ldots, w_d$ such that

$$dist(y^{(i)}, w_0 + \sum_j w_j \cdot x_j^{(i)}) \tag{10.2}$$

is as small as possible for all $i = 1, \ldots, n$.

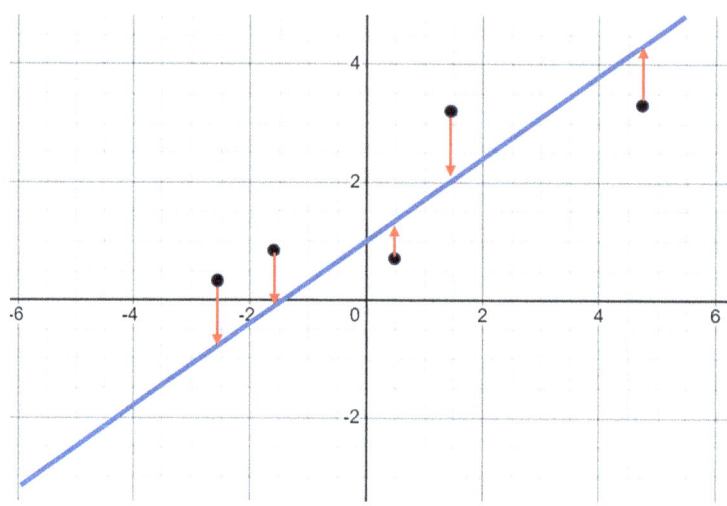

Fig. 10.2 A one-dimensional LR, relating $y = w_1 x + w_0$. The dots represent the pairs $(x^{(i)}, y^{(i)})$ and the red arrows the Euclidean distance between the labels $y^{(i)}$ and the predicted labels $w_1 x^{(i)} + w_0$

Giving a geometric interpretation we think of each pair $(x^{(i)}, y^{(i)})$ as a $(d+1)$-dimensional point. The goal is then to find a hyperplane

$$H : y = \sum_{i=1}^{d} w_i \cdot x_i + w_0 \qquad (10.3)$$

that minimizes Eq. 10.2 over all the points. See Fig. 10.2 for an example of a one-dimensional LR.

For conciseness, we set $x_0^{(i)} = 1$ for all $i = 1, \ldots, n$. Then Eq. 10.3 becomes

$$H : y = \sum_{i=0}^{d} w_i \cdot x_i. \qquad (10.4)$$

Geometrically, this lifts each point $x^{(i)}$ from $\mathbb{R}^d$ to $\mathbb{R}^{d+1}$, and specifically onto the surface defined by $x_0 = 1$, and looking for a hyperplane H that passes through the origin. We now use vector notation to state the LR loss function:

$$\arg\min_{w} \, dist(y, X \cdot w), \qquad (10.5)$$

10.2 Machine Learning Models

where $y = (y^{(1)}, \ldots, y^{(n)})$, $w = (w_0, w_1, \ldots, w_d)$ and

$$X = \begin{pmatrix} 1 & x_1^{(1)} & & x_d^{(1)} \\ \vdots & & \ddots & \\ 1 & x_1^{(n)} & & x_d^{(n)} \end{pmatrix}. \tag{10.6}$$

Commonly, the distance function $dist$ is taken to be the Euclidean distance squared. Since $\frac{1}{n}$ is constant within an instantiation of a training process, we can multiply by a factor of $\frac{1}{n}$ without affecting the w that achieves the minimum. This is useful to keep the intermediate values small when there are many records to train on. We therefore set the cost:

$$E = \frac{1}{n} dist(y, X \cdot w) = \frac{1}{n} \|y - X \cdot w\|_2^2. \tag{10.7}$$

The minimum is achieved when $\frac{\partial E}{\partial w_i} = 0$ for $0 \leq i \leq n$. Skipping the gory details we can rearrange these equations as $Aw = b$, where $A = X^T X$ and $b = X^T y$. Solving for w in plaintext is then easy: $w = A^{-1} b$. However, under FHE inverting A is a challenging task.

Training Using SGD

Another way to train an LR model is using SGD. We start from a random or arbitrary hyperplane

$$H^{(0)} : y = w_0^{(0)} + w_1^{(0)} \cdot x_1 + \cdots + w_d^{(0)} \cdot x_d, \tag{10.8}$$

and iteratively update w to minimize the distance from the labels. Specifically, at the ith iteration, we have $H^{(i)} : y = \sum_j w_j^{(i)} x_j$. The partial derivatives are then:

$$\frac{\partial E}{\partial w_j} = \frac{-2}{n} \sum x_j \cdot (y^{(i)} - X \cdot w^{(t)}). \tag{10.9}$$

The parameters w_i are then updated:

$$w_j^{(t+1)} = w_j^{(t)} - \alpha \cdot \frac{\partial E}{\partial w_j} = w_j^{(t)} + \alpha \cdot \frac{2}{n} \sum x_j \cdot (y^{(i)} - X \cdot w^{(i)}), \tag{10.10}$$

where α is a parameter called the *learning rate*.

Putting it together with the syntax introduced in Chap. 8, we assume a tile size of $t_1 \times t_2$ and set:

- The input matrix X be of shape $[\frac{n}{t_1}, \frac{d+1}{t_2}]$, that is, X is a $n \times (d+1)$.

- The label vector y be of shape $[\frac{n}{t_1}, \frac{*}{t_2}]$, that is, y is a vector of size n where the values are duplicated along the second dimension.
- The model vector w be of shape $[\frac{*}{t_1}, \frac{d+1}{t_2}]$, that is, w is a vector of size $d+1$ where the values are duplicated along the first dimension.

Using these shapes Eq. 10.10 can be implemented with this code (Line 3 in Algorithm 16) which we explain below.

$$w^{(t)} = w^{(t-1)} + \frac{2\alpha}{n} \cdot \text{Sum}^1(X \cdot \text{Sum}^2(X \cdot w))$$

- $X \cdot w$ is a multiplication between a $[\frac{n}{t_1}, \frac{d+1}{t_2}]$ tile tensor and a $[\frac{*}{t_1}, \frac{d+1}{t_2}]$ tensor. As explained in Chap. 8 the output has a shape of $[\frac{n}{t_1}, \frac{d+1}{t_2}]$, effectively multiplying every sample of X with the weights w of the model.
- $\text{Sum}^2(\cdot)$ sums over the second dimension. This results in a tensor with shape $[\frac{n}{t_1}, \frac{*}{t_2}]$ and effectively performs the sum $\sum x_i \cdot w_i$ for all samples in parallel.
- Multiplying $X \cdot \text{Sum}^2(\cdot)$ involves tiles of shape $[\frac{n}{t_1}, \frac{d+1}{t_2}]$ and $[\frac{n}{t_1}, \frac{*}{t_2}]$. The output is again a tile with shape $[\frac{n}{t_1}, \frac{d+1}{t_2}]$. This effectively multiplies the ith coordinate x_i with the ith coordinate of $\text{Sum}^2(\cdot)$ for all coordinates i and all the samples in X in parallel.
- Finally, $\text{Sum}^1(\cdot)$ sums over the first dimension and results in a tensor of shape $[\frac{*}{t_1}, \frac{d+1}{t_2}]$. This has the effect of summing each coordinate over all the samples in X.
- Eventually, the output is scaled and added to modify the weights w.

We conclude the discussion on LR with Algorithm 16 that implement linear regression training using the tile tensor notation from Chap. 8 and function sum over dimension, $\text{Sum}^i(T)$, mentioned in Sect. 8.9

Algorithm 16: Linear regression training

Input: n points in $\mathbb{R}^{d+1}$ given as a tensor with shape $shape(X) = [\frac{n}{t_1}, \frac{d+1}{t_2}]$, their labels given as a tensor with shape $shape(y) = [\frac{n}{t_1}, \frac{*}{t_2}]$.

Output: The model weights given as a tensor with shape $shape(w) = [\frac{*}{t_1}, \frac{d+1}{t_2}]$ that minimize (approximately) the cost in Eq. 10.7.

1 Initialize $w^{(0)}$ with an arbitrary value
   // Run τ SGD iterations
2 **for** $t = 1, \ldots, \tau$ **do**
3    $w^{(t)} = w^{(t-1)} + \frac{2\alpha}{n} \cdot \text{Sum}^1(X \cdot \text{Sum}^2(X \cdot w))$
4 **end**
5 **return** $w^{(\tau)}$

10.2.1.2 Training Linear Regression Under FHE

As we have covered above, there are two ways to train an LR model. The first is by solving $w = A^{-1}b$, which gives the exact solution that minimizes the cost but in HE is limited to low dimensions. Solving $w = A^{-1}b$ under FHE involves inverting the $d \times d$ matrix A. This is a hard task under FHE because it involves a deep computation that includes branches and computing reciprocal inverses of numbers.

The second is using SGD which is efficient for high dimensions but gives a solution that only approximates the cost-minimization problem. In addition, when training using SGD there is the issue of stopping condition as mentioned in Chap. 4 and at the beginning of this chapter.

In [11] the authors proposed to train an LR model using a protocol that involved two non-colluding servers as we explain below. They used the Paillier encryption scheme, which is an additive HE. The authors used the fact that the plaintext space F in Paillier, which is a *mathematical ring*, has a (very) large size emanating from its security requirement. Given, $a \cdot b^{-1}$, where $a, b < \sqrt{|F|}$ and b^{-1} is the reciprocal inverse of b in F, they used *rational reconstruction* [10, 23] to decode the rational number $\frac{a}{b}$.

Later, in [1] the authors showed how that can be done using BGV. Here the authors used multiple keys with different (co-prime) ring sizes together with the CRT to achieve a large ring size. In addition they showed how to pack the messages efficiently into ciphertexts and use the SIMD feature of BGV. In what follows, we shortly describe their protocol to invert a matrix A using two non-colluding servers. (Please note: obtaining the model $w = A^{-1}b$ using only additive HE (and two non-colluding servers) requires more details but is similar. The more detailed scenario will not be described here.)

- **Settings.** The protocol includes a client and two servers: $S1$ and $S2$.
- **Uploading.** The client encrypts the matrix A, sends $\text{Enc}(A)$ to $S1$, and sends the secret key to $S2$.
- **Masking A.** $S1$ draws a regular matrix R, computes $\text{Enc}(A \cdot R)$, and sends $\text{Enc}(A \cdot R)$ to $S2$.
- **Inverting $A \cdot R$.** $S2$ decrypts $A \cdot R$, inverts it to get $(A \cdot R)^{-1} = R^{-1} \cdot A^{-1}$, encrypts it, and sends $\text{Enc}(R^{-1} \cdot A^{-1})$ to $S1$.
- **Unmasking A^{-1}.** $S1$ multiplies by R to get $\text{Enc}(R \cdot R^{-1} \cdot A^{-1}) = \text{Enc}(A^{-1})$.

This protocol is secure since the view of $S1$ includes only encrypted messages, and the view of $S2$ includes only messages that are indistinguishable from random.

10.2.1.3 Logistic Regression

LoR is a model whose goal is to classify data points as one of two values (e.g., "1" or "0"). The regression model tries to fit the hyperplane of Eq. 10.3, so points above H are classified "1" and the points below H are classified as "0"; see Fig. 10.3. Using this distinctive hyperplane we can classify new points by checking their position

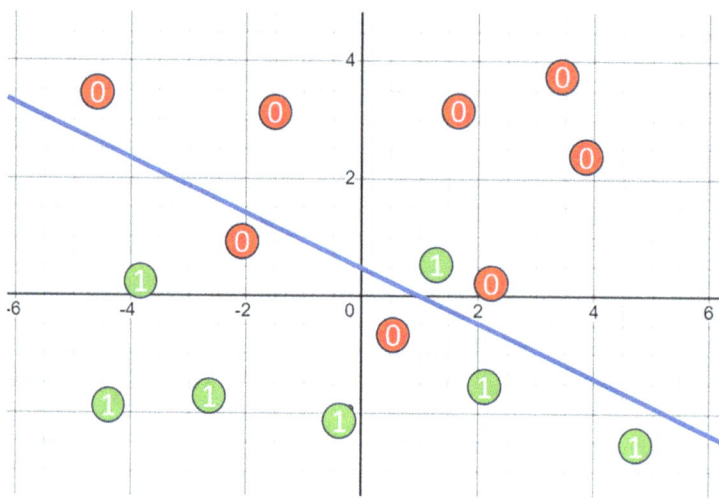

Fig. 10.3 A 2D LoR, classifying (x_1, x_2) as "0" if it is above the line $w_2 \cdot x_2 + w_1 \cdot x_1 + w_0 = 0$ or "1" otherwise. The figure presents three prediction mistakes, two 0s identified as 1 and one 1 that is identified as 0

relative to H. Furthermore, the "deeper" a point is on one side of H, the more confident the model is at its prediction.

More specifically, given a point $x = (x_1, \ldots, x_d)$ the model assigns a likelihood score $p(x) \in [-1, 1]$, where a higher value of $p(x)$ indicates a higher confidence that x is classified as "1," where

$$Pr[y = 1 \mid x] = \frac{1}{1 + e^{-(w_0 + \sum w_i \cdot x_i)}}. \tag{10.11}$$

Similar to the LR model, given n points $x^{(1)}, \ldots, x^{(n)}$ and their labels $y^{(1)}, \ldots, y^{(n)}$, we wish to find the parameters $w = (w_0, \ldots, w_d)$ then maximize the likelihood score:

$$\arg\max_{w} \prod_{y^{(j)}=1} Pr[y=1 \mid x^{(j)}] \prod_{y^{(j)}=0} \left(1 - Pr[y=1 \mid x^{(j)}]\right) =$$

$$\arg\max_{w} \prod_{y^{(j)}=1} \frac{1}{1 + e^{-w_0 - \sum w_i \cdot x_i^{(j)}}} \prod_{y^{(j)}=0} \left(1 - \frac{1}{1 + e^{-w_0 - \sum w_i \cdot x_i^{(j)}}}\right) = \tag{10.12}$$

Also here, we lift the problem to $d + 1$ dimensions by mapping $(x_1, \ldots, x_d)$ to $(1, x_1, \ldots, x_d)$. That is, we add a new coordinate x_0^j to each point and set it to 1. Abusing the notation for x and w which are now $d + 1$-dimensional vectors, we can

10.2 Machine Learning Models

rewrite the loss function as

$$\arg\max_{w} \prod_{y^{(j)}=1} \frac{1}{1+e^{-\sum w_i \cdot x_i^{(j)}}} \prod_{y^{(j)}=0} \left(1 - \frac{1}{1+e^{-\sum w_i \cdot x_i^{(j)}}}\right) = \quad (10.13)$$

since $1 - \frac{1}{1+e^{-t}} = \frac{1}{1+e^{t}}$ we can rewrite this product as

$$\arg\max_{w} \prod_{y^{(j)}=1} \frac{1}{1+e^{-\sum w_i \cdot x_i^{(j)}}} \prod_{y^{(j)}=0} \frac{1}{1+e^{\sum w_i \cdot x_i^{(j)}}} = \quad (10.14)$$

Setting $y'^{(j)} = 2y^{(j)} - 1 \in \{-1, 1\}$ and using the monotonicity of $\log(\cdot)$ we write

$$\begin{aligned}
&= \arg\max_{w} \log \prod_{j} \frac{1}{1+e^{-y'^{(j)} \cdot \sum w_i \cdot x_i^{(j)}}} \\
&= \arg\min_{w} \sum_{j} 1 + e^{-y'^{(j)} \cdot \sum w_i \cdot x_i^{(j)}}.
\end{aligned} \quad (10.15)$$

As before, we add a multiplicative factor of $\frac{1}{n}$ to keep the intermediate values small and set

$$J(w) = \frac{1}{n} \sum_{j} 1 + e^{-y'^{(j)} \cdot \sum w_i \cdot x_i^{(j)}} \quad (10.16)$$

and try to minimize $J(w)$ using SGD. Deriving by w_i we have

$$\frac{\partial J}{\partial w_i} = \frac{1}{n} \sum_{j} \text{Sigmoid}(-y^{(j)} \cdot \sum w_i x_i^{(j)}) \cdot y^{(j)} \cdot x_i^{(j)}. \quad (10.17)$$

The SGD algorithm therefore starts with a random value $w_i^{(0)}$ and iteratively updates

$$w_i^{(t+1)} = w_i^{(t)} + \alpha \cdot \frac{1}{n} \sum_{j} \text{Sigmoid}(-y'^{(j)} \cdot \sum w_i x_i^{(j)}) \cdot y'^{(j)} \cdot x_i^{(j)}, \quad (10.18)$$

where α is the learning factor.

The first FHE training procedure is due to [13] who showed to train a LoR model using SGD with Nesterov's acceleration [17]. Later in [4] the authors showed how to train 30,000 models in parallel using SIMD.

Approximating Sigmoid

The Sigmoid function used in the training is non-polynomial and to use under FHE it needs to be approximated by a polynomial, e.g., as discussed in Chaps. 5–6. In [14] the authors used least squares approximation to approximate Sigmoid over the input range $[-8, 8]$. They showed approximations of degree 3, 5, and 7. Later, the authors of [13] used these approximations to train a LoR model. They first set $y = \frac{x}{8}$ and then compute

$$\text{Sigmoid}_3(y) = -0.81562y^3 + 1.20096y + 0.5 \tag{10.19}$$

$$\text{Sigmoid}_5(y) = 1.3511295y^5 - 2.3533056y^3 + 1.53048y + 0.5 \tag{10.20}$$

$$\text{Sigmoid}_7(y) = -2.50739y^7 + 5.43402y^5 - 4.19407y^3 + 1.73496y + 0.5 \tag{10.21}$$

Figure 10.4 shows the different approximations.

Putting it together with the syntax introduced in Chap. 8, we assume a tile size of $t_1 \times t_2$ and set:

- The input matrix X be of shape $[\frac{n}{t_1}, \frac{d+1}{t_2}]$, that is, X is a $n \times (d+1)$.

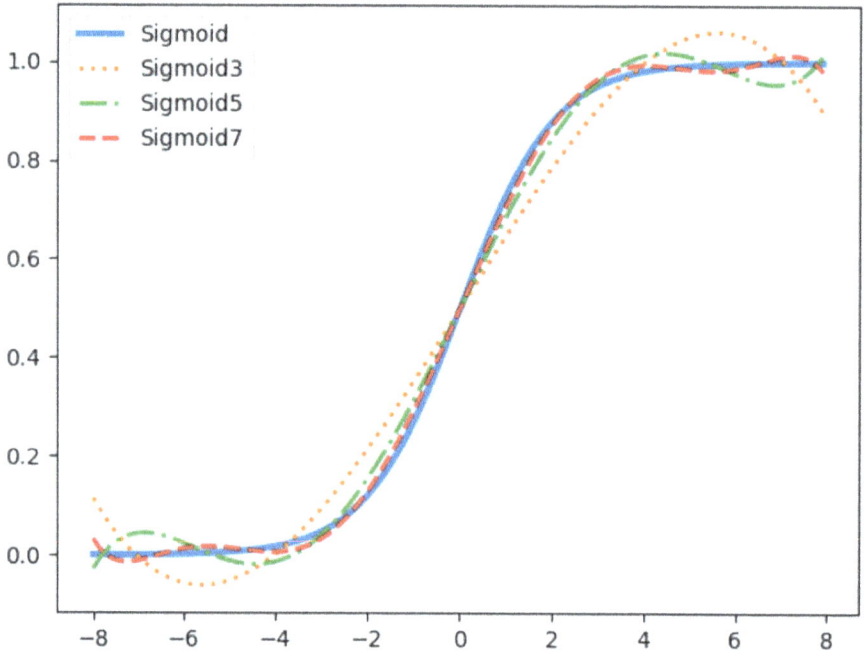

Fig. 10.4 Polynomial approximations of the Sigmoid function over $y = x/8$, for $x \in [-8, 8]$ using polynomials of degree 3, 5, and 7 from [14]

- The label vector y be of shape $[\frac{n}{t_1}, \frac{*}{t_2}]$, that is, y is a vector of size n where the values are duplicated along the second dimension.
- The model vector w be of shape $[\frac{*}{t_1}, \frac{d+1}{t_2}]$, that is, w is a vector of size $d+1$ where the values are duplicated along the first dimension.

Using these shapes Eq. 10.10 can be implemented with this code (Line 3 in Algorithm 17) which we explain below (the explanation is similar to that of LR but we repeat it with the necessary adjustments).

$$w^{(t)} = w^{(t-1)} + \frac{\alpha}{n} \text{Sum}^1(X \cdot y' \cdot \text{Sigmoid}(y' \cdot \text{Sum}^2(X \cdot w))).$$

- $X \cdot w$ is a multiplication between a $[\frac{n}{t_1}, \frac{d+1}{t_2}]$ tile tensor and a $[\frac{*}{t_1}, \frac{d+1}{t_2}]$ tensor. As explained in Chap. 8 the output has a shape of $[\frac{n}{t_1}, \frac{d+1}{t_2}]$, effectively multiplying every sample of X with the weights w of the model.
- $\text{Sum}^2(\cdot)$ sums over the second dimension. This results in a tensor with shape $[\frac{n}{t_1}, \frac{*}{t_2}]$ and effectively performs the sum $\sum x_i \cdot w_i$ for all samples in parallel.
- Multiplying $y' \cdot \text{Sum}^2(\cdot)$ involves a multiplication between two tensors of shape $[\frac{n}{t_1}, \frac{*}{t_2}]$.
- Applying Sigmoid($\cdot$) effectively applies the #Sigmoid function of all samples in parallel.
- Multiplying $X \cdot y'\text{Sigmoid}(\cdot)$ involves a tile of shape $[\frac{n}{t_1}, \frac{d+1}{t_2}]$ and two tiles of shape $[\frac{n}{t_1}, \frac{*}{t_2}]$. The output is again a tile with shape $[\frac{n}{t_1}, \frac{d+1}{t_2}]$. This effectively multiplies the ith coordinate x_i with the ith coordinate of Sigmoid($\cdot$) for all coordinates i and all the samples in X in parallel.
- Finally, $\text{Sum}^1(\cdot)$ sums over the first dimension and results in a tensor of shape $[\frac{*}{t_1}, \frac{d+1}{t_2}]$. This has the effect of summing each coordinate over all the samples in X.
- Eventually, the output is scaled and added to modify the weights w.

We conclude the discussion on LoR with Algorithm 17 that implements logistic regression training using the tile tensor notation from Chap. 8 and function sum over dimension, $\text{Sum}^i(T)$, mentioned in Sect. 8.9

10.2.2 Decision Trees

A decision tree (DT) is a tree where each inner node v is associated with a condition C_v and each leaf l is associated with a label L_l; see Fig. 10.5. The height of a node is its distance from the leaves in its subtree. When using a DT for prediction (in plaintext), we start at the root and evaluate in recursion. When visiting node v we evaluate the condition C_v. If C_v is met, we continue into the right child; otherwise, we continue into the left child. When we reach a leaf l, we output L_l. For example, in Fig. 10.5 a person over 40 with LDL lower than 180 and no family history of heart

Algorithm 17: Logistic regression training

Input: n points in $\mathbb{R}^{d+1}$ given as a tensor with shape $shape(X) = [\frac{n}{t_1}, \frac{d+1}{t_2}]$, their labels given as a tensor with shape $shape(y') = [\frac{n}{t_1}, \frac{*}{t_2}]$.

Output: The model weights given as a tensor with shape $shape(w) = [\frac{*}{t_1}, \frac{d+1}{t_2}]$ that minimize (approximately) the cost in Eq. 10.15.

1 Initialize $w^{(0)}$ with an arbitrary value
   // Run τ SGD iterations
2 **for** $t = 1, \ldots, \tau$ **do**
3 $\quad\;\; w^{(t)} = w^{(t-1)} + \frac{\alpha}{n} \text{Sum}^1(X \cdot y' \cdot \text{Sigmoid}(y' \cdot \text{Sum}^2(X \cdot w)))$
4 **end**
5 **return** $w^{(\tau)}$

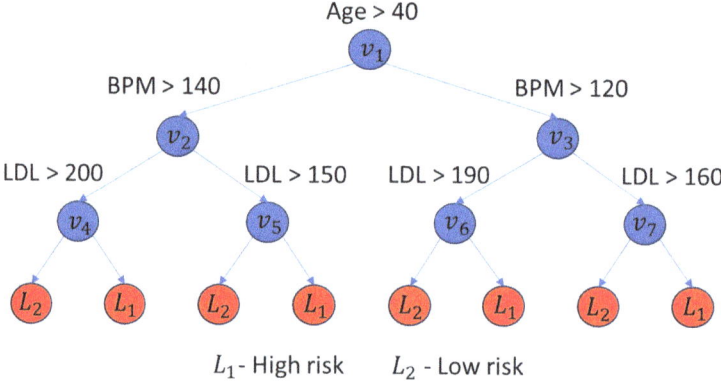

Fig. 10.5 A DT for determining the risk of heart disease. The tree considers parameters such as age, weight, LDL level, whether the person is a smoker, and whether they have a family history of heart diseases. Blue nodes are intermediate nodes, while red nodes are decision nodes with associated labels

diseases will start at the root then continue to the right child, then the left child, and finally, the left child and reach a leaf whose label is "Low Risk."

In the privacy-preserving setting, the tree (i.e., the conditions at the inner nodes and the labels at the leaves) and the query parameters are encrypted. In this case, training and traversing a DT has some challenges emanating from FHE. In what follows we first discuss the challenges of prediction using a DT.

As mentioned in Chap. 4, it is impossible to branch on a condition that depends on an encrypted value. Instead, we can compute both branches and use the condition as a multiplexer. This naïve method leads to computing the entire tree. Unfortunately, there is no way to avoid linear time processing when traversing a DT. That can be shown by a reduction from the PIR and remembering there's a lower bound $\Omega(n)$ for PIR shown in [3].

Since we have a lower bound of $\Omega(n)$, the naïve algorithm (Algorithm 18) seems quite efficient.

10.2 Machine Learning Models

Algorithm 18: Naive decision tree prediction

Input: A tree T where an inner node v is associated with a condition C_v and a leaf l is associated with a label L_l.
Output: The label L_l of the leaf l reached when traversing the tree.

1 set $\chi_{root} = 1$
2 for *all inner nodes v in BFS order* **do**
3   Compute under FHE c_v, indicating whether C_v is met.
4   set $\chi_{Right(v)} = \chi_v \cdot c_v$, where $Right(v)$ is the right son of v
5   set $\chi_{Left(v)} = \chi_v \cdot (1 - c_v)$, where $Left(v)$ is the left son of v
6 end
7 return $\sum_l \chi_l \cdot L_l$

This algorithm traverses the tree in a breadth-first-search (BFS) order starting at the root. The algorithm also maintains indicators χ_v indicating whether the plaintext traversal had reached the node v. For the root, χ_{root} is trivially set to 1. At a node v that the algorithm reaches, we compute c_v indicating whether the condition C_v is met. Then we set $\chi_{Right(v)} = \chi_v \cdot c_v$ and $\chi_{Left(v)} = \chi_v \cdot (1 - c_v)$. At the end, the algorithm multiplies the L_l by χ_l for each leaf l. This eliminates all the labels and the only surviving label is that of the node the plaintext traversal reached.

It is easy to see that the naïve algorithm visits every node of the tree for a total of $O(n)$ which does not contradict the $\Omega(n)$ lower bound. However, the naïve algorithm evaluates $O(n)$ conditions C_v (Line 3). Since this evaluation is in many cases significantly more expensive than computing χ_v (Lines 4 and 5), it is interesting to ask whether we can evaluate conditions only sublinear times. The answer to that is affirmative, as we next discuss.

In [2] the author suggested a way to traverse a DT under FHE in a way that requires the evaluation of only the conditions along the path the plaintext traversal would have taken. Their intuition was to keep an array of all nodes $N[1], \ldots, N[2n-1]$, where the $n-1$ are inner nodes and n are leaves. The node structure also includes the index of its right and left children; denote them as v_R and v_L, for the node v. Their algorithm starts at the root node and then acts as follows. For each node v they reach, they evaluate c_v (that indicates whether C_v is met) and then use PIR to copy the node whose index is $v_R \cdot c_v + v_L \cdot (1 - c_v)$. Their algorithm terminates after $\log n$ step when we reach a leaf. This indeed traverses only a single path, thus evaluating only $\log n$ conditions. We note that the overall algorithm is still linear because fetching $N[v_R \cdot c_v + v_L \cdot (1 - c_v)]$ takes linear time.

A more general approach called *copy and recurse* was introduced in [19], also mentioned in Chap. 4. Their approach can be applied for any r-ary tree (i.e., a tree where every inner node has r children) that the traversal algorithm (in plaintext) continues only into $\xi = r^c < r$ of them. Using *copy and recurse* the number of conditions that are evaluated is comparable to that of the plaintext algorithm. For the case of a binary tree, the running time then is $O(t \cdot n^{\log_r 2} + r \cdot n)$ for any $2 \leq r \leq n$ and t the time to compute c_v.

10.2.3 Time-Series Analysis

A time series is a sequence X of numeric values over time, and a time-series model such as ARIMA [21] attempts to fit a given time series so as to enable the prediction of future or intermediate unknown values. The model can also serve to detect an anomaly within the time series if the predicted value is too far from the observed future value when it actually occurs. For example, if we expect the signal measured in a time series to behave linearly, then we can model the time series as a straight line using LR. We can then extrapolate this line to a future time in order to predict a future value of the time series.

Suppose that the data of the data series is private, and that the user wishes for the server to fit the model in order to predict a future value. The user can encrypt the time series in one or more ciphertexts and send them to the server, which will proceed to fit the model and compute the prediction under FHE. For example, if we assume a simple linear model then, we can use LR training and inference under FHE as described in Sect. 10.2.1. Here, we show how to train and use a specific time-series model, namely, *ARIMA(p,q,d)*, from encrypted time-series data under HE. In this process we also discuss how to compute useful statistics over encrypted data (like variance, covariance, and correlations) and how to solve a system of linear equations under HE.

An autoregressive (AR) time-series model with a parameter p assumes a linear relationship between the value X_t of the time series at time t and its p preceding values. That is,

$$X_t = \mu + \sum_{i=1}^{p} \varphi_i X_{t-i} + \epsilon_t \qquad (10.22)$$

where φ_i are the model coefficients and ϵ_t is an additional error at time t that is usually assumed to be an independent, identically distributed variable sampled from a normal distribution with zero mean.

Example 10.1 (AR($p=2$)) The formula for an AR($p=2$) model where the value linearly depends on the previous two values is:

$$X_t = \mu + \varphi_1 X_{t-1} + \varphi_2 X_{t-2} + \epsilon_t. \qquad (10.23)$$

We can fit the values of μ, φ_1, and φ_2 by training a multiple LR model in which the values X_t correspond to the dependent variable and the corresponding values X_{t-1} and X_{t-2} are the two explanatory variables. Fitting AR(2) is actually a bit simpler than fitting a general LR model with two explanatory variables because the statistical properties needed for the fitting process such as the mean and variance are the same for all the variables (i.e., for X_t, X_{t-1}, and X_{t-2}), assuming the time series is long enough.

In a moving average (MA) time-series model with parameter q, the deviation of the values relative to the mean depends linearly on the q preceding errors, i.e., on

10.2 Machine Learning Models

the q preceding residues of the same model:

$$X_t = \mu + \sum_{i=1}^{q} \theta_i \epsilon_{t-i} + \epsilon_t. \tag{10.24}$$

The rationale for the dependence of the current value on previous errors is that the errors may occur due to some past "disturbance" that caused a deviation from the basic model and which takes some time ($\sim q$ time units) to wear out.

An autoregressive moving average (ARMA) time-series model with parameters p and q combines the behavior of an AR(p) model and an MA(q) model. In addition, it extends the AR model by adding a linear dependence of the value at time t and the q preceding errors, i.e., the q preceding residues of the same model:

$$X_t = \mu + \sum_{i=1}^{p} \varphi_i X_{t-i} + \sum_{i=1}^{q} \theta_i \epsilon_{t-i} + \epsilon_t. \tag{10.25}$$

An ARMA(p, q) time-series model can be used to predict up to q future values beyond the last value of the modeled time series.

Example 10.2 (ARMA($p = 2, q = 1$)) The following formula for the ARMA($p = 2, q = 1$) model assumes that the value linearly depends on the previous two values and also on the previous single error.

$$X_t = \mu + \varphi_1 X_{t-1} + \varphi_2 X_{t-2} + \theta_1 \epsilon_{t-1} + \epsilon_t. \tag{10.26}$$

The training of an ARMA model includes the computation (under FHE) of several statistical properties of the time series and its shifted versions (see below). For that we use the following notations: for a time series X of N elements, X_t is the value of the time series X at time $0 \leq t < N$ and c_x is the encryption of X under FHE using a tile tensor of shape $\left[\frac{N}{s}\right]$, where $s \geq$ where $s \geq N$. X^i is the time series X shifted i time slots into the past. That is, $X_j^i = X_{j-i}$ and X^i is defined only from time $j \geq i$, i.e.,

$$X^i = \left(\text{NA}, \text{NA}, \ldots, \text{NA}, X_i^i, X_{i+1}^i, X_{i+2}^i, \ldots, X_{N-i}^i\right) \tag{10.27}$$
$$= (\text{NA}, \text{NA}, \ldots, \text{NA}, X_0, X_1, X_2, \ldots, X_{N-i}).$$

The encryption of X^i is $c_{x^i} = \text{ShiftR}_i(c_x)$ (see Definition 2.9).

Example 10.3 (A Shifted Time Series) Consider a series X of $N = 5$ elements encrypted in ciphertexts of $s = 8$ slots as follows:

$$X = (1, 2, 3, 4, 5) \qquad c_x = (1, 2, 3, 4, 5, 0, 0, 0) \tag{10.28}$$
$$X^2 = (\text{NA}, \text{NA}, 1, 2, 3) \qquad c_{x^2} = (\text{NA}, \text{NA}, 1, 2, 3, 0, 0, 0). \tag{10.29}$$

It can be seen that the jth value of X^2 is *two* time steps into the past relative to the jth value of X (except for the NA values). Note that under encryption we often replace the NA values with zeroes by using plaintext masks.

Remark 10.4 Storing a series of N values in a tile tensor with shape $\left[\frac{N}{s}\right]$, where $s \geq N$, basically means to use a single ciphertext and place the series in its first N slots. Actually, tile tensors support more sophisticated features relevant for ARMA models. Namely, using interleaved packing, $\left[\frac{N\sim}{s}\right]$ (see Definition 9.1), we can drop the constraint $s \geq N$ and use an arbitrarily large series. Whenever we need to shift or rotate the series, we can use pseudo-rotations (Definition 9.8), which are efficiently supported by the interleaved packing method of tile tensors. For simplicity, we continue with the simpler packing method for the rest of this section.

10.2.3.1 Generating Stationary Time Series by Using a Differencing Process

The AR and ARMA models assume that a data series is stationary, meaning that its mean and variance are more or less constant over time, and that there is no seasonality in the data, i.e., no behavior that repeats at regular intervals such as at yearly holidays, weekends, or nightly. For example, the time series of Fig. 10.6 is *not* stationary because its mean rises over time. It is often possible to construct a stationary time series from a non-stationary one by differencing the original non-stationary series. That is, if a non-stationary time series X has $N = 1000$

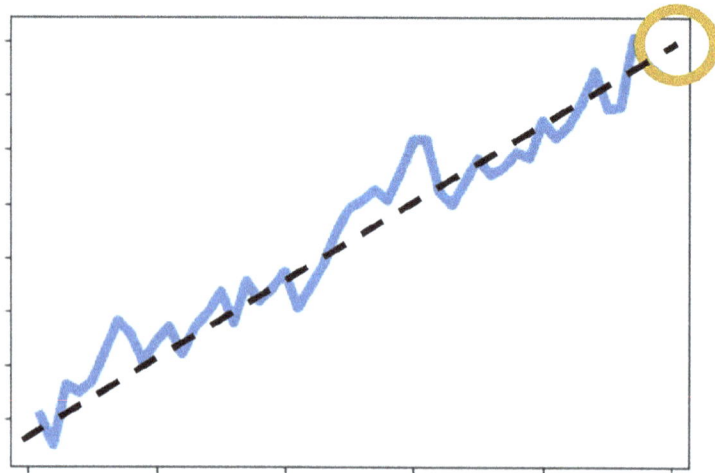

Fig. 10.6 A time series (blue) is modeled as a straight (black) line, e.g., using LR. The model predicts the next future value (gold circle). The x-axis represents the time, and the y-axis represents the measured value

10.2 Machine Learning Models

values, then it is possible to create a new time series Z with 999 values by setting $Z_i = X_{1+1} - X_i$ for $0 \leq i < N - 1$. If Z is still not stationary, we can repeat the differencing process until hopefully we get a stationary sequence of differences with enough structure left in the data. Such a model is called an *ARIMA(p, q, d)* time-series model, which compared to *ARMA(p, q)* also has an *integrated* part d. This part refers to the initial step of the model, where the original time series is differenced d times to get a stationary ARMA(p, q) time series.

The FHE implementation for time-series analysis described in this chapter assumes an HE *approximated scheme* such as CKKS, since the model parameters and the intermediate computations involve real values. In addition, we assume here that the entire time series fits within one ciphertext of the FHE scheme, which is often the case. For example, when using ciphertexts of 32K slots, a time series of up to 32K values is being considered. Handling larger series is clearly doable using the tile tensor methods learned in Part III.

If the d parameter of an ARIMA(p, d, q) model is greater than 0, then the time series must first be differenced d times before the training begins on the differenced stationary series. Each differencing can be carried out by rotating the ciphertext containing the time series by one slot to the left and then subtracting it from the original unrotated series. The differenced time series is then located in the first $N-d$ slots and the remaining slots should be zeroed out as described in Algorithm 19 and shown in Fig. 10.7a.

Algorithm 19: Differencing the input time series d times

Input: c_x a tile tensor of shape $\left[\frac{N}{s}\right]$ that encrypts a time series X of length N. d the number of required differencing steps according to the ARIMA(p, d, q) model.
Output: $cd_x[0:d]$: An array of $d+1$ encrypted tile tensors of shape $\left[\frac{N-i}{s}\right]$, where $cd_x[i]$ contains the i'th differencing of X.

1 $cd_x[0] \leftarrow c_x$
2 **for** *For i in 1..d* **do**
3 $\quad$ $cd_x[i] \leftarrow \text{Rot}_{-1}(cd_x[i-1])$
4 $\quad$ $cd_x[i] \leftarrow cd_x[i] - cd_x[i-1]$
5 **end**
6 $P\left[\frac{N-d}{s}\right] \leftarrow$ a tile tensor that encodes the vector $(1, 1, \ldots, 1) \in R^{N-d}$
7 $cd_x[d] \leftarrow cd_x[d] \odot P$
8 **return** $cd_x[0:d]$

The Algorithm 19's output, $cd_x[d]$, can now be trained as a stationary time series, learning the associated φ and θ parameters (see Eq. 10.25). The resulting model can then be used to predict the future q values of the time series $cd_x[d]$, which is currently of length $N-d$ and its values are values are

$$cd_x[d][i] = cd_x[d-1][i+1] - cd_x[d-1][i], \text{ for } i \in [0, N-d-1]. \quad (10.30)$$

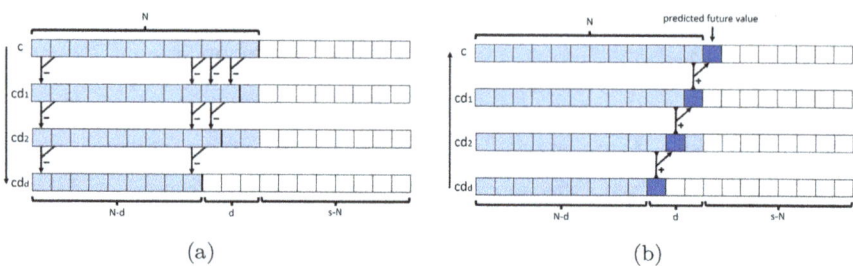

Fig. 10.7 Handling differencing during training and prediction. (**a**) Each differencing involves a subtraction of the rotated time series from the previous time series, and a reduction of the length of the time series by 1. The final time series $cd_x[d]$ is presumably stationary and can be trained as an ARMA(p, q) series. (**b**) The trained ARMA(p, q) model of $cd_x[d]$ is used to predict the next value of $cd_x[d]$. This predicted value is then used to predict the next value of $cd_x[d-1]$, $cd_x[d-2]$, etc. until the next value of the original time series C is computed

When $q = 1$, the predicted value corresponds to $cd_x[d][N - d]$, so we can also predict the next value of $cd_x[d - 1]$ as

$$cd_x[d - 1][N - d + 1] = cd_x[d][N - d] + cd_x[d - 1][N - d]. \quad (10.31)$$

We can continue and use $cd_x[d-1][N-d+1]$ to similarly predict $cd_x[d-2][N-d+2]$, etc. until we compute $cd_x[0][N] = c_x[N]$, i.e., the next value for the original input time series in c_x. These computations can all be carried out with FHE by repeatedly adding $cd_x[i]$ and rotating the result once to the right. The final prediction can then be extracted from the $N+1$'th slot of the final time series. See Algorithm 20 and Fig. 10.7b

---

Algorithm 20: Compute the predicted value for the original time series from the predicted value of the differenced series

Input: $cd_x[0 : d]$: An array of $d + 1$ encrypted tile tensors of shape $\left[\frac{N-i}{s}\right]$ from Algorithm 19, where $cd_x[d][N - d]$ contains the predicted future value of X in $cd_x[d]$.
Output: c_x a ciphertext in which $c_x[N]$ encrypts the predicted future value for X.
1 **for** *For i in d..0* **do**
2 $\quad cd_x[i - 1] \leftarrow cd_x[i - 1] \oplus cd_x[i]$
3 $\quad cd_x[i - 1] \leftarrow \text{Rot}_1(cd_x[i - 1])$
4 **end**
5 **return** $cd_x[0]$

10.2 Machine Learning Models

10.2.3.2 Fitting ARIMA or ARMA Time-Series Models Under FHE

The objective of training an ARIMA(p, d, q) or an ARMA(p, q) model for a given time series is to learn the φ and θ coefficients of Eq. 10.25 and the variance of the error ϵ in that equation. If the time series is not stationary, then an ARIMA model is required and the time series is differenced d times as described before.

Next, we explain how to train a differenced time series as an ARMA series. To this end, we define the following statistics over the time series, in which we rely on the Sum(c_x) operator that returns a tile tensor of shape $\left[\frac{*}{s}\right]$ with the encryption of $\sum_{i=1}^{N} X_i$ duplicated in all slots. It is computed using the using the RotateAndSum Algorithm of Sect. 7.2 over the s slots of c_x. The reason why the sum value is duplicated in all slots is that c_x contains the N values of the time series X and zeroes otherwise. In practice, Sumc_x does *not* need to assume that the N values of X are placed in the first N slots of c. The methods described below for computing the covariance and correlation rely on this fact.

- Mean$_{N(c_x)}$: returns a tile tensor of shape $\left[\frac{*}{s}\right]$ with the encryption of $\frac{1}{N}$Sum(c_x) duplicated in all slots. It is computed using a simple scalar-ciphertext product over the Sum($\cdot$) operator (recall that we are assuming an HE *approximated scheme*).
- Var$_N$ (C_X): the variance of the N elements of X computed by

$$\text{Var}_N(c_x) = \text{Mean}_N(c_x{}^2) - \text{Mean}_{N(c_x)}{}^2, \quad (10.32)$$

where the square operation is the FHE elementwise product of a ciphertext with itself.
- Cov$_{N-i}$ (c_x, c_{x^i}): the covariance of the $N - i$ elements of the time series X and X^i encrypted in c_x and c_{x^i}, respectively, starting from slot i in both cases. To achieve this, the first i elements of c_x are first zeroed out using a mask P, i.e., $\tilde{c}_x = c_x \odot P$ and the covariance is computed by

$$\text{Cov}_{N-i}(c_x, c_{x^i}) = \text{Mean}_{N-i}(\tilde{c}_x \odot c_{x^i}) - \text{Mean}_{N-i}(\tilde{c}_x) \odot \text{Mean}_{N-i}(c_{x^i}). \quad (10.33)$$

When N is large enough, the means of X and X^i can be assumed to be equal, which simplifies the computation to

$$\text{Cov}_{N-i}(\tilde{c}_x, c_{x^i}) = \text{Mean}_{N-i}(\tilde{c}_x \odot c_{x^i}) - \text{Mean}_{N-i}(c_{x^i})^2. \quad (10.34)$$

- Corr$_{N-i}$ (c_x, c_{x^i}): the Pearson correlation of the $N - i$ elements of the time series X and X^i encrypted in ciphertext c_x and c_{x^i}, respectively, starting from slot i in both cases. Denote by SD (c_x) the encryption of the standard deviation of X duplicated in all slots. Also here, the first i elements of c_x are first zeroed out using a mask P, i.e., $\tilde{c}_x = c_x \odot P$, and the correlation is computed by

$$\text{Corr}_{N-i}\left(\tilde{c}_x, c_{x^i}\right) = \frac{\text{Cov}_{N-i}\left(\tilde{c}_x, c_{x^i}\right)}{\text{SD}_{N-i}\left(c_x\right)\text{SD}_{N-i}\left(c_{x^i}\right)}. \tag{10.35}$$

When N is large enough, c_x and c_{x^i} are assumed to have the same standard deviation and the above equation is written as

$$\text{Corr}_{N-i}\left(\tilde{c}_x, c_{x^i}\right) = \frac{\text{Cov}_{N-i}\left(\tilde{c}_x, c_{x^i}\right)}{\text{Var}_{N-i}\left(c_{x^i}\right)}. \tag{10.36}$$

In any case, the resulting correlation is duplicated in all the slots of output ciphertext.

10.2.3.3 Fitting an ARMA($p > 1, q = 1$) Time Series

An ARMA($p > 1, q = 1$) model is

$$X_t = \mu + \sum_{i=1}^{p} \varphi_i X_{t-i} + \theta_1 \epsilon_{t-1} + \epsilon_t, \tag{10.37}$$

where its coefficients can be computed by following the steps of Algorithm 21, upon receiving an encryption of a time series X as input. The algorithm's steps may seem at first complex, and the next section explains how each of them was derived.

Remark 10.5 The division and square root operations need to be computed under FHE using polynomials that estimate these functions. The estimating polynomials can be found using the methods described in Chaps. 5 and 6 and then evaluated using the methods described in Sect. 4.8. The main difficulty is that when looking for a good estimating polynomial for these functions, for example, using the Remez algorithm, one must first determine the range of inputs for which the estimation is optimized. One could ask the user to provide the expected range for the values in the denominator and inside the square root, but this may be considered to violate the privacy of the data. In our case, however, the correlations and the θ parameter are guaranteed to have an absolute value less than 1. There is no similar limit to the covariance, but one can ask the user to normalize the time series before sending its encryption to the server for analysis so that the covariance of the series and its past values also remains small. In such a case the user would need to transform the normalized prediction back to the pre-normalized domain.

10.2 Machine Learning Models

Algorithm 21: Fitting an ARMA($p > 1, q = 1$) time series

1 Compute the φ_i parameters by solving a system of p linear equations. See the text.
2 $\mu = \text{Mean}_N(c_x)(1 - \sum_{i=1}^p \varphi_i)$
3 $c_{rx} = (\mu + \sum_{i=1}^p \varphi_i c_{x^i}) - c_x$
4 $\theta_1 = \dfrac{1 - \sqrt{1 - 4\text{Corr}_{N-p-1}(c_{rx}, c_{rx^1})^2}}{2\text{Corr}_{N-p-1}(c_{rx}, c_{rx^1})}$
5 $cov_i = \text{Cov}_{N-i}(c_x, c_{x^i})$ for $i \in i \in [1, p]$
6 $\text{Var}(\epsilon_t) = \dfrac{cov_1 - (\varphi_1 \text{Var}_N(c_x) + \sum_{i=2}^p \varphi_i cov_{i-1})}{\theta}$.

Computing the μ and φ_i Parameters

Step 1 of Algorithm 21 computes the φ_i parameters and uses these to compute μ on Step 2. We start with describing the latter computation. When the length of the time series N is large enough, we can assume that all the rotated and truncated time series X^i for $i \in [1, p]$, encrypted as the corresponding ciphertexts c_{x^i} used in the algorithm, have the same mean $m = \text{Mean}(X_{t-1})$ and that it is equal to the expectation of X_t. In addition, by definition, the errors ϵ_t and ϵ_{t-1} are assumed to be sampled from a normal distribution with zero mean. Thus, taking the expectation of X_t from Eq. 10.37 results with

$$m = \mathbb{E}(X_t) = \mathbb{E}(\mu) + \mathbb{E}\left(\sum_{i=1}^p \varphi_i X_{t-i}\right) + \theta_1 \mathbb{E}(\epsilon_{t-1}) + \mathbb{E}(\epsilon_t)$$

(10.38)

$$= \mu + \sum_{i=1}^p \varphi_i \mathbb{E}(X_{t-i}) = \mu + \sum_{i=1}^p \varphi_i m$$

from which we get

$$\mu = m(1 - \sum_{i=1}^p \varphi_i).$$

(10.39)

Going back to Step 1, our goal is to compute the φ_i coefficients. We do that in two steps: (1) generating a system of linear equations and (2) solving the system. We start with generating the system, where for brevity we consider the case $p = 4$ and subsequently generalize it for any $p > 1$. Specifically, the model for ARIMA($p = 4, q = 1$) is:

$$X_t = \mu + \sum_{j=1}^4 (\varphi_j X_{t-j}) + \theta_1 \epsilon_{t-1} + \epsilon_t$$

(10.40)

and the linear equations require computing different covariance values. For these computations, we make the assumption that the mean of the series m and hence μ are 0 (since the formulas for the covariance should be the same even if we consistently shift all the time-series values to a different mean).

Let $var = \text{Var}(X)$ and $cov_i = \text{Cov}(X^i, X) = \text{Cov}(X^i_j, X_j)$ for all j in which the two series are defined. Because the means of X and of X^i are assumed to be 0, we get that

$$cov_j = \mathbb{E}(X_{t-j} X_t)$$

$$= \mathbb{E}\left(X_{t-j}\left(\sum_{i=1}^{4}(\varphi_j X_{t-i}) + \theta_1 \epsilon_{t-1} + \epsilon_t\right)\right)$$

$$= \sum_{i=1}^{4}\varphi_i \mathbb{E}\left(X_{t-j} X_{t-i}\right).$$

The final step is valid because there is no dependence or covariance between a time series value (X_{t-j} considering cases where $j \geq 2$) and future errors (like ϵ_{t-1} or ϵ_t).

Placing $i = 2, 3, 4, 5$ in Eq. 10.41 leads to

$$\begin{aligned} cov_2 &= \varphi_1 cov_1 + \varphi_2 var + \varphi_3 cov_1 + \varphi_4 cov_2 \\ cov_3 &= \varphi_1 cov_2 + \varphi_2 cov_1 + \varphi_3 var + \varphi_4 cov_1 \\ cov_4 &= \varphi_1 cov_3 + \varphi_2 cov_2 + \varphi_3 cov_1 + \varphi_4 var \\ cov_5 &= \varphi_1 cov_4 + \varphi_2 cov_3 + \varphi_3 cov_2 + \varphi_4 cov_1. \end{aligned} \quad (10.41)$$

After computing the variance var and the covariances cov_j above under FHE, as described in Sect. 10.2.3.2, we hold a system of *four* linear equations for the variables $\varphi_1, \varphi_2, \varphi_3, \varphi_4$ which can be written as

$$\begin{bmatrix} cov_1 & var & cov_1 & cov_2 \\ cov_2 & cov_1 & var & cov_1 \\ cov_3 & cov_2 & cov_1 & var \\ cov_4 & cov_3 & cov_2 & cov_1 \end{bmatrix} \begin{bmatrix} \varphi_1 \\ \varphi_2 \\ \varphi_3 \\ \varphi_4 \end{bmatrix} = \begin{bmatrix} cov_2 \\ cov_3 \\ cov_4 \\ cov_5 \end{bmatrix} \quad (10.42)$$

For the general case where $p > 1$, the above can be generalized so that

$$cov_j = \sum_{i=1}^{p} \varphi_i cov_{|j-i|} \quad \text{for} \quad j \in [2, p+1] \quad (10.43)$$

where $cov_0 = var$. This facilitates a system of p linear equations for the variables φ_i where $i \in [1, p]$, which can be computed under FHE.

10.2 Machine Learning Models

Solving a System of Linear Equations Under FHE

Eventually, to estimate φ_i, we need to solve a system of linear equations under FHE, i.e., a system where all the equations' coefficients are encrypted. Denote by M, X, and V the matrix and the two vectors of Eq. 10.42, respectively, such that $MX = V$. For every $i \in [1, p]$ let M_i be the matrix M where column i is replaced by V. According to Cramer's rule, for every $i \in [1, p]$

$$X_i = \frac{det(M_i)}{det(M)} \qquad (10.44)$$

where $det(M)$ is the determinant of the matrix M. The division of Cramer's rule can be carried out using a polynomial estimation of the function $\frac{1}{x}$. Computing a determinant of a matrix involves just multiplications, additions, and subtractions, which can all be carried out directly under FHE on the encrypted matrices M and M_i. Multiplications are costly under FHE, so it makes sense to optimize the process of computing the determinant as much as possible. This may include, for example, parallelizing the computations of the determinants of sub-matrices and caching such determinants to avoid repeated computations.

Computing θ_1 (Algorithm 21, Step 4)

Step 3 of Algorithm 21 computes the encryption c_{rx} of the sequence of residues of predictions of the model AR(p) relative to the actual values of ARMA(p, q). In other words

$$RX_t = (\mu + \sum_{i=1}^{p} \varphi_i X_{t-i} + \theta_1 \epsilon_{t-1} + \epsilon_t) - (\mu + \sum_{i=1}^{p} \varphi_i X_{t-i}) = \theta_1 \epsilon_{t-1} + \epsilon_t. \qquad (10.45)$$

Thus, the time series RX behaves according to an MA($q = 1$) model with $\mu = \mathbb{E}(RX) = 0$.

Recall that the ϵ_t and ϵ_{t-1} values are assumed to be independent, identically distributed variables sampled from a normal distribution with zero mean, so their covariance is also zero.

$$\text{Var}(RX) = \text{Var}(\theta_1 \epsilon_{t-1} + \epsilon_t)$$
$$= \mathbb{E}\left((\theta_1 \epsilon_{t-1} + \epsilon_t)^2\right) - (\mathbb{E}(\theta_1 \epsilon_{t-1} + \epsilon_t))^2 = \mathbb{E}\left((\theta_1 \epsilon_{t-1} + \epsilon_t)^2\right)$$
$$= \mathbb{E}\left((\theta_1 \epsilon_{t-1})^2 + \epsilon_t^2\right) = (\theta_1^2 + 1)\mathbb{E}(\epsilon_t^2) = (\theta_1^2 + 1)\text{Var}(\epsilon_t).$$
$$(10.46)$$

The covariance of consecutive values of RX is

$$\text{Cov}\left(RX, RX^1\right) = \mathbb{E}(RX_t RX_{t-1})$$
$$= \mathbb{E}((\theta_1 \epsilon_{t-1} + \epsilon_t)(\theta_1 \epsilon_{t-2} + \epsilon_{t-1})) \quad (10.47)$$
$$= \mathbb{E}(\theta_1 \epsilon_{t-1}^2) = \theta_1 \text{Var}(\epsilon_t)$$

and the Pearson correlation between pairs of consecutive values of RX is

$$cor_1 = \text{Corr}\left(RX, RX^1\right) = \frac{\text{Cov}\left(RX, RX^1\right)}{\text{Var}(RX)}$$
$$= \frac{\theta_1 \text{Var}(\epsilon_t)}{(\theta_1^2 + 1)\text{Var}(\epsilon_t)} = \frac{\theta_1}{\theta_1^2 + 1}. \quad (10.48)$$

From Eq. 10.48 we get that θ_1, and in fact also $\frac{1}{\theta_1}$, are the solutions to the quadratic equation,

$$cor_1 \theta_1^2 - \theta_1 + cor_1 = 0. \quad (10.49)$$

Because the θ parameter of MA/ARMA/ARIMA models must be smaller than 1, we choose it to be the smaller root

$$\theta_1 = \frac{1 - \sqrt{1 - 4cor_1^2}}{2cor_1}. \quad (10.50)$$

This expression can be evaluated under FHE using approximating polynomials for the $\sqrt{x}$ function in the range $[0, 1]$ and for the $\frac{1}{x}$ function in the range $[-2, 2]$ (since correlations are always in the range $[-1, 1]$).

Computing $Var(\epsilon_t)$ (Algorithm 21, Step 5)

We show here how to derive the variance of the errors ϵ_t observed at different times t, i.e., the variance of the set of differences between the model's predictions and the actual values. We refer to this variance (somewhat informally) as $Var(\epsilon_t)$. Note that we can do this without deriving the errors themselves. Section 10.2.3.6 will later show how knowing this variance can help us perform anomaly detection on the time series under HE, i.e., to determine if the difference between the actual and predicted values is "too large."

The following derivation for cov_1 holds, where ϵ is the time series of ϵ_t values, and where $\mathbb{E}(S_1, S_2)$ is the expected value of the product of corresponding values from the two series. Note that the errors are independent and that while the values of X are dependent on past errors, they are *not* dependent on future errors, and thus

10.2 Machine Learning Models

the covariance between X^1 and ϵ is 0, and $\mathbb{E}(X^1, \epsilon) = 0$ (because the means of X^1 and of ϵ are also assumed to be 0).

$$\begin{aligned}
cov_1 &= \mathbb{E}(X^1 X) = \mathbb{E}(X_{t-1} X_t) \\
&= \mathbb{E}\left(X_{t-1} \left(\sum_{i=1}^{p} \varphi_i X_{t-i} + \theta_1 \epsilon_{t-1} + \epsilon_t\right)\right) \\
&= \sum_{i=1}^{p} (\varphi_i \mathbb{E}(X_{t-1} X_{t-i})) + \theta_1 \mathbb{E}(X_{t-1} \epsilon_{t-1}) \\
&= \varphi_1 var + \sum_{i=2}^{p} (\varphi_i cov_{i-1}) + \theta_1 \mathbb{E}(X_t \epsilon_t) \\
&= \varphi_1 var + \sum_{i=2}^{p} (\varphi_i cov_{i-1}) + \theta_1 \text{Var}(\epsilon_t)
\end{aligned} \qquad (10.51)$$

where the last equation holds because

$$\mathbb{E}(\epsilon_t X_t) = \mathbb{E}\left(\epsilon_t \left(\sum_{i=1}^{p} \varphi_i X_{t-i} + \theta_1 \epsilon_{t-1} + \epsilon_t\right)\right) = \mathbb{E}(\epsilon_t \epsilon_t) = \text{Var}(\epsilon_t) \qquad (10.52)$$

thus,

$$\text{Var}(\epsilon_t) = \frac{cov_1 - (\varphi_1 var + (\sum_{i=2}^{p} \varphi_i cov_{i-1}))}{\theta_1}. \qquad (10.53)$$

10.2.3.4 Fitting an ARMA($p = 1, q = 1$) Time Series

Algorithm 21 can be simplified to Algorithm 22 when fixing $p = 1$ to get an ARMA($p = 1, q = 1$) model.

$$X_t = \mu + \varphi_1 X_{t-1} + \theta_1 \epsilon_{t-1} + \epsilon_t. \qquad (10.54)$$

Here, we only need to estimate one parameter φ_1 instead of p, which does not involve solving a system of linear equations under FHE. Also, the sequence of residues c_{rx} computed in Step 3 is now defined with respect to the model $M = \text{AR}(p = 1)$ rather than $M = \text{AR}(p = 1)$.

Algorithm 22: Fitting an ARMA($p = 1, q = 1$) time series

Input: c_x: An encrypted time series.
Output: The coefficients of an ARMA(1, 1) model.

1 $\varphi_1 = \dfrac{\text{Cov}_{N-2}\left(c_x, c_{x^2}\right)}{\text{Cov}_{N-1}\left(c_x, c_{x^1}\right)}$

2 $\mu = (1 - \varphi_1)\text{Mean}_N(c_x)$

3 $c_{rx} = (\mu + \varphi_1 * c_x{}^1) - c_x$

4 $\theta = \dfrac{1 - \sqrt{1 - 4\text{Corr}_{N-p-1}\left(c_{rx}, c_{rx^1}\right)^2}}{2\text{Corr}_{N-p-1}\left(c_{rx}, c_{rx^1}\right)}$ // In ARMA(1, 1): $p = 1$.

5 $\text{Var}(\epsilon_t) = \dfrac{\text{Cov}_{N-1}\left(c_x, c_{x^1}\right) - \varphi_1}{\theta}$

6 return $\varphi, \theta, \text{Var}(\epsilon_t)$

10.2.3.5 Predicting Future Values with an Encrypted ARMA($p, q = 1$) Time-Series Model

We saw how an ARMA model can be trained under FHE so as to learn the encrypted coefficients of the model, where the goal of the training is to use the trained model to predict a future value. In general, an ARMA(p, q) model can be used to predict q future value(s) that are expected to follow the final value of the time series. In our case $q = 1$ so we can predict only one future value.

After the training we know the (encrypted) values of the parameters μ, φ_i, and θ_1, and we also know all the (encrypted) values c_x of X. However, we do not know the values of the ϵ_t errors for all the time slots t. These errors are the residues of the model predictions relative to the series, but the model predictions depend in turn on past errors, so there is a circular dependency which needs to be resolved somehow.

One way to handle this is to guess an initial error of 0 for the first p time points of the series and then continue the computation of the predictions and corresponding errors from that point onward based on that assumption. If the series indeed corresponds to an ARIMA model, then the resulting ϵ_t values should eventually stabilize to error values that resolve the circular dependency. For example, suppose we have an ARMA($p = 1, q = 1$) series of length N. We can assume that $\hat{\epsilon}_0 = 0$ for time $t = 0$ and can then predict

$$\hat{X}_1 = \mu + \varphi_1 X_0 + \theta_1 \hat{\epsilon}_0. \tag{10.55}$$

Now, we can predict $\hat{\epsilon}_1$ as

$$\hat{\epsilon}_1 = X_1 - \hat{X}_1 \tag{10.56}$$

10.2 Machine Learning Models

and continue to compute all the $\hat{X}_t$ and corresponding errors $\hat{\epsilon}_t$ in the same manner, until we reach the end of the series and predict

$$\hat{X}_{N-1} = \mu + \varphi_1 X_{N-2} + \theta_1 \hat{\epsilon}_{N-2} \tag{10.57}$$

$$\hat{\epsilon}_{N-1} = X_{N-1} - \hat{x}_{N-1}. \tag{10.58}$$

We are now at last able to predict the future value

$$\hat{X}_N = \mu + \varphi_1 X_{N-1} + \theta_1 \hat{\epsilon}_{N-1}. \tag{10.59}$$

If N is large enough, the circular dependency of the sequence of ϵ_t predictions will be resolved before the end of the series so that the final prediction would be accurate. Luckily, if the time series corresponds well to an ARIMA model, then the ϵ_t predictions usually stabilize quickly enough, say within ~20 iterations of applying Eqs. 10.57 and 10.58.

The above predictions of the $\hat{\epsilon}_t$ values of the time series involve only additions, subtractions, and multiplications, so they can all be done directly under FHE. The problem is that the product depth of the computation is as deep as the length of the time series, which could be very long indeed and thus impractical for FHE. However, as stated above, the predictions of the $\hat{\epsilon}_t$ values do not really need the full length of the series in order to stabilize. For example, if we estimate that the $\hat{\epsilon}_t$ predictions will stabilize after ~20 iterations, then we can start the chain of $\hat{\epsilon}_t$ predictions at the $N - 20$'th time slot, instead of at slot 0. We start by assuming that $\hat{\epsilon}_{N-20} = 0$ and then compute the corresponding values of $\hat{X}_{N-19} \Rightarrow \hat{\epsilon}_{N-19} \Rightarrow \hat{X}_{N-18} \Rightarrow \hat{\epsilon}_{N-18} \Rightarrow \ldots \hat{\epsilon}_{N-1} \Rightarrow \hat{X}_N$. The product depth now is just ~20 which is reasonable for an FHE computation.

10.2.3.6 Anomaly Detection Based on an ARIMA Model

We can also use a trained model for detecting an anomaly in a time series. Using the model we predict a future value and then check if the difference between the actual and predicted values is "too large." In the case of ARIMA, the errors ϵ_t (i.e., the differences between the model's predictions and the actual values) are assumed to be independent, identically distributed variables sampled from a normal distribution with zero mean. We can thus use the cumulative density function (CDF) for the normal distribution in order to determine at least how many standard deviations from the mean we need to be in order to detect an anomalous event that should occur with probability p or less. For example, suppose that the absolute difference between the actual and predicted value is err and that we want to report an anomaly when the probability p of this event is 1 in 100,000 or less. According to the CDF of the normal distribution, this probability corresponds to an observed error that is

$z_p = 4.417$ standard deviations or more from the mean of 0. Thus,

$$|err - 0| \geq z_p \text{SD}(\epsilon_t). \tag{10.60}$$

Squaring both sides leads to

$$err^2 \geq z_p^2 \text{Var}(\epsilon_t) \tag{10.61}$$

and we report an anomaly when the following expression is positive

$$err^2 - z_p^2 \text{Var}(\epsilon_t). \tag{10.62}$$

We saw in Sect. 10.2.3.3 how to estimate $\text{Var}(\epsilon_t)$ under FHE. We can thus send the value of expression 10.62 as the anomaly report. The user holding the FHE secret key would decrypt the report and declare an anomaly when the resulting plain value is positive.

The above expression assumes that we know the variance of ϵ_t, though in our case we actually just have an estimation for it based on samples from the time series. In general, some corrections are required when using an estimated variance to compute confidence intervals. However, when the time series is long enough, we can safely assume that our variance estimation is accurate.

10.2.4 K-NNs

In a simple k nearest neighbors (kNN) classifier, we are given a dataset of n points $P \subset \mathbb{R}^d$ where each point $p_i \in P$ also has a label $l_i \in \{0, 1\}$, then for a query $q \in \mathbb{R}^d$ we predict the label of q by taking the majority of labels of the k points in P whose distance to q is smallest. Distance here can be any metric but most commonly the L_1 and L_2 norms are used. The advantage of kNN is its simplicity. Specifically, kNN does not require a training phase. Prediction is also simple. See, for example, Algorithm 23

Algorithm 23: k nearest neighbor classifier

Input: $P \subset \mathbb{R}^d$: a set of n points.
$L \in \{0, 1\}^d$: a binary label, where l_i is the label of $p_i \in P$
$q \in \mathbb{R}^d$: a query.
Output: l_q: the predicted label for q.
1 **for** *Each point $p_i \in P$* **do**
2 $\quad$ compute $d_i = dist(p_i, q)$
3 **end**
4 Find the k smallest distances $d_{i_1}, \ldots, d_{i_k}$.
5 **return** $majority(l_{i_1}, \ldots, l_{i_k})$

10.2 Machine Learning Models

In the privacy-preserving setting, P belongs to one party, q belongs to a second party, and the code runs on the cloud (a third party). The privacy requirement is that only the label of q becomes known (usually to the party owning q). The privacy-preserving version of the problem involves finding the top-k values which is a hard problem under FHE. Solutions to the privacy-preserving problem use a mixture of tools. In [20] the authors used FHE to compute the distances of q from the points of P and then used *oblivious transfer* to read the labels of the k nearest points. In [8] the authors used kd-tree to find the k nearest neighbors more efficiently. The tree was encrypted with searchable encryption to allow for a fast execution. However, using searchable encryption leaks information. For example, the access patterns leaks. In [7] the authors use additive HE together with garbled circuits and distributed ORAM. In [9] the authors introduced the first protocol that was secure against the data owner and against the querier. However, their protocol required the parties to stay active during the entire protocol.

To improve the efficiency of kNN under FHE, the authors of [18] proposed to use approximation and randomization. First, they relaxed the classifier to k-ish nearest neighbors. They showed that the accuracy of the classifier does not change much when considering $\frac{1}{2}k < \kappa < 2k$ neighbors (see the accuracy as a function of k in Fig. 10.8). Then they showed a randomized algorithm over FHE that uses this relaxation to achieve a better running time.

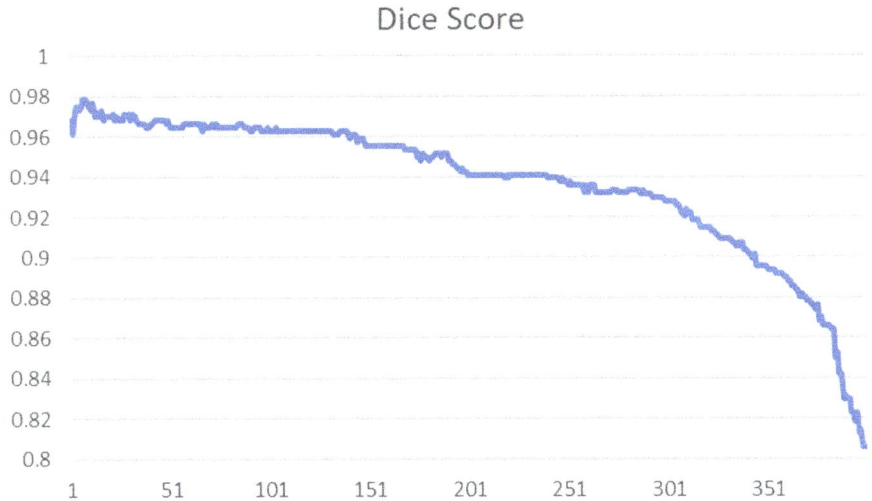

Fig. 10.8 The dice (F_1) score of a kNN classifier as it changes as a function of k as calculated on a breast cancer database of 569 samples

10.3 Lab Exercises

10.3.1 Statistical Metrics for Time-Series Analysis

Create a CKKS ciphertext c_1 in which the first 500 slots contain 500 values rising from 0.0 to 1. Set ciphertext c_2 to be the rotation of c_1 one slot to the left. Now use the methods described in Sect. 10.2.3.2 to compute the following statistics:

1. Compute the sum of the first 500 elements of c_1.
2. Compute the mean of the first 500 elements of c_1.
3. Compute the variance of the first 500 elements of c_1.
4. Compute the covariance of the first 499 elements of c_1 and c_2.
5. Compute the Pearson correlation of the first 499 elements of c_1 and c_2.
   Note that c_1 and c_2 practically contain the same elements, so the denominator of the Pearson correlation (i.e., the product of the standard deviations of c_1 and c_2) can safely be estimated to be the variance of c_1.
   Also, note that the variance in the denominator in this case is within the convergence range of Goldschmidt's polynomials (i.e., in the range (0, 2); see Sect. 6.3.2). Thus, you may use a Goldschmidt polynomial, say $G_{d=10}$, to estimate the reciprocal of the variance.

10.3.1.1 Solving a System of Linear Equations Under FHE

Consider the following system of three linear equations with three unknowns:

$$\begin{bmatrix} 1 & 0 & 5.4 \\ 2.5 & 7 & -2 \\ 0 & -4.1 & 3.6 \end{bmatrix} \begin{bmatrix} x_1 \\ x_2 \\ x_3 \end{bmatrix} = \begin{bmatrix} 51.6 \\ -38.5 \\ 48.8 \end{bmatrix}$$

6. Implement FHE code for computing the determinant of a 3 by 3 matrix, where each of the nine entries is encrypted in a separate ciphertext. Then use this code to compute the determinant of the left-hand-side matrix of the above system of equations.
7. Compute the multiplicative inverse of the above determinant. Note that the determinant's value is in the range where $SC_{d=10, s=40}$ estimates $\frac{1}{x}$ with good accuracy under FHE (i.e., $[-56.5308, -4.4019] \cup [4.4019, 56.5308]$; see Sect. 6.3.2).
8. Solve the above system of equations by computing the three determinants for the respective numerators of Cramer's rule and then multiply them with the multiplicative inverse of the denominator computed above.

10.3 Lab Exercises

10.3.1.2 LR and Solving a System of Linear Equations Under FHE

1. Consider the following database with six features $(X_1, \ldots, X_6)$ and a label y.

$$\begin{bmatrix} X_1 & X_2 & X_3 & X_4 & X_5 & X_6 & y \\ 1 & 5 & 6 & -3 & 2 & -2 & 23.1 \\ 0 & 0 & -3 & 4 & 6 & -3 & -3.1 \\ 5 & -3 & -3 & 3 & -2 & 1 & 4.05 \\ 1 & 7 & 2 & 6 & 0 & -8 & -46.04 \\ 0 & -3 & 9 & 0 & 4 & 9 & 136.02 \\ 4 & 2 & -4 & 5 & 5 & 9 & 110.3 \\ 3 & 2 & 6 & -7 & 3 & -8 & -26.97 \\ 1 & -1 & 8 & 7 & 5 & 6 & 121.8 \end{bmatrix}$$

Use SGD to find the linear regression model, i.e., the weights $w_1, \ldots, w_6$, $bias$ that minimize $bias - y + \sum_{i=1}^{6} X_i \cdot w_i$.

2. Estimate the time it would take to find the model by solving $w = A^{-1}b$, where $A = X^T X$ and $b = X^T b$.

3. Consider the matrices

$$X = \begin{bmatrix} 1 & 0 & 5.4 \\ 2.5 & 7 & -2 \\ 0 & -4.1 & 3.6 \end{bmatrix} \qquad y = \begin{bmatrix} 51.6 \\ -38.5 \\ 48.8 \end{bmatrix}.$$

Let $A = X^T X$ and $b = X^T y$, and use SGD to solve for $w = A^{-1}b$.

- Compare the accuracy of the answer with the accuracy of your answer in the previous question.
- Compare the time to compute your answer with the time it took to compute the answer of the previous question.

10.3.1.3 Regression Training Using SGD

1. Change Algorithm 16 to train a batch of b different LR models.
2. Change Algorithm 17 to train a batch of b different LR models, and use different Sigmoid approximations.
3. What is the shape that achieves the fastest latency?
4. What is the shape that achieves the fastest amortized time?

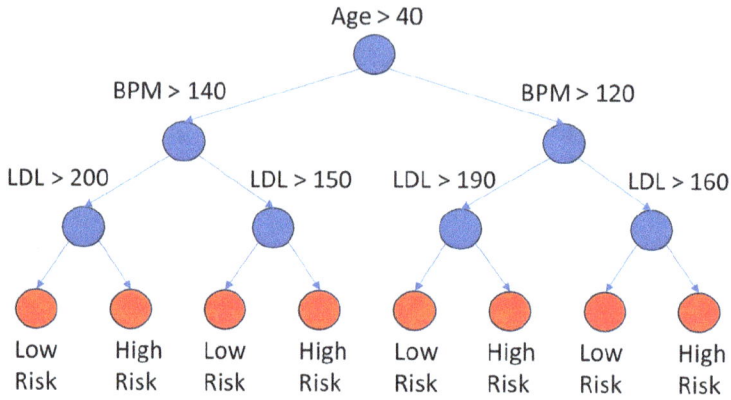

Fig. 10.9 A (made-up) decision tree for determining risk for heart failure

10.3.1.4 Decision Trees

Consider the decision tree depicted in Fig. 10.9. Notice that each level compares the same variable (age at the root, BPM at the second level, and LDL at the third level). Assume age, BPM, and LDL are all integers and $0 \leq age \leq 100$, $40 \leq BPM \leq 200$, and $100 \leq LDL \leq 300$ and label High Risk $= 1$ and Low Risk $= 0$. Use CKKS to encrypt the values in the tree and the values in inference query (age, BPM, LDL) and implement

$$isGreater(x, y) = \begin{cases} 1 + \beta & \text{if } x - y > \alpha \\ 0 + \beta & \text{if } x - y < -\alpha \end{cases} \qquad (10.63)$$

$(isGreater(x, y)$ may have any value when $|x - y| < \alpha$) using the approximations in Sect. 6.1.

In all subsequent questions we assume the output of the decision tree needs to be accurate up to a noise of β', that is, the output of the decision tree needs to be $0 + e$ or $1 + e$, where $|e| < \beta'$.

1. Compute the value β when using the naïve algorithm (Algorithm 18).
2. Compute the value β when using copy and recurse.
3. Compare the value β to guarantee a noise bound of β'.

References

1. Akavia, A., Shaul, H., Weiss, M., Yakhini, Z.: Linear-regression on packed encrypted data in the two-server model. In: Proceedings of the 7th ACM Workshop on Encrypted Computing & Applied Homomorphic Cryptography, WAHC'19, pp. 21–32. Association for Computing Machinery, New York (2019). https://doi.org/10.1145/3338469.3358942

2. Azogagh, S., Delfour, V., Gambs, S., Killijian, M.O.: PROBONITE: private one-branch-only non-interactive decision tree evaluation. In: Proceedings of the 10th Workshop on Encrypted Computing & Applied Homomorphic Cryptography, WAHC'22, p. 23–33. Association for Computing Machinery, New York (2022). https://doi.org/10.1145/3560827.3563377
3. Beimel, A., Ishai, Y., Malkin, T.: Reducing the servers computation in private information retrieval: PIR with preprocessing. In: Bellare, M (ed.) Advances in Cryptology—CRYPTO 2000, pp. 55–73. Springer, Berlin (2000). https://doi.org/10.1007/3-540-44598-6_4
4. Bergamaschi, F., Halevi, S., Halevi, T.T., Hunt, H.: Homomorphic training of 30,000 logistic regression models. In: Applied Cryptography and Network Security: 17th International Conference, ACNS 2019, Bogota, Colombia, June 5–7, 2019, Proceedings, pp. 592–611. Springer, Berlin (2019). https://doi.org/10.1007/978-3-030-21568-2_29
5. Boemer, F., Lao, Y., Cammarota, R., Wierzynski, C.: nGraph-HE: a graph compiler for deep learning on homomorphically encrypted data. In: Proceedings of the 16th ACM International Conference on Computing Frontiers, CF '19, p. 3–13. Association for Computing Machinery, New York (2019). https://doi.org/10.1145/3310273.3323047
6. Bonawitz, K., Ivanov, V., Kreuter, B., Marcedone, A., McMahan, H.B., Patel, S., Ramage, D., Segal, A., Seth, K.: Practical secure aggregation for privacy-preserving machine learning. In: Proceedings of the 2017 ACM SIGSAC Conference on Computer and Communications Security, CCS '17, p. 1175–1191. Association for Computing Machinery, New York (2017). https://doi.org/10.1145/3133956.3133982
7. Chen, H., Chillotti, I., Dong, Y., Poburinnaya, O., Razenshteyn, I., Riazi, M.S.: SANNS: scaling up secure approximate k-nearest neighbors search. In: Proceedings of the 29th USENIX Conference on Security Symposium, SEC'20. USENIX Association, USA (2020). https://www.usenix.org/conference/usenixsecurity20/presentation/chen-hao
8. Du, J., Bian, F.: A privacy-preserving and efficient k-nearest neighbor query and classification scheme based on k-dimensional tree for outsourced data. IEEE Access **8**, 69333–69345 (2020). https://doi.org/10.1109/ACCESS.2020.2986245
9. Elmehdwi, Y., Samanthula, B.K., Jiang, W.: Secure k-nearest neighbor query over encrypted data in outsourced environments. In: 2014 IEEE 30th International Conference on Data Engineering, pp. 664–675 (2014). https://doi.org/10.1109/ICDE.2014.6816690
10. Fouque, P.A., Stern, J., Wackers, J.G.: Cryptocomputing with rationals. In: Financial Cryptography, 6th International Conference, FC 2002, Southampton, Bermuda, March 11–14, 2002, Revised Papers, pp. 136–146 (2002). https://doi.org/10.1007/3-540-36504-4_10
11. Giacomelli, I., Jha, S., Joye, M., Page, C.D., Yoon, K.: Privacy-preserving ridge regression with only linearly-homomorphic encryption. In: Preneel, B., Vercauteren, F. (eds.) Applied Cryptography and Network Security - 16th International Conference, ACNS 2018, Leuven, Belgium, July 2–4, 2018, Proceedings, Lecture Notes in Computer Science, vol. 10892, pp. 243–261. Springer, Berlin (2018). https://doi.org/10.1007/978-3-319-93387-0_13
12. Hatamizadeh, A., Yin, H., Roth, H., Li, W., Kautz, J., Xu, D., Molchanov, P.: GradViT: gradient inversion of vision transformers. In: 2022 IEEE/CVF Conference on Computer Vision and Pattern Recognition (CVPR), pp. 10011–10020 (2022). https://doi.org/10.1109/CVPR52688.2022.00978
13. Kim, A., Song, Y., Kim, M., Lee, K., Cheon, J.H.: Logistic regression model training based on the approximate homomorphic encryption. BMC Med. Genom. **11**(4), 83 (2018). https://doi.org/10.1186/s12920-018-0401-7
14. Kim, M., Song, Y., Wang, S., Xia, Y., Jiang, X.: Secure logistic regression based on homomorphic encryption: design and evaluation. JMIR Med. Inform. **6**(2), e19 (2018). https://doi.org/10.2196/medinform.8805
15. Lemaréchal, C.: Cauchy and the gradient method. Doc Math. Extra **251**(254), 10 (2012)
16. McMahan, B., Moore, E., Ramage, D., Hampson, S., Arcas, B.A.Y.: Communication-efficient learning of deep networks from decentralized data. In: Singh, A., Zhu, J. (eds.) Proceedings of the 20th International Conference on Artificial Intelligence and Statistics, Proceedings of Machine Learning Research, vol. 54, pp. 1273–1282. PMLR (2017). https://proceedings.mlr.press/v54/mcmahan17a.html

17. Nesterov, Y.E.: A method of solving a convex programming problem with convergence rate $o\left(\frac{1}{k^2}\right)$. Dokl. Akad. Nauk SSSR **269**(3), 543–547 (1983). http://mi.mathnet.ru/dan46009
18. Shaul, H., Feldman, D., Rus, D.: Secure k-ish nearest neighbors classifier. Proc. Privacy Enhancing Technol. **2020**(3), 42–61 (2020). https://doi.org/10.2478/popets-2020-0045
19. Shaul, H., Kushnir, E., Moshkowich, G.: Secure range-searching using copy-and-recurse. In: Proceedings of the Privacy Enhancing Technologies Symposium (PETS '24) (2024)
20. Sun, M., Yang, R.: An efficient secure k nearest neighbor classification protocol with high-dimensional features. Int. J. Intell. Syst. **35**(11), 1791–1813 (2020). https://doi.org/10.1002/int.22272
21. The Pennsylvania State University - Eberly College of Science: STAT 510 Applied Time Series Analysis (2023). https://online.stat.psu.edu/stat510/
22. Truex, S., Baracaldo, N., Anwar, A., Steinke, T., Ludwig, H., Zhang, R., Zhou, Y.: A hybrid approach to privacy-preserving federated learning. In: Proceedings of the 12th ACM Workshop on Artificial Intelligence and Security, AISec'19, p. 1–11. Association for Computing Machinery, New York (2019). https://doi.org/10.1145/3338501.3357370
23. Wang, P.S., Guy, M.J.T., Davenport, J.H.: P-adic reconstruction of rational numbers. ACM SIGSAM Bull. **16**(2), 2–3 (1982). https://doi.org/10.1145/1089292.1089293
24. Zhang, C., Li, S., Xia, J., Wang, W., Yan, F., Liu, Y.: BatchCrypt: efficient homomorphic encryption for Cross-Silo federated learning. In: 2020 USENIX Annual Technical Conference (USENIX ATC 20), pp. 493–506. USENIX Association (2020). https://www.usenix.org/conference/atc20/presentation/zhang-chengliang
25. Zhu, L., Liu, Z., Han, S.: Deep leakage from gradients. In: Advances in Neural Information Processing Systems, vol. 32. Curran Associates (2019). https://proceedings.neurips.cc/paper_files/paper/2019/file/60a6c4002cc7b29142def8871531281a-Paper.pdf

Chapter 11
Case Study: Neural Networks

Abstract This chapter aims to combine all the previous techniques for implementing a large NN model (e.g., ResNet50). The reader first learns the limitations of such an implementation and the ML techniques to overcome them.

11.1 Neural Networks

For many years cryptographers have viewed homomorphically encrypted NNs as a distant mountain. The complexity of the calculations, the computation depth, and the required practical performance figures made that mountain seem insurmountable. Today, this situation starts to change, as we will see in this chapter.

NNs are comprised of various combinations of basic components, each with its own HE-related challenges. In this section we list several commonly used layer types that make up NNs, and we discuss the challenges associated with implementing them under HE as well as ways to handle these challenges.

NNs are powerful and robust ML models, and the intuition for their structure comes from an attempt to imitate brain operations. Consequently, the basic building block of an NN is called a neuron. Basic neurons receive a set of inputs $x_1, \ldots, x_n$ with some associated weights $w_1, \ldots, w_n$. A simple output of a neuron is the value $f(\sum x_i w_i + b)$, where the nonlinear function f is often called an activation function and is usually chosen from a small set of popular nonlinear functions that have been proven to be effective (see Sect. 11.1.2). The motivation for calling these activation functions comes again from biological neurons that are either activated or not, depending on their inputs. See an example of a neuron in Fig. 11.1.

What gives the NN its power is the topology in which the neurons are connected together. The network is partitioned into layers. Each layer arranges neurons and their connections in a different structure and the output of a layer is the input of the next layer. The training phase adjusts the input weights of all neurons of a given network topology while minimizing some loss function, but not changing the neuron connection themselves.

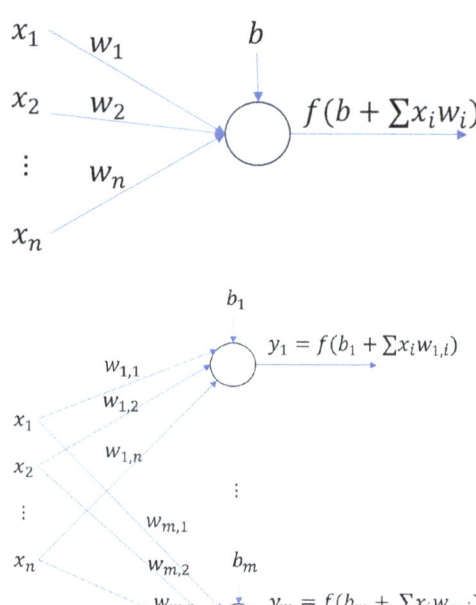

Fig. 11.1 A single neuron with inputs $x = (x_1, \ldots, x_n)$ with weights $w = (w_1, \ldots, w_n)$, a bias b, and an output $y = f\left(b + \sum_i x_i w_i\right)$. The function f is called the activation function

Fig. 11.2 An FC layer with $x = (x_1, \ldots, x_n)$ inputs and $y = (y_1, \ldots, y_m)$ outputs. The layer involves a weight matrix W of shape $m \times n$ and a bias vector b of m elements

11.1.1 Fully Connected Layers

In an FC layer every input neuron is connected to every output neuron. The number of input neurons (which is the number of neurons in the output of the previous layer) and the number of output neurons are parameters of the layers. For a layer with n input neurons $x = (x_1, \ldots, x_n)$ and m output neurons $y = (y_1, \ldots, y_m)$ the value of the jth output neuron is given by $y_j = f(\sum x_i w_{j,i} + b_j)$. More compactly we write,

$$y = f(W \cdot x + b), \tag{11.1}$$

where x and y are vectors as shown above, W is an $m \times n$ matrix, b is a vector of m elements, and f is some activation function as discussed in Sect. 11.1.2. See Fig. 11.2 for an example.

An FC layer is intuitive and general as every neuron in the input may affect every neuron in the output. Theoretically, any layer type such as the convolution and average pooling layers (which we discuss below) can be implemented as an FC layer. However, using a more specific layer reduces the number of parameters that needs to be trained. This reduces the training time and the amount of data on which the network needs to be trained.

11.1 Neural Networks

Under HE, an FC layer is often translated into three high-level operations, namely, a vector-matrix multiplication of the inputs vector with the weights matrix and two additional steps of adding the bias vector b to the above result and applying an activation function. Chapter 9.8 compares different vector-matrix multiplication methods. Each method uses a different number of homomorphic rotations and multiplications, and the preferred method depends on the context of the computation, i.e., on the packing scheme used for the inputs and outputs of the multiplication, on whether further matrix-vector multiplications are required etc.

The step of adding the bias b does not pose a significant challenge when implementing an HE version of the NN. The reason is that HE addition operations are considered fast and they also do not impact on the chain index or increase the accumulated error much as compared to multiplication and rotation operations.

11.1.2 Activation Functions

As mentioned above, the output of a neuron is set to be $y_j = f(\sum x_i \cdot w_i + b_j); f(\cdot)$ is a nonlinear function which is referred to as the *activation function*. It is essential to have a nonlinear function at each layer; otherwise, a multiple layer NN would be trivialized and could be represented by a single layer.

Two important properties of activation functions are their output range and being continuously differentiable. An activation function with a limited range could still support gradient-based training methods but may induce the *vanishing gradients phenomenon* as values at the extremes would map to the same values. An activation function with an infinite range on the other hand usually requires smaller learning rates. It is not contingent that an activation function would be continuously differentiable; however, when training is performed using gradient-based methods, a continuously differentiable function would facilitate the analysis of the gradients and might be better suited to certain optimization methods.

Activation functions can be considered as a part of the network topology (sometimes they are described as a separate layer) and are usually not changed during the training phase. However, looking ahead, we note that under FHE, there are training techniques that involve training the parameters of the activation functions.

There are several different activation functions that have been found to be particularly effective and we illustrate the most popular ones: ReLU, Sigmoid, GELU, and tanh in Fig. 11.3. These activation functions must be approximated when used with HE, since they are all non-polynomial. ReLU is defined as ReLU$(x) = max(0, x)$, has an output range of $[0, \inf)$, and is not continuously differentiable. Nevertheless, it is one of the most commonly used activation functions in NNs. It has been shown in numerous papers that it can be approximated or simply replaced with a simple square function or some other low-degree polynomial [2, 11, 15, 20, 32, 41] that are easily implemented with HE operations (see Sect. 6.2). While successful in shallow networks over small images, the square activation suffers from accuracy

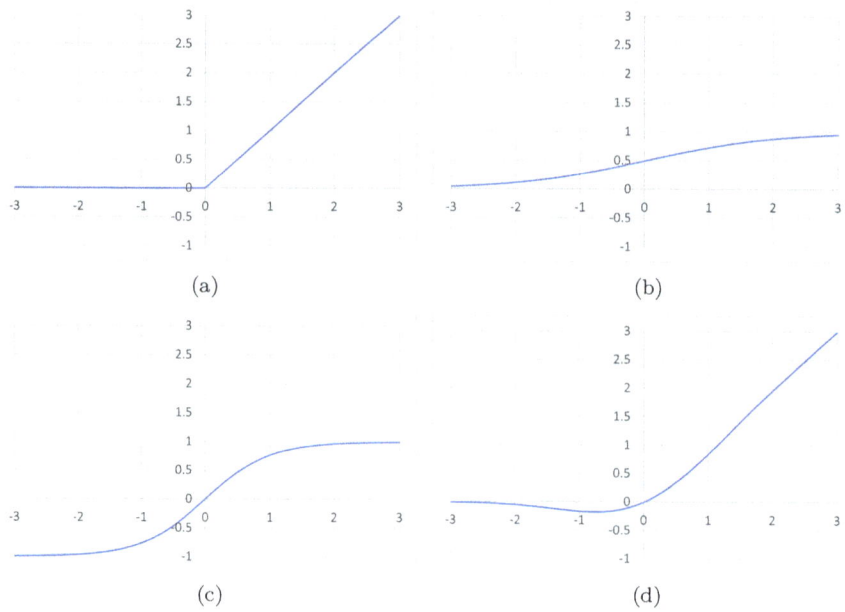

Fig. 11.3 Popular activation functions. (**a**) ReLU(x) = max$(0, x)$. (**b**) Sigmoid$(x) = \frac{1}{1+e^{-x}}$. (**c**) tanh(x). (**d**) GELU$(x) = \frac{1}{2}x \left(Erf \left(1 + \frac{x}{\sqrt{2}} \right) \right)$

degradation when used in deep networks. In addition, due to accuracy concerns, some works prefer to replace the ReLU activation with a higher-degree polynomial function [5, 23, 28, 42]. Another approach suggested by [35] leverages Hermite polynomials to more accurately approximate the ReLU activation function using a low-degree polynomial. More discussion on ways to approximate the ReLU activation function is found in Part II and specifically Sect. 6.2.

Another commonly used activation is the Sigmoid activation function, which has a limited range since $f(x) \in (0, 1)$, and is continuously differentiable. Since the Sigmoid contains both a division operation and an exponent operation, it is not easily implemented using HE operations and would require a high-degree polynomial to be accurately approximated. Sect. 10.2.1.3 reported several approximations to the Sigmoid function.

For very deep NNs and also for transformers, it is often better to use the GELU activation function whose approximations tend to support a wider input range. The GELU activation function is defined as GELU$(x) = \frac{x}{2}(1 + \text{erf}(\frac{x}{\sqrt{2}}))$. It has a range of $(-0.17, \inf)$ and is continuously differentiable. We leave the approximation of GELU(x) and tanh(X) as exercises.

Another popular non-polynomial function is Softmax, which is used either as an activation function or at the final layer of the NN when assigning probabilities to the inferred classes. What makes Softmax different from the other above activations is that it is a multivariate activation, while the other activations are univariate. Specifically, it is defined as

$$\text{Softmax}(x) = \frac{e^{x_i}}{\sum_j^n e^{x_j}}, \qquad (11.2)$$

where $x \in \mathbb{R}^n$. The Softmax activation function has a limited range since $f(x) \in (0, 1)$ and is continuously differentiable. The Softmax function can be viewed as a smooth approximation of $argmax$, taking n real numbers and normalizing them to n probabilities where the highest input corresponds to the highest probability.

Note that the Softmax function includes non-polynomial components, like division and exponents. One can use polynomial approximations as described in previous chapters. However, as these approximations entail a higher multiplication depth, other practical approaches have been suggested. These include replacing the multivariate Softmax function with a univariate ReLU or simply relying on the client side to perform the Softmax operation on the final output values of the network.

11.1.2.1 Trainable Activation Functions

Polynomial activation functions may be used in HE NNs as an approximation for some popular non-polynomial activation function, but they can in fact also be considered directly on their own merit, even for non-HE NNs. In [31] the authors considered the computational complexity of training with polynomial activation functions and specifically with square activation functions. Another approach was taken in [36, 37, 40] where the authors had a different activation function for each neuron whose parameters were set during the training phase. For large NNs this quickly becomes infeasible to train. In contrast, the suggestion in [39] was to train a single polynomial for each NN layer. A similar approach was also used later in [6].

11.1.3 Convolution Layers

To understand the intuition behind convolution layers, imagine you are given a black-and-white image (i.e., each pixel is 0 or 1, respectively) and are looking for a horizontal line that is three pixels long. Now consider the 3×3 kernel in Table 11.1 (see Sect. 9.4 for the terminology). It is easy to see that applying this kernel on a 3×3 block with a horizontal line in the middle line (i.e., 0's in the first and last lines and 1's in the middle line) gives the maximal value. This kernel, therefore, can be thought of as giving a score to how similar a 3×3 block is to a horizontal line. Stacking multiple convolution layers in a row can be thought of as giving a score to how similar is a large block to a complex structure, for example, the digit "3."

Convolution layers are not restricted only to dealing with images, but in the context of images, the input to the convolution layer is a 3D array of values. The first

Table 11.1 A kernel that computes a score of how similar a 3 × 3 block contains a horizontal line

−1	−1	−1
1	1	1
−1	−1	−1

7	54	5	7	0	2	1	3
3	1	53	8	7	6	4	5
9	3	2	8	5	6	7	8
6	4	60	1	0	9	7	3
9	7	26	2	14	38	4	91
5	3	71	41	35	6	74	51
32	5	6	54	9	8	23	65
4	2	45	7	8	1	42	53

⟹

54	53	7	5
9	60	9	8
8	71	38	91
32	54	9	65

Fig. 11.4 Max-pooling an input matrix of size [8, 8] with a filter of size [2, 2] and strides of 2, resulting with an output of size [4, 4], where each output pixel is the maximal value in the corresponding filtered section of data

two dimensions are height and width and the third is the channels. The parameters of the layer that are changed during training are the values in the kernels. The output is another 3D array, where again the first two dimensions are height and width and the third is the channels.

Similar to the FC layers, one would often aspire to reduce the number of costly rotation and multiplication operations required for performing the convolution layers using various packing schemes and optimization algorithm as described in Sect. 9.4.

11.1.4 Sum, Mean, and Max-Pooling

Pooling layers are generally placed after one or more convolutional layers in order to downsample their output. This is useful in avoiding overfitting by abstracting away the precise location of features and reducing the number of learned features. The pooling operation is similar to convolution, with a typically small square filter that is applied to contiguous non-overlapping areas of the two-dimensional data. Applying the filter downsamples the corresponding square data section into a single value. This value is the sum, mean, or maximum of all values in this data section for sum, mean, and max-pooling layers, respectively (Fig. 11.4). As with convolution, pooling can also be further customized with different strides and padding settings. When applied to multiple input channels, the pooling is done on each channel separately.

11.1 Neural Networks

Sum pooling can be thought of as a special case of convolution where the filter has 1 in all its elements. Let us start with the case where the strides is set to 1. This setting is not normally used, but it offers a simple starting point. We use Eq. 9.6, which because of all the values being constants (equal to 1) the equation reduces to summing over pseudo-rotations of the input T_I. A more efficient version of the same computation utilizes the RotateAndSum technique, as in Algorithm 10, where we first sum over rows using $\log(h_F)$ pseudo-rotation-and-sum iterations of T_I, then $\log(w_F)$ iterations for summing over columns.

Strides can be handled using the same techniques explained in Sect. 9.4.3, either adjusting the resulting tile tensor shape to reflect that only some of the slots contain the output, or discarding whole tiles where needed. Padding can be handled as in Sect. 9.4.2. If masks are required for non-trivial padding, then summation can no longer be fully performed using RotateAndSum, and the masked tiles need to be added separately. Regarding multiple channels, since pooling layers work on each channel separately, there is no need to sum over the channel dimension (i.e., no need for Eq. 9.16).

Mean-pooling is the same as sum pooling except that the factor $\frac{1}{h_F w_F}$ needs to be multiplied either before or after the computation, or inside the masks if masks are needed. Max-pooling is also similar, except that the RotateAndSum is replaced with the rotate-and-max algorithm. See more about this and additional techniques related to max-pooling in Sect. 6.1.3.

11.1.5 Normalization

Normalization layers are used between layers to normalize function input values. Choosing the right normalization technique to use depends on the NN architecture, use case, and data types. For example, for a batch B of m inputs $x_0, \ldots, x_m$ with k features each, i.e., the j feature of input x_i is $x_{i,j}$. A batch-normalization layer is computed for each feature j independently by

$$\text{BatchNorm}(x, j) = \gamma \left(\frac{x_{i,j} - \mu_{B,j}}{\sqrt{\sigma_{B,j}^2 + \epsilon}} \right) + \beta, \tag{11.3}$$

where

$$\mu_{B,j} = \frac{1}{m} \sum_{i=1}^{m} x_{i,j}, \tag{11.4}$$

$$\sigma_{B,j} = \sqrt{\frac{1}{m} \sum_{i=1}^{m} (x_{i,j} - \mu_{B,j})^2}, \tag{11.5}$$

and γ and β are scaling factors learned during training to shift and scale the output range. In contrast, *layer-normalization* looks at feature statistics across a single sample and adjusts each feature so that overall there would be a zero mean and unit variance. Here,

$$\text{LayerNorm}(x_i)_j = \gamma \left(\frac{x_{i,j} - \mu_j}{\sqrt{\sigma_j^2 + \epsilon}} \right) + \beta, \quad (11.6)$$

where

$$\mu_i = \frac{1}{m} \sum_{j=1}^{m} x_{i,j}, \quad (11.7)$$

$$\sigma_i = \sqrt{\frac{1}{m} \sum_{j=1}^{m} (x_{i,j} - \mu_i)^2}. \quad (11.8)$$

Both methods involve arithmetic operations that might be costly to implement when using FHE, such as division and square root, so one would need to employ appropriate approximations like for square root inverse, which was discussed in Example 5.4.

11.1.5.1 Dropout

Dropout [38] is a regularization method that helps to combat overfitting. It is implemented using a vector of independent random variables with a probability parameter $p \in (0, 1)$ that is applied at the outputs of each neuron. The random variables reduce the number of handled values and outputs of each layer and thus reduce the reliance on a particular node or set of nodes, increasing the robustness and decreasing overfitting. Typically, the dropout layer is more relevant to the training phase, while during inference all p values are set to 1. The dropout technique can be leveraged to also accelerate FHE training operations by dropping entire tiles. This could be done using a dedicated dropout mechanism.

11.2 Training FHE-Friendly Networks

By design, most NNs are not FHE-friendly because they include non-polynomial components, such as max-pooling or the ReLU activation function. In an attempt to bridge this gap, early works combined HE and MPC, by relying on the client to do some of the heavy-lifting (see Sect. 3.3). Whenever a NN component seemed too complex for HE, client-aided solutions sent the intermediate values of the NN

11.2 Training FHE-Friendly Networks

as ciphertexts to the client. The client would then decrypt the results, run the non-polynomial operation in the clear, re-encrypt the results, and send them for further processing back to the server.

Another reason to use client-aided designs is that bootstrapping was considered a time-heavy operation that many early FHE implementations did not support. Consequently, various solutions included sending the ciphertext back to the client for re-encryption whenever the HE chain index was getting too low. As sending such interim results to the client might reveal confidential information regarding the model, some of these early works relied on MPC to provide the necessary mitigation. Among these are SecureML [34], MiniONN [30], and GAZELLE [21].

While these works provided a fast and at times accurate solution, communication bandwidth overheads and questions regarding the real-life practicality of asking the client to stay online for the interactions, and the security and privacy implications of such solutions led researchers to attempt to implement a solution that fully resides at the server side and does not rely on mid-processing client interactions.

CryptoNets [15] was the first work to provide a shallow non-interactive NN inference over HE encrypted data that exclusively uses HE. The activation functions were replaced by a square function as it was both efficient in terms of multiplication depth and sufficient in terms of prediction accuracy as an approximation. [41] used a polynomial approximation rather than a simple square function to approximate the activation function. In 2019 CHET [11] was able to implement SqueezeNet-CIFAR [10], and in 2021 HElayers [2] implemented AlexNet [25] over encrypted images of the COVIDx CT-2A [17] dataset and later on also on the ImageNet [12] dataset.

The use of bootstrapping in FHE-based solution was done, e.g., in [8], allowing for a non-interactive and a more layer-depth-scalable solution (i.e., facilitating NNs of deeper layers). As the performance of CKKS bootstrapping improved, more works incorporated bootstrapping to support bigger NNs. [29] implemented ResNet-20 using CKKS in about 176 minutes (but with only 111.6 bits of security). Aharoni et al. [4] implemented a polynomial version of a full SqueezeNet model [19] which requires 40 sequential multiplication operations in about 4 minutes and 128-bit security. Finally, works such as [5, 23, 26, 28] implemented a bigger non-interactive CKKS-based ResNet44, ResNet50, ResNet101, and ResNet152.

One main difference between [23, 26, 28] and [5] is that the former replaced activation functions with their approximations without further retraining. This simplifies the model preparation phase for a pre-trained model, but forces the solution to use high-degree polynomials to support the large input range for the replaced activations. In contrast, [5] advocated a solution where further training (fine-tuning) or training a model from scratch is allowed. In these cases, lower degree polynomials can achieve similar levels of accuracy. Next, we discuss some training methods where the network is trained from scratch or when it is allowed to be fine-tuned.

11.2.1 Training Methods

Training methods for polynomial NNs attempt to imitate the prediction accuracy produced by their non-polynomial NN counterparts while minimizing training resources and processing duration. The simplest approach is to train the NNs from scratch while disregarding the original pre-trained non-polynomial NN. However, getting the network to converge could be difficult when working with deep NNs. A two-phase training approach was demonstrated in [6], but to discuss it, we first need to mention a commonly used ML concept called knowledge distillation (KD).

11.2.1.1 Knowledge Distillation (Transfer Learning)

KD [16, 18, 33] is the process of transferring knowledge from a complex (teacher) model to a simpler (student) model to improve efficiency. This can be implemented by adding a model fidelity distance as a loss term to the loss function, measuring the extent with which the student model predictions match those of the teacher model and "penalizing" differences. Rather than using hard predictions, one can use probabilities like in the Softmax function. These prediction goals are termed *soft targets* and can be calculated for the ith class as

$$Q^\tau[i] = \frac{\exp(\frac{y_i}{\tau})}{\sum_j \exp(\frac{y_j}{\tau})}, \tag{11.9}$$

where y_i denotes the teacher model prediction logits (i.e., classification model outputs prior to some normalization such as by a Softmax function) and τ is a hyperparameter that controls the probability distribution over the classes. Given a student soft target prediction vector $Q_s^\tau[i]$ and a teacher soft target prediction vector $Q_t^\tau[i]$, the loss term that embodies the KD can be described as

$$L_{KD} = w\tau^2 \cdot \text{CrossEntropy}(Q_s^\tau[i], Q_t^\tau[i]). \tag{11.10}$$

Once again the loss term should be weighted with a hyper-parameter w to fine-tune its significance over the whole objective function. In our case, KD will transfer the "knowledge" from a stronger (non-polynomial) teacher model to a weaker (polynomial) student model.

11.2.2 A Two-Phase FHE-Friendly Training

We can now describe the a general recipe from [6] for making a given (plaintext) NN more HE-friendly; see an illustration in Fig. 11.5.

11.2 Training FHE-Friendly Networks

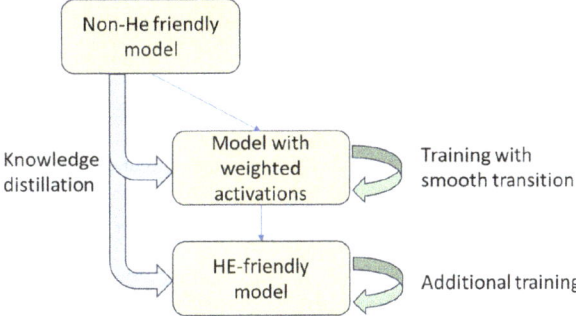

Fig. 11.5 A general recipe from [6] to transform a non-HE-friendly model to be HE-friendly. The recipe starts by smoothly replacing the activation functions with HE-friendly activation functions and then using KD to train the HE-friendly model

The recipe starts by replacing the activation functions with HE-friendly activation functions. The replacement happens in a smooth way using the following weighted activations:

$$\text{weightedAct}_e(x) = (1 - \lambda_e) \cdot \text{nonPolyAct}(x) + \lambda_e \cdot \text{polyAct}(x). \qquad (11.11)$$

Here, nonPolyAct(x) is the original non-FHE-friendly activation function, and polyAct(x) is the learned polynomial activation function (e.g., $ax^2 + bx$ was used in [6] with parameters a and b), and for a fixed parameter d, at every fine-tuning epoch e we have $\lambda_e(x) = \min(1, \frac{e}{d})$.

At the second step, the model is trained using KD: using the original model as the "teacher" and the HE-friendly model as the "student" while using the teacher's soft labels when training the student model.

11.2.3 A Three-Phase FHE-Friendly Training

A three-phase approach for gradually shifting from a pre-trained model to a polynomial one for DNNs is described in [5] and summarized in Algorithm 24. It receives a pre-trained NN model denoted by M with $|M|$ non-polynomial layers, and a training set TS and a smaller disjoint set RS, where TS and RS comprise together the original training set of M.

In the initial phase of the algorithm the original loss function of the model M is updated to include *range loss* regularization. The goal of this new loss term is to help limit the input ranges of the layers and center them around 0. Given an input vector to the $i - th$ layer x_i, the range loss term rl can be described as

$$rl = \|(\|\mathbf{x}^i\|_p)_{0 \leq i < |M|}\|_q, \qquad (11.12)$$

with p usually chosen to be $p = \infty$ and $q \in \{1, 2, \infty\}$. The loss term should be weighted with a hyper-parameter w to fine-tune its significance over the whole objective function. Through the training algorithm, the input ranges for each layer would be minimized while the overall accuracy of the non-polynomial model is minimally degraded.

Algorithm 24: Training polynomial NNs, [5]

1. **Input:** A pre-trained NN model (M), a training set (TS), a small disjoint set (RS) and a hyper-parameter to represent the confidence level (α).
2. **Output:** A trained HE-friendly polynomial model (M_{HE}).
3. Add a regularization range loss term rl to loss(M).
4. Fine-tune M over TS until input ranges to the layers are small enough and accuracy is satisfying. The resulting model is M'.
5. Evaluate M' over RS and compute the pairs (min $\mathbf{x}^i$, max $\mathbf{x}^i$)$_{0 \leq i < |M|}$ per sample. x_i denotes the input vector to the ith layer.
6. Estimate the range $([x_{min}^i, x_{max}^i])_{0 \leq i < |M|}$ for values of $\mathbf{x}^i$ with confidence level α.
7. Replace the functions $f_i(x)$ of the non-polynomial model layers with polynomial approximations $P_i(x)$ over the estimated ranges $[x_{min}^i, x_{max}^i]$. The new model is M_{HE}. Note that the replacement isn't and shouldn't necessarily be abrupt. It is better at this point to use more gradual methods like the gradual shifting from non-polynomial to polynomial activation functions, e.g. using changing weights, coefficients training, or other gradual methods.
8. Fine-tune M_{HE} over TS until convergence.
9. return M_{HE}.

In the second phase of the algorithm, the input ranges per layer are estimated with confidence level α. The estimation can be performed using the sampled training set RS (that would of course not be used during the training or validation phases).

In the third and last phase the original non-polynomial activation functions are replaced with polynomial approximations. This can be done using the various methods described in Sect. 11.1.2 given the range estimations that were calculated in the previous phase. Additional fine-tuning is usually necessary after replacing the activation functions.

11.3 The LeNet NN

Putting the layers above together we can describe the LeNet network that was trained on the MNIST [13] dataset to identify a digit written in an image. A compact description of LeNet is given in Fig. 11.6. The network gets an input RGB image of 32×32 pixels as its input. It first runs a convolution layer with 6 kernels of size 5×5, the output of which is 6 channels of 28×28 neurons. This is reduced to 6 channels of 14×14 neurons using average pooling. Subsequently, there is another convolutional layer with 16 kernels of size 5×5, with an output of 16 channels of size 10×10, which are reduced in size again using average pooling into 16 channels of size 5×5. The resulting 400 neurons serve as the input of an FC layer with 120 neurons in the output which are fed to another FC layer with only 84 neurons in the output. Finally, these neurons are the input to the last FC layer that outputs only 10 values that represent the confidence *score* of the network for each digit. The digit

11.3 The LeNet NN

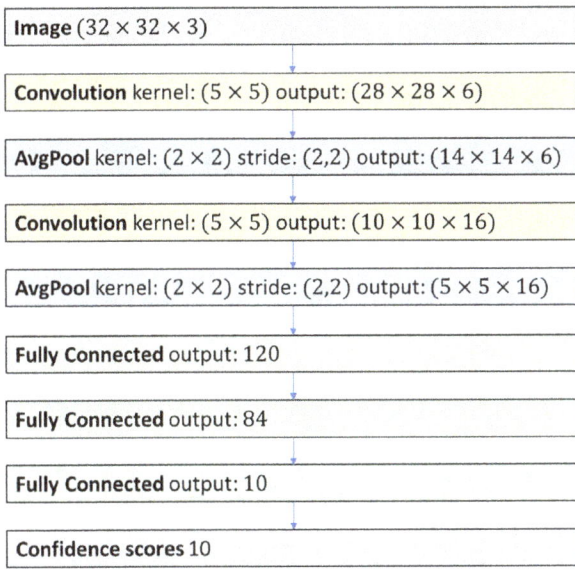

Fig. 11.6 An illustration of the LeNet NN for classifying an image of a digit from the MNIST dataset [13]

with the highest score is the one that the network is most confident about as being the digit that is written in the image.

Below is a suggested way (not necessarily the most efficient) to pack the input and the layers of the network

- **Input** The input to the network is an image with 32×32 pixels and 3 color channels, or a tensor of shape [3, 32, 32]. This is packed in a tile tensor of shape $\left[\frac{32\sim}{32}, \frac{3}{4}, \frac{32\sim}{16}, \frac{*}{8}\right]$. This follows the usage of interleaved packing for convolution as explained in Sect. 9.4 and specifically the handling of the multiple channels and filters as explained in Sect. 9.4.4.
- **Convolution Layer 1**
  - **Kernel:** 5×5 kernel packed with shape $\left[5, 5, \frac{*}{32}, \frac{3}{4}, \frac{*}{16}, \frac{6}{8}\right]$.
  - **Bias:** with shape $\left[\frac{*}{32}, \frac{*}{4}, \frac{*}{16}, \frac{6}{8}\right]$.
  - **Output:** of shape $\left[\frac{28?\sim}{32}, \frac{1?}{4}, \frac{28?\sim}{16}, \frac{6}{8}\right]$. Note that the channel dimension has moved from being second to being fourth. This is because the multiple channels and filters are handled as matrix-vector multiplication (see Sect. 8.10) which is causing the dimensions to alternate. The first and third dimensions are still marked as interleaved, though with these sizes they are already equivalent to basic tiling.
- **Activation Layer** Apply the polynomial activation $A1(x) = 0.061395x^2 + 0.134343x$; see Fig. 11.7. Here and in the next activation layers, the coefficients of the activation polynomials were set during the training step.
- **Average Pooling 1:** Average 2×2 values into one value.

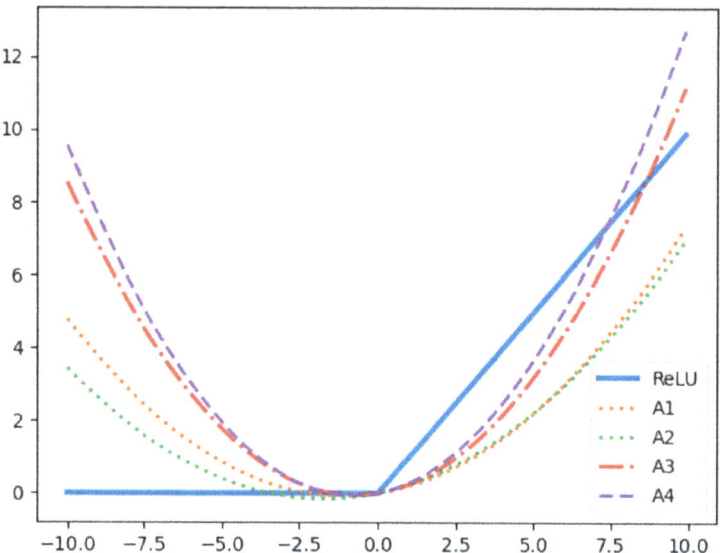

Fig. 11.7 Graphs of the activation functions used by our LeNet5 design. Specifically four quadratic trainable functions were used: $A1$, $A2$, $A3$, and $A4$; while not mimicking ReLU, they still allow the network to yield good prediction results

- **Output:** of shape $\left[\frac{14\sim}{16}, \frac{1?}{8}, \frac{14\sim}{16}, 1, \frac{6}{8}\right]$. Note that this layer splits the first dimension from being $\left[\frac{14\sim}{32}\right]$ to $\left[\frac{14\sim}{16}, \frac{1?}{2}\right]$ as explained in Sect. 9.4.3. Then 2 dimensions $\left[\frac{1?}{2}, \frac{1?}{4}\right]$ are merged into $\left[\frac{1?}{8}\right]$ since in a tile tensor shape, consecutive dimensions $\left[\frac{1?}{a}, \frac{1?}{b}\right]$ can be merged to single dimension $\left[\frac{1?}{ab}\right]$.

- **Convolution Layer 2**
  - **Kernel:** 5×5 kernel packed with shape $\left[5, 5, \frac{*}{16}, \frac{16}{8}, \frac{*}{16}, 1, \frac{6}{8}\right]$.
  - **Bias:** with shape $\left[\frac{*}{16}, \frac{16}{8}, \frac{*}{16}, 1, \frac{*}{8}\right]$.
  - **Output:** of shape $\left[\frac{10?\sim}{16}, \frac{16}{8}, \frac{10?\sim}{16}, \frac{1?}{8}\right]$. Note that the channel dimension has moved back from being fourth to being second.

- **Activation Layer 2** Apply the polynomial activation $A2(x) = 0.052587x^2 + 0.181897x$; see Fig. 11.7.
- **Average Pooling 2:** Average 2×2 values into one value.
  - **Output:** of shape $\left[\frac{5\sim}{8}, \frac{1}{2}, \frac{16}{8}, \frac{5\sim}{8}, \frac{1?}{16}\right]$. Again, the first dimension was split from being $\left[\frac{10?\sim}{16}\right]$ to $\left[\frac{5\sim}{8}, \frac{1}{2}\right]$ and the third dimension $\left[\frac{10?\sim}{16}\right]$ was split to $\left[\frac{5\sim}{8}, \frac{1}{2}\right]$ where the second part was then merged into the following dimension.

11.3 The LeNet NN

- **Flatten.** The flattening layer is a logical layer that does not change the packing or the values of the neurons.
- **Fully Connected 1.**
  - **Weights:** of shape $\left[\frac{5\sim}{8}, \frac{1}{2}, \frac{16}{8}, \frac{5\sim}{8}, \frac{120}{16}\right]$. The weight logical tensor has a shape [120, 400]. Since we did not flatten the input image, we split the dimension of size 400 to [5, 16, 5] to match the input. The matrix multiplication will then sum over all three dimensions.
  - **Bias:** of shape $\left[\frac{*}{8}, \frac{1}{2}, \frac{*}{8}, \frac{*}{8}, \frac{120}{16}\right]$.
  - **Output:** of shape $\left[\frac{*}{8}, \frac{1}{2}, \frac{1?}{8}, \frac{1?}{8}, \frac{120}{16}\right]$. Dimensions 0, 2, 3 have been summed over. The first one results in duplications, and the other two with unknowns, as explained in Definition 8.33.
- **Activation Layer 3** Apply the polynomial activation $A3(x) = 0.0996568x^2 + 0.142005x$; see Fig. 11.7.
- **Fully Connected 2.**
  - **Weights:** of shape $\left[\frac{84}{8}, \frac{*}{2}, \frac{*}{8}, \frac{*}{8}, \frac{120}{16}\right]$. Here we multiply with the logical matrix [84, 120]. We place the 84 dimensions in the first dimension, which happens to be unused in the input. A better strategy could have been to merge all the input unused dimensions to one, and place our matrix first dimension there.
  - **Bias:** of shape $\left[\frac{84}{8}, \frac{*}{2}, \frac{*}{8}, \frac{*}{8}, \frac{*}{16}\right]$.
  - **Output:** of shape $\left[\frac{84}{8}, \frac{1?}{2}, \frac{1?}{8}, \frac{1?\sim}{8}, \frac{1?}{16}\right]$. The bias is a logical tensor of shape [6], one value for each kernel. It is packed in this particular way for compatibility with the convolution output; see Sect. 8.8 regarding elementwise operators.
- **Activation Layer 4** Apply the polynomial activation $A4(x) = 0.112863x^2 + 0.170549x$; see Fig. 11.7.
- **Fully Connected 3.**
  - **Weights:** of shape $\left[\frac{84}{8}, \frac{10}{2}, \frac{*}{8}, \frac{*}{8}, \frac{*}{16}\right]$.
  - **Bias:** of shape $\left[\frac{*}{8}, \frac{10}{2}, \frac{*}{8}, \frac{*}{8}, \frac{*}{16}\right]$.
  - **Output:** of shape $\left[\frac{*}{8}, \frac{10}{2}, \frac{1?}{8}, \frac{1?}{8}, \frac{1?}{16}\right]$.
- **Output.** The output of the network is ten values (each value duplicated multiple times) where the ith value is the confidence score that the image encodes the ith digit.

11.4 Obstacles with Implementing ResNet50 NNs and Even Deeper DNNs

Implementing DNNs such as ResNet50 requires resolving issues that one might have ignored in shallower NNs. These challenges can be roughly categorized to performance- and accuracy-related challenges. These challenges are inherently intertwined since performance and accuracy are often traded-off. While higher-degree polynomial approximations and an abundance of skip connections may improve accuracy, their effect on performance in DNNs can be drastic. Reducing the number of costly bootstrap operations, architecture-aware ciphertext packing, and removal of skip connections that might not make sense in terms of chain index and tensor shape may help to reduce latency. In addition, since polynomials are used throughout the NN, certain phenomena unique to deep NNs may be witnessed, and other undesired phenomena may be exaggerated. For example, one major challenge is that polynomial approximations are only accurate in a certain input range. As the vectors propagate through more and more layers (and polynomials), their value variance's becomes bigger to the point where some of them deviate from the valid approximation domain. At that point due to the high value gradient nature of the high-order polynomial approximations, one might witness imprecise results at best and output value explosions at worst. Other phenomena such as vanishing and exploding gradients are exaggerated during the training phase and should be addressed properly. One may alleviate such problems by, for example, increasing the degree of the approximating polynomial, or by adapting to the limited range during training by using a loss function that favors valid ranges, or by introducing more frequent normalization steps.

11.5 Reducing Computations Under Encryption

11.5.1 Pruning

A common practice to improve inference running time of NNs and decrease overfitting is to reduce the number of parameters (neurons or weights). The process of reducing the number of neurons is called *pruning* and it is a commonly used practice in plaintext NNs as well. In this process, neurons, weights, or both are iteratively removed from the network until the accuracy of the network drops below an accepted level. In each iteration, the goal is to remove the neurons or weights that contribute the least to the output of the network. For weights, this can be done by removing the weights that are close to 0. For neurons, these can be the neurons whose output was close to 0 on a set of predefined benchmark inputs (see Fig. 11.8 for an example). After pruning the network, it should be fine-tuned. See Fig. 11.9.

11.5 Reducing Computations Under Encryption

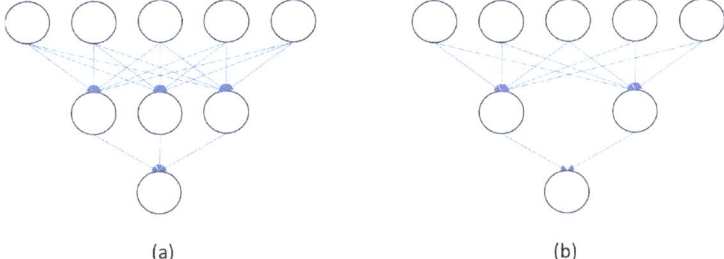

Fig. 11.8 The effect of pruning. (**a**) is a NN and (**b**) is the pruned network, where the middle neuron in the middle layer was removed

Fig. 11.9 Pruning process

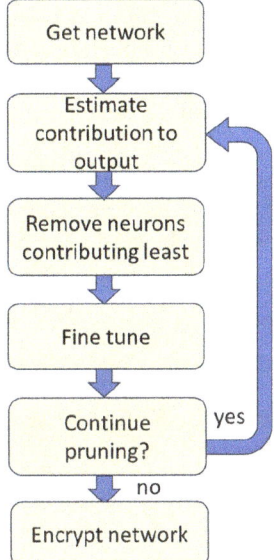

11.5.1.1 Pruning Under FHE

As mentioned, the goal of pruning is to reduce latency when running a network. This is also true when the pruning process happens before the network is encrypted and the pruned network runs under FHE. See Fig. 11.9. When encrypting a pruned network data leakage and packing should be considered as we now explain.

Data Leakage Even in encrypted networks, the topology (i.e., the number of neurons and their connections) is still revealed to some extent, which may leak private information. Even the single piece of information that a network was pruned leaks the information that the data it was trained on was such that it could produce a smaller (pruned) network. This seems to be unavoidable as the purpose of the pruning process is to decrease the size of the network.

Packing We recall that a ciphertext in FHE holds multiple values in a SIMD manner (see Chaps. 7–9). This means that removing only part of the neurons mapped to the slots of a single ciphertext will not have any effect on the running time performance of the network. Instead, to leverage the pruning capability, one needs to prune the network in a way that eventually removes entire ciphertexts. For example, [9] removed entire ciphertexts including all the neurons that were mapped to them. The ciphertexts that were removed were those whose neurons had the least effect on the network. They consider the diagonalization packing used by GAZELLE's [22] (see also Sect. 9.7.1). In this case, when computing matrix vector multiplication $M \cdot v$, removing a diagonal in M saves not only a multiplication operation but also a rotation operation on v. In another work [3] the authors applied permutation on the neurons so neurons with little effect on the network were packing in the same ciphertext, thus causing smaller degradation in accuracy when pruning the same number of ciphertexts. The implementation they provided used the matrix vector multiplication shown in Sect. 8.10. In this case the number of multiplications indeed drops but the number of rotations (coming from the sum-over-dimension operation) does not change. Permuting the order of neurons is easily done for fully connected layers that are implemented by a matrix-vector multiplication. It is harder to do, for example, for convolution layers whose implementation is optimized to a specific packing. In the LeNet example mentioned above, the neurons in each of the last three fully connected layers can be permuted so the neurons with least effect on the network will be mapped to the same ciphertext.

11.5.2 Split Networks

Another method to improve the overall performance of a NN's processing of encrypted data is to split the NN model to a public NN and a private NN. The public model can be a non-polynomial model that is trained over public data and used as a sort of feature extractor at the client side. After the features are extracted they would be homomorphically encrypted and sent to the server side to be analyzed using a smaller, private polynomial NN or ML model (e.g., LoR as used by [27]). This method has the advantage of simplifying the training process as the depth of the polynomial model decreases. One should note however that this method comes with some drawbacks, including the extra work required at the client side and the inability to operate on raw data that hasn't gone through the feature extraction at the server side.

11.5.3 Neural Architectural Search (NAS)

Neural architecture search (NAS) is an umbrella term for various methods of automating NN architecture engineering [14]. NAS methods differ by the *Search*

Space i.e., which architectural building blocks will be used in the search and consequently by the space of possible architectures. The *Search Strategy* is defined as the method of performing the exploration, i.e., of finding the best NN architecture in the Search Space. The *Performance Estimation Strategy* is defined as the method being used to estimate the performance of each explored potential architecture.

Note that the trivial estimation method of simply training each explored model architecture and then evaluating them using a test data set is probably too costly to be impractical. In the context of training HE-friendly polynomial NNs, NAS has been suggested as a way to automate the transformation and design of HE-friendly NNs [7, 32]. Regarding the Search Space, one can explore different polynomial alternatives with different corresponding mixtures of cryptographic primitives (additions, multiplications, and rotations), as well as varying multiplication depths and tile tensor shapes, each with its own performance and accuracy estimates. Similarly, one can also explore various cryptographic parameters (such as the number of slots, scale, etc.) while adhering to security, memory, and performance constraints.

11.6 Lab Exercise: Train Large HE-Friendly NNs

The following lab exercise requires a programming environment such as Python. This environment should include libraries that can handle NN inference and training such as [1]:

1. Use the techniques learned in Part II to approximate the GELU and tanh functions using the Remez algorithm.
2. Use the techniques learned in Part II to approximate the GELU and tanh functions using a piecewise function approximation, using quadratic approximation at every function range.
3. Use the techniques learned in Part II to suggest approximations for the Softmax function with n inputs. What assumptions have you made on the input range?

In the following questions use some reference AlexNet [25] implementation trained on CIFAR-10 [24]:

4. Use the techniques learned in Part II to approximate the GELU and tanh functions.
5. Modify the activation layers with different non-polynomial functions such as ReLU, GELU, tanh, and Sigmoid and retrain the model. Compare the resulting prediction accuracy as well as the time it took to train the models.
6. Approximate the activation functions using different polynomial approximation methods (e.g., Taylor polynomial Sect. 5.2 and Remez 5.3) and compare the approximation accuracy.
7. Using a polynomial approximation of a high degree (e.g., 48), use the weights of a pre-trained non-polynomial AlexNet architecture and only replace the non-

polynomial functions without any re-training or KD. Compare the prediction accuracy to the retrained version. How does the accuracy change as the degree of the polynomial approximations increases?

8. Attempt the above experiments with a ResNet-50 architecture trained on CIFAR-10 and then decrease the number layers to 18 and repeat the previous experiments.

In the following questions use LeNet5 and the CIFAR-10 dataset.

9. Train a polynomial (HE-friendly) model and compare the resulting prediction accuracy with the accuracy of the original non-polynomial model.
10. Run an inference operation under FHE on a batch of 1, 2, 4, 8 inputs. For this, you can use an FHE SDK such as pyhelayers to convert a network from, e.g., an ONNX representation to a FHE-based format. Extract the tile tensor shapes, e.g., using the operation

    ```
    pyhelayers.set_neural_net_verbosity_level(
            pyhelayers.VERBOSITY_REGULAR)
    ```

    when using pyhelayers. What differences do you observe?

References

1. Abadi, M., Agarwal, A., Barham, P., Brevdo, E., Chen, Z., Citro, C., Corrado, G.S., Davis, A., Dean, J., Devin, M., Ghemawat, S., Goodfellow, I., Harp, A., Irving, G., Isard, M., Jia, Y., Jozefowicz, R., Kaiser, L., Kudlur, M., Levenberg, J., Mané, D., Monga, R., Moore, S., Murray, D., Olah, C., Schuster, M., Shlens, J., Steiner, B., Sutskever, I., Talwar, K., Tucker, P., Vanhoucke, V., Vasudevan, V., Viégas, F., Vinyals, O., Warden, P., Wattenberg, M., Wicke, M., Yu, Y., Zheng, X.: TensorFlow: large-scale machine learning on heterogeneous systems (2015). https://www.tensorflow.org/. Software available from tensorflow.org
2. Aharoni, E., Adir, A., Baruch, M., Drucker, N., Ezov, G., Farkash, A., Greenberg, L., Masalha, R., Moshkowich, G., Murik, D., et al.: HElayers: a tile tensors framework for large neural networks on encrypted data. PoPETs (2023). https://doi.org/10.56553/popets-2023-0020
3. Aharoni, E., Baruch, M., Bose, P., Buyuktosunoglu, A., Drucker, N., Pal, S., Pelleg, T., Sarpatwar, K., Shaul, H., Soceanu, O., Vaculin, R.: Efficient pruning for machine learning under homomorphic encryption. In: Computer Security – ESORICS 2023: 28th European Symposium on Research in Computer Security, The Hague, The Netherlands, September 25–29, 2023, Proceedings, Part IV, pp. 204–225. Springer, Berlin (2024). https://doi.org/10.1007/978-3-031-51482-1_11
4. Aharoni, E., Baruch, M., Drucker, N., Ezov, G., Kushnir, E., Moshkowich, G., Soceanu, O.: Poster: Secure SqueezeNet inference in 4 minutes. In: 43rd IEEE Symposium on Security and Privacy (2022). https://www.ieee-security.org/TC/SP2022/downloads/SP22-posters/sp22-posters-50.pdf
5. Baruch, M., Drucker, N., Ezov, G., Kushnir, E., Lerner, J., Soceanu, O., Zimerman, I.: Training large scale polynomial CNNs for E2E inference over homomorphic encryption (2023). arXiv preprint arXiv:2304.14836. https://arxiv.org/abs/2304.14836
6. Baruch, M., Drucker, N., Greenberg, L., Moshkowich, G.: A methodology for training homomorphic encryption friendly neural networks. In: Applied Cryptography and Network Security Workshops, pp. 536–553. Springer, Cham (2022). https://doi.org/10.1007/978-3-031-

References

16815-4_29

7. Bian, S., Jiang, W., Lu, Q., Shi, Y., Sato, T.: Nass: Optimizing secure inference via neural architecture search. ECAI 2020 (2020). https://doi.org/10.3233/FAIA200288
8. Bourse, F., Minelli, M., Minihold, M., Paillier, P.: Fast homomorphic evaluation of deep discretized neural networks. In: CRYPTO. Springer, Cham (2018)
9. Cai, Y., Zhang, Q., Ning, R., Xin, C., Wu, H.: Hunter: HE-friendly structured pruning for efficient privacy-preserving deep learning. In: Proceedings of the 2022 ACM on Asia Conference on Computer and Communications Security, ASIA CCS '22, pp. 931–945. Association for Computing Machinery, New York (2022). https://doi.org/10.1145/3488932.3517401
10. Corvoysier, D.: Experiment with SqueezeNets, commit:2619f730b4e91313057039feb81788c5648e3951 (2017). https://github.com/kaizouman/tensorsandbox/tree/master/cifar10/models/squeeze
11. Dathathri, R., Saarikivi, O., Chen, H., Laine, K., Lauter, K., Maleki, S., Musuvathi, M., Mytkowicz, T.: CHET: an optimizing compiler for fully-homomorphic neural-network inferencing. In: Proceedings of the 40th ACM SIGPLAN Conference on Programming Language Design and Implementation, PLDI 2019, pp. 142–156. New York (2019). https://doi.org/10.1145/3314221.3314628
12. Deng, J., Dong, W., Socher, R., Li, L.J., Li, K., Fei-Fei, L.: Imagenet: a large-scale hierarchical image database. In: 2009 IEEE Conference on Computer Vision and Pattern Recognition, pp. 248–255. IEEE, Piscataway (2009).
13. Deng, L.: The mnist database of handwritten digit images for machine learning research. IEEE Signal Process. Mag. **29**(6), 141–142 (2012)
14. Elsken, T., Metzen, J.H., Hutter, F.: Neural architecture search: a survey. J. Mach. Learn. Res. **20**(1), 1997–2017 (2019)
15. Gilad Bachrach, R., Dowlin, N., Laine, K., Lauter, K., Naehrig, M., Wernsing, J.: Cryptonets: applying neural networks to encrypted data with high throughput and accuracy. In: International Conference on Machine Learning, pp. 201–210 (2016). http://proceedings.mlr.press/v48/gilad-bachrach16.pdf
16. Gou, J., Yu, B., Maybank, S.J., Tao, D.: Knowledge distillation: a survey. Int. J. Comput. Vis. **129**, 1789–1819 (2021)
17. Gunraj, H., Sabri, A., Koff, D., Wong, A.: Covid-net ct-2: enhanced deep neural networks for detection of covid-19 from chest ct images through bigger, more diverse learning. Front. Med. **8** (2022). https://doi.org/10.3389/fmed.2021.729287
18. Hinton, G., Vinyals, O., Dean, J.: Distilling the knowledge in a neural network (2015). arXiv preprint arXiv:1503.02531
19. Iandola, F.N., Moskewicz, M.W., Ashraf, K., Han, S., Dally, W.J., Keutzer, K.: SqueezeNet: AlexNet-level accuracy with 50x fewer parameters and <1MB model size (2016). arXiv preprint arXiv:1602.07360
20. Jiang, X., Kim, M., Lauter, K., Song, Y.: Secure outsourced matrix computation and application to neural networks. In: Proceedings of the 2018 ACM SIGSAC Conference on Computer and Communications Security, pp. 1209–1222 (2018)
21. Juvekar, C., Vaikuntanathan, V., Chandrakasan, A.: GAZELLE: a low latency framework for secure neural network inference. In: 27th USENIX Security Symposium (USENIX Security 18), pp. 1651–1669 (2018)
22. Juvekar, C., Vaikuntanathan, V., Chandrakasan, A.: GAZELLE: a low latency framework for secure neural network inference. In: 27th USENIX Security Symposium (USENIX Security 18), pp. 1651–1669. USENIX Association, Baltimore (2018). https://www.usenix.org/conference/usenixsecurity18/presentation/juvekar
23. Kim, D., Park, J., Kim, J., Kim, S., Ahn, J.H.: HyPHEN: a hybrid packing method and optimizations for homomorphic encryption-based neural networks (2023). arXiv preprint arXiv:2302.02407
24. Krizhevsky, A., Hinton, G., et al.: Learning multiple layers of features from tiny images (2009). http://www.cs.utoronto.ca/~kriz/learning-features-2009-TR.pdf

25. Krizhevsky, A., Sutskever, I., Hinton, G.: ImageNet classification with deep convolutional neural networks. Neural Inform. Process. Syst. **25** (2012). https://doi.org/10.1145/3065386
26. Lee, E., Lee, J.W., Lee, J., Kim, Y.S., Kim, Y., No, J.S., Choi, W.: Low-complexity deep convolutional neural networks on fully homomorphic encryption using multiplexed parallel convolutions. In: Chaudhuri, K., Jegelka, S., Song, L., Szepesvari, C., Niu, G., Sabato, S. (eds.) Proceedings of the 39th International Conference on Machine Learning, Proceedings of Machine Learning Research, vol. 162, pp. 12403–12422. PMLR (2022). https://proceedings.mlr.press/v162/lee22e.html
27. Lee, G., Kim, M., Park, J.H., Hwang, S.w., Cheon, J.H.: Privacy-preserving text classification on bert embeddings with homomorphic encryption (2022). arXiv preprint arXiv:2210.02574
28. Lee, J., Lee, E., Lee, J.W., Kim, Y., Kim, Y.S., No, J.S.: Precise approximation of convolutional neural networks for homomorphically encrypted data. IEEE Access (2023). https://doi.org/10.1109/ACCESS.2023.3287564
29. Lee, J.W., Kang, H., Lee, Y., Choi, W., Eom, J., Deryabin, M., Lee, E., Lee, J., Yoo, D., Kim, Y.S., No, J.S.: Privacy-preserving machine learning with fully homomorphic encryption for deep neural network. IEEE Access **10**, 30039–30054 (2022)
30. Liu, J., Juuti, M., Lu, Y., Asokan, N.: Oblivious neural network predictions via MiniONN transformations. In: Proceedings of the 2017 ACM SIGSAC Conference on Computer and Communications Security, CCS '17, pp. 619–631. Association for Computing Machinery, New York (2017). https://doi.org/10.1145/3133956.3134056
31. Livni, R., Shalev-Shwartz, S., Shamir, O.: On the computational efficiency of training neural networks. In: Proceedings of the 27th International Conference on Neural Information Processing Systems - Volume 1, NIPS'14, pp. 855–863. MIT Press, Cambridge (2014)
32. Lou, Q., Jiang, L.: Hemet: A homomorphic-encryption-friendly privacy-preserving mobile neural network architecture. In: International Conference on Machine Learning, pp. 7102–7110. PMLR (2021)
33. Mirzadeh, S.I., Farajtabar, M., Li, A., Levine, N., Matsukawa, A., Ghasemzadeh, H.: Improved knowledge distillation via teacher assistant. Proc. AAAI Conf. Artif. Intell. **34**, 5191–5198 (2020). https://doi.org/10.1609/aaai.v34i04.5963
34. Mohassel, P., Zhang, Y.: SecureML: a system for scalable privacy-preserving machine learning. In: 2017 IEEE Symposium on Security and Privacy (SP), pp. 19–38 (2017). https://doi.org/10.1109/SP.2017.12
35. Park, J., Kim, M.J., Jung, W., Ahn, J.H.: Aespa: accuracy preserving low-degree polynomial activation for fast private inference (2022). arXiv preprint arXiv:2201.06699
36. Piazza, F., Uncini, A., Zenobi, M.: Neural networks with digital LUT activation functions. In: Proceedings of 1993 International Conference on Neural Networks (IJCNN-93-Nagoya, Japan), vol. 2, pp. 1401–1404 (1993). https://doi.org/10.1109/IJCNN.1993.716806
37. Scardapane, S., Scarpiniti, M., Comminiello, D., Uncini, A.: Learning Activation Functions from Data Using Cubic Spline Interpolation, pp. 73–83. Springer, Cham (2019). https://doi.org/10.1007/978-3-319-95098-3_7
38. Srivastava, N., Hinton, G., Krizhevsky, A., Sutskever, I., Salakhutdinov, R.: Dropout: a simple way to prevent neural networks from overfitting. J. Mach. Learn. Res. **15**(1), 1929–1958 (2014)
39. Wu, W., Liu, J., Wang, H., Tang, F., Xian, M.: PPolyNets: achieving high prediction accuracy and efficiency with parametric polynomial activations. IEEE Access **6**, 72814–72823 (2018). https://doi.org/10.1109/ACCESS.2018.2882407
40. Zhang, M., Xu, S., Fulcher, J.: Neuron-adaptive higher order neural-network models for automated financial data modeling. IEEE Trans. Neural Netw. **13**(1), 188–204 (2002). https://doi.org/10.1109/72.977302
41. Zhang, Q., Yang, L.T., Chen, Z.: Privacy preserving deep computation model on cloud for big data feature learning. IEEE Trans. Comput. **65**(05), 1351–1362 (2016). https://doi.org/10.1109/TC.2015.2470255
42. Zimerman, I., Baruch, M., Drucker, N., Ezov, G., Soceanu, O., Wolf, L.: Converting transformers to polynomial form for secure inference over homomorphic encryption (2023). arXiv preprint arXiv:2311.08610. https://arxiv.org/abs/2311.08610

Glossary

Activation function A nonlinear transformation that is applied to the weighted sum of neuron inputs in an NN.

Analytic function A function $f(x)$ is analytic in some domain D if for any x_0 in D, its Taylor series about x_0 converges to the value of f in some neighborhood of x_0. Examples include any polynomial function, e^x, $log_a(x)$, and standard trigonometric functions.

Approximated scheme An HE scheme, where $Dec(f(Enc(x))) = f(x) + \epsilon$ for some small ϵ.

Batch normalization A normalization technique in NNs, it normalizes values across a batch of inputs per a single neuron in order to achieve better training stability.

Bootstrapping An FHE technique that enables further computation over ciphertexts, by for example, reducing noise in BGV or enabling further multiplications on a ciphertext in CKKS.

Chebyshev polynomials A set of polynomials for minimizing the error in approximation of functions over a given interval.

Chebyshev's Equioscillation theorem A theorem that states that there exists a unique optimal approximation function that minimizes the maximum error of the approximation on some predefined interval.

Ciphertext An encrypted plaintext.

Circuit partitioning The process of dividing complex hardware circuits into smaller and manageable sub-circuits for the sake of analysis, optimization, or parallel processing.

Circuit model A computation model in which the behavior of algorithms does not depend on the input.

Classification models ML methods that map input data into predefined categories or classes based on some learned patterns.

Cleartext Data represented in its native domain, e.g., floating-point numbers, boolean data, or strings.

Column-major flattening A flattening convention where the elements along the first dimension are flattened consecutively, see Definition 7.10.

Comparison model A computation model in which the behavior of algorithms depends only on comparison of values that depend on the input. Also referred as comparison-based model.

Computationally bounded adversaries Adversaries with bounded computational resources. For example, PPT adversaries in some security parameter λ.

Copy and recurse A technique to transform a plaintext algorithm based on tree-traversal into an algorithm running under FHE. The FHE algorithm has a similar running time as the plaintext algorithm (any processing done at the nodes the plaintext algorithm visits) with an additional overhead proportional to the tree size.

Cyclotomic polynomial The nth cyclotomic polynomial for $n \in \mathbb{Z}_+$, is the unique irreducible polynomial with integer coefficients that divides $x^n - 1$ but does not divide $x^k - 1$ for any $k < n$.

Equidistant points A set of equally spaced points at consistent intervals along a given range.

Exact HE scheme The opposite of an approximated scheme. An HE scheme, where $\text{Dec}(f(\text{Enc}(x))) = f(x)$.

Exponentiation by squaring An efficient method for exponentiation that reduces the number of required multiplications.

Function domain The set of input values where a function is defined.

Function range The set of all possible output values of a function when operating on a given domain.

Greedy algorithm An algorithm that at each decision point takes the action with the highest immediate value without planning ahead.

Horner's method An efficient method to perform polynomial evaluation.

Hybrid encryption See transciphering.

Hyper-parameters A set of parameters that influence the performance of an ML model, which are not learned directly from the input data to the model.

Layer normalization A normalization technique in NNs, it normalizes values withing the scope of a layer of a NN in order to achieve better training stability.

Malleable scheme An encryption scheme, where ciphertexts can be manipulated to produce a valid decrypted plaintext without the knowledge of the decryption key.

Mathematical ring A mathematical structure, usually a set with two operations on it addition and multiplication. A ring is a generalization of a mathematical field, where operations are commutative and every element in the set has an inverse.

Minimax poly-approximation A minimax polynomial approximation $P(X)$ with degree at most n of a function $f(x)$ in the domain $[a, b]$ is the polynomial that minimizes the maximal approximation error across the entire domain among all polynomials of degree at most n. There exists a unique minimax polynomial of degree at most n for every continuous function in a closed range.

Non-scalar product The operation or the result of multiplying two ciphertexts. Note that both ciphertext-ciphertext and ciphertext-plaintext products can be considered non-scalar because they involve element-wise operations on vectors. However, in this book we generally use the term non-scalar multiplication or non-scalar product to refer to the slower operation of multiplying two ciphertexts.

Oblivious transfer A cryptographic protocol between two parties A and B, where A holds two values v_0 and v_1, and B holds a bit $b \in \{0, 1\}$. The object of the protocol is to transfer v_b to B without revealing b to A or v_{1-b} to B.

Orthonormal basis A vector space with mutually perpendicular vectors with unit length.

Packing scheme A way data is packed within a list of ciphertexts or plaintext elements.

Plaintext Cleartext data encoded to a specific format defined by the HE scheme but still unencrypted.

Polynomial ring A mathematical structure (a ring) that is constructed from the set of polynomials with coefficients in another ring (e.g., $\mathbb{Z}_q$).

Post-quantum secure A property of cryptographic schemes that means that they stay secure even in the presence of a quantum computer.

Pruning The process of removing redundant connections or neurons to improve efficiency while preserving performance of an NN.

Pseudo-rotation Rotation of the elements of a tensor, where some elements shift out on one side, and arbitrary elements shift in on the other side, see Definition 7.16.

Rational reconstruction A method for recovering a rational number from its value modulo a sufficiently large integer.

Rectangular wide matrices Matrices with more columns than rows.

Regression models ML methods for estimating continuous numerical values based on input features.

Reinforcement learning An ML for making decisions by interacting with an environment while maximizing the cumulative rewards.

Relinearization In FHE schemes like CKKS, a ciphertext is represented by a pair of polynomials (b, a) of a polynomial ring, decrypted as $b + a \cdot s$ in the modular arithmetic of the ring where s is the secret key. A multiplication of two such ciphertexts naturally results in a product P that is represented by 3 polynomials (c, b, a) decrypted as $c + b \cdot s + a \cdot s^2$ in the modular arithmetic of the ring. This representation is more costly to handle than the original representation in terms of both time and space. A Relinearize operation on P modifies its internal representation and brings it back to the more efficient form of a pair of polynomials. It is therefore common practice to perform the Relinearize operation following a multiplication.

Remez algorithm An iterative algorithm that searches for the minimax polynomial approximation $P(X)$ for a given function $f(x)$, a required maximal degree n of $P(X)$ and a given input domain in which the estimation is required $[a, b]$.

Rescaling In approximate FHE schemes such as CKKS the input is multiplied by a scale during encoding and then divided back by the scale at the final decoding. The correct scale of a product of two ciphertexts is the product of the scales of the multiplicands. It is often advisable not to diverge too far from the original smaller scale (e.g., so as not to overflow beyond the ring moduli), and it is therefore common practice to rescale the product back to the original scale following a multiplication. This operation is called Rescaling.

Rotatable dimension A dimension of a tensor stored flattened in a ciphertext such that it can be rotated using a single HE rotation operation, see Definition 7.20.

Row-major flattening A flattening convention where the elements along the last dimension are flattened consecutively, see Definition 7.11.

Runge's phenomenon The unexpected oscillations that occur when interpolating a function using high-degree polynomials with equidistant points.

Same padding (convolution) Zero padding added to the input image of a 2D convolution such that the output will have the same size as the input.

Scalar product The operation or the result of multiplying a ciphertext by a scalar (i.e. a single plain value, either integer, real, or complex).

Semantic security A cryptographic property, where ciphertext leaks no information about the encrypted plaintext beyond what is possible to learn from random guessing.

Sorting networks An algorithmic arrangement for efficiently sorting data through parallel comparison operations.

Stationary A stationary time series has statistical properties like mean and variance that remain constant over time.

Supervised learning An ML approach where models are trained using labeled data.

Taylor expansion A Taylor expansion of a function f(x) at a point a is an infinite sum of terms expressed in terms of the function's derivatives at a,

$$\sum_{n=0}^{\infty} \frac{f^n(a)}{n!}(x-a)^n .$$

Taylor series A Taylor expansion of a function f(x) at some point - see *Taylor expansion*.

Taylor polynomial A finite prefix of some infinite Taylor series - see *Taylor series*.

Tile tensors A data structure containing the data of tensor stored within multiple tiles alongside metadata specifying how the data is packed inside the tiles, see Definition 8.5.

Transciphering Transforming data encrypted under one cipher scheme to another while still encrypted.

Turing machine An abstract mathematical model of computation capable of executing algorithms by manipulating symbols on a strip of tape according to a table of rules.

Turing complete A computational system that is capable of simulating any *Turing machine*, indicating full computational power of a computing device.

Unsupervised learning An ML approach where models infer patterns from unlabeled data.

Valid padding (convolution) In contrast with same padding, valid padding means no padding is added to the input image of a 2D convolution, hence the filters will be placed only where they can cover valid pixels of the input image.

Vanishing gradients phenomenon A phenomenon that apply mostly to DNNs training, where the backpropagation process encounters small gradients, which yield a slow learning rate for the early layers.

Index

A
Activation function, 133–134
Advanced encryption standard, 80
Advanced vector extensions, 84
Ajtai, Koml'os, and Szemer'edi, 72
Analytic, 45, 46, 114, 393
Anomaly detection, 61–62, 65, 263–264
Application interface, vii, 18, 21
Approximated scheme, 17, 255, 293
Approximation, approximation theory, viii, 6, 7, 21–22, 57, 80, 97, 111–123, 125–146, 235, 236, 246, 265, 267, 268, 274–276, 279, 282, 286, 289, 290
Artificial intelligence, 5, 136
Attribute inference, 58
Autoregressive (AR), 250
Autoregressive integrated moving average (ARIMA), 235, 250, 260
Autoregressive moving average (ARMA), 251, 260

B
Basic tiling, 180–181, 183, 202, 230, 283
Batch normalization, 277
Benaloh scheme, 15
Bitwise representation, 19, 75, 127, 129, 145
Blind data linkage (BDL), 55
Boolean circuits, 19, 32, 149
Bootstrap, 27, 74, 90, 112, 145, 208, 228, 286
Bootstrapping, 21, 24, 56, 57, 73, 74, 80, 89, 97, 194, 229, 233, 279, 293

Brakerski/Fan and Vercauteren (B/FV), 16, 31, 32, 151
Brakerski, Gentry, Vaikuntanathan (BGV), 16, 31, 32, 129, 151
Branches, 71, 79, 84, 126, 248
Breadth-first-search (BFS), 249
Broadcasting, 174–176, 186, 187, 192
Bytewise representation, 19

C
Central processing unit (CPU), 10, 38, 87, 88, 208
Channel, 49, 174, 210, 216–219, 227, 228, 276, 277, 282–284
Chebyshev expansion, 139–141
Chebyshev nodes, 113, 137–143, 146
Chebyshev polynomials, 118, 136–137, 146, 293
Chebyshev's Equioscillation theorem, condition, 115, 293
Cheon, Kim, Kim, and Song (CKKS), vii, 16, 18–21, 23, 24, 27, 28, 31–33, 40, 45, 54, 71, 72, 74, 79–81, 84 , 87–89, 93, 99, 101–103, 126, 129, 130, 135, 136, 142, 145–146, 151, 164, 165, 171, 197, 208, 231, 236, 237, 253, 266, 279, 295
Chief information security officer (CISO), vii
Chillotti-Gama-Georgieva-Izabachene (CGGI), 16
Chinese remainder theorem (CRT), 20, 243

Ciphertext, 14, 17, 18, 22–28, 31–33, 38–42, 47, 49, 50, 54–56, 63, 70–72, 74, 80, 82, 85–87, 89, 92, 93, 97–99, 101–103, 121, 125, 129, 130, 132, 133, 135, 145, 146, 151–173, 176–179, 182, 191, 193, 196, 197, 201, 206–208, 211, 212, 223, 243, 250–257, 266, 279, 286, 288, 293
Circuit model, 69–73, 82–83
Circuit partitioning, 98, 293
Circuits, 16, 19, 32, 50, 69–74, 87, 89, 90, 96–99, 101, 102, 152, 178, 194, 227, 230, 265
Classification models, 6, 280
Cleartext, 9, 17–21, 24–26, 55, 171, 197, 230, 235, 293, vii
Client, 38, 41–44, 50, 56, 57, 63, 64, 133, 237, 238, 243, 278, 279, 288
Closest vector problem (CVP), 29, 30
Cloud, 13, 15, 18, 42, 45, 47, 49, 50, 55, 57, 265
Column-major, 157, 158, 294
Communication, 9, 41, 44, 49–51, 65, 87, 168, 279
Comparison (of ciphertexts), 72, 102
Comparison model, 69–73, 294
Compatible shapes, 174, 175, 186
Compatible tile tensor shapes, 187, 188, 205
Complex packing, 164–166, 171, 220–223, 230, 231
Composed polynomial, 118, 120, 131–134
Computationally bounded adversaries, 44, 294
Conditional branches, 70, 71, 126
Confidential computing (CC), 10
Conjugation, 89, 161, 165
Convolution, 152, 156, 158, 173, 174, 210–220, 227–228, 231, 272, 275–277, 282, 283, 285, 288, 296
Convolutional neural network (CNN), ix, 174
Copy and recurse, 77–79, 102, 249, 268, 294
Correlation (Pearson), 255–256, 260, 266
Cos, cosine, 115, 136–144, 146
Covariance, 250, 255, 256, 258–261, 266
Covert adversaries, 44
Cramer's rule, 259, 266
Cumulative density function (CDF), 263
Cyclotomic polynomial, 31, 294

D
Data owner, 37, 41–43, 45, 51–54, 239, 265
Data science (DS), viii, ix, 3–10, 16, 125
Decision tree (DT), 75, 76, 152, 247–249, 268
Deep neural network (DNN), 42, 281, 286

Degree of a polynomial, 94, 104, 105, 120, 128, 135, 136, 140
Depth (of computation, circuit, product), 27, 73–74, 84, 89–91, 93–96, 99, 101, 104, 105, 112, 120, 123, 127–135, 137, 140, 144, 145, 235, 263, 271
Determinant, 259
Diagonalized tiling, 225, 226
Diagonal packing, 168–171, 219, 223–228, 230
Differential privacy (DP), 43
Directed a-cyclic (hyper) graph (DAG), 73, 97
Division, 21, 56, 89, 94, 105, 118, 134, 135, 256, 259, 274, 275, 278
Double angle formula, 142
Ducas-Micciancio (DM), 16
Duplication, 58, 183–186, 191, 192, 228–230, 285

E
Economization (of a power series), 140
Electrical power and energy system (EPES), 64
Elementwise operators, 161, 174–176, 186–188, 205, 207, 228
ElGamal, 15
Encoded tile tensors, 196–197
Encrypted tile tensors, 231, 253, 254
Entity resolution (ER), 52, 55
Equality, 26, 90, 91, 111, 117, 125–132
Equidistant points, 137, 138, 146
Euler's formula, 142
Evaluation key, 28
Exact scheme, 17, 21, 24, 26, 41
Exponentiation by squaring, 91–92, 99, 101, 103, 120, 135, 142, 155, 156
External tensor, 179–181, 183, 186–189, 202–204, 210, 215, 220–222, 225, 226
Extrapolation, 88

F
Fast Fourier transform (FFT), 84
Federated learning (FL), 15, 51–52, 238–239
Fermat's little theorem, 130, 134
FHE-friendly, 121, 278–282
FHE over the torus (TFHE), 16, 18, 20, 22, 24, 32, 41, 74, 81, 87–89, 102, 130, 149
FHEW, 24, 80
Filter, 79, 211–214, 217, 219, 276
Finite field, 18, 25, 134
Floating-point arithmetic, 20, 90, 129, 130
Fraud detection, 59–60
Fully connected (FC), 195, 272–273, 285, 288

Index

Fully homomorphic encryption (FHE), 9, 14, 16, 17, 19, 21, 27, 29, 37–55, 60, 70, 72, 79, 80, 82, 84, 89, 95, 98, 111, 121, 129, 144, 145, 151, 159, 197, 212, 236, 246, 256, 258, 264, 265, 287
Function domain, 128
Function owner, 37, 41–43
Function range, 128, 289
Fundamental theorem of algebra, 93

G
Galois keys, 27
Game-of-life, 88
Garbled circuits (GC), 9, 50, 265
General data protection regulation (GDPR), 4, 13, 51, 237
Giant step baby step (GSBS), 28, 195, 229
Goldschmidt's division algorithm, 118
Goldwasser Micali scheme, 15
Graphical processing unit (GPU), 38
Greater than, 71, 130, 131, 142, 143, 145, 253
Greedy algorithm, 74

H
Haar's condition, 117, 118
HeaaN, 142, 143
Health insurance portability and accountability act (HIPAA), 61, 62
Homomorphic encryption (HE), 3–33, 73, 151–153
Homomorphic noise, 21, 26, 90, 112, 115, 136
Honest but curious, 44, 45
Horner's-method (Horner's-rule, polynomial evaluation method), 90–91, 103
Hybrid bitwise, 129, 145–146
Hybrid encryption, 87
Hyper-parameters, 7, 280–282

I
IEEE standard, 112
Image-to-columns, 230
Indistinguishability under chosen plaintext attack (IND-CPA), 38–41
Inequality, 117, 125–132
Inference, 7–8, 45, 47–50, 57, 58, 195, 250, 279, 286, 290
Intellectual property (IP), 8, 42–43, 62
Interleaved complex packing, 221, 222
Interleaved tiling, 202, 203, 205, 206, 209, 230
Interpolation, 113, 125, 137–141, 143

K
Key generation, 14, 31, 49
Knowledge distillation, 280

L
Lattice-based problems, 16
Layer normalization, 277, 278
League, 132
Learning with errors (LWE), 29–31, 134
Least-significant bit (LSB), 74
Levelled homomorphic encryption (LHE), 16, 74
Linear programming (LP), 97
Linear regression (LR), 81, 236, 239–243, 267
Linear secret sharing scheme (LSSS), 50
Local sensitive hash (LSH), 55
Logical-and, 19, 32
Logical tensor, 178, 181, 183, 184, 186, 188, 189, 201–203, 205, 210, 220, 222, 223, 225, 226, 229
Logical-xor, 18, 19
Logistic regression (LoR), 239–247
Lookup table (LUT), 80
Low-density lipoprotein (LDL), 77, 78, 248, 268

M
Machine learning (ML), vii, 5, 7–8, 37, 125, 235–268
Malicious adversaries, 44, 45, 50
Malleable (scheme), 44, 56, 294
Mathematical ring, 20, 243, 294
Matrix multiplication, 152, 161, 173, 174, 176, 188, 192, 195, 219, 222, 223, 228–229, 231, 273, 285
Maximum (Max), 8, 73, 83–85, 89, 127, 131, 132, 196, 276
Max pooling, 125, 131, 276–278
Mean, 64, 99, 101, 111, 113, 119, 138, 140, 141, 144–146, 158, 192, 208, 236, 250, 252, 255, 257–259, 261–264, 266, 276–278
Mean time to detect (MTTD), 64
Membership inference attacks, 58
Minimax polynomial, 115–119, 122, 127–129, 133, 134, 141, 144, 264, 295
Minimum (Min), 8, 64, 196, 207, 236, 341
Model extraction attack, 43, 58
Model inversion attack, 58
Modulus chain index, 27
Most-significant bit (MSB), 129

Moving average (MA), 250–251
Multi-key FHE (MK-FHE), 47, 54
Multi-party computation (MPC), viii, 9, 10, 45, 47, 50, 239, 278, 279
Multi-party FHE (MP-FHE), 43, 45–51
Multiple inputs multiple outputs (MIMO), 210, 217
Multiplication chain index, 27, 73, 89
Multiplication depth, 57, 128, 163, 184, 186, 197, 198, 201, 228, 229, 289

N
Naïve polynomial evaluation method, 90, 91, 99, 103
Neural architecture search (NAS), 288–289
Neural network (NN), ix, 17, 122, 152, 195–196, 238, 271–290
Newton–Raphson method, 120, 121
Noise, 10, 17, 20–28, 32, 41, 47, 48, 54, 55, 57, 73, 74, 80, 90, 112, 115, 136, 155, 184, 236–237, 268
Nondeterministic polynomial time (NP), 74, 97
Non-scalar product, 89–96, 99, 101, 104, 105, 118, 120, 127–129, 131, 133, 135, 294
Normalization, 113, 277–278, 280, 286
Number theoretic transform (NTT), 84
Numerical noise, 21
Numerical stability, 22, 90, 95–96, 103, 112, 118, 119, 135

O
Oblivious transfer (OT), 265, 295
Orthonormal basis, 139, 295

P
Packing scheme, ix, 86, 152, 153, 179, 206, 207, 273, 276, 295
Padding, 54, 178, 185, 195, 205, 210, 213–214, 226, 276, 277, 296
Pafnuty Chebyshev, 137
Paillier scheme, 14–16, 81, 243
Partially homomorphic encryption (PHE), 15, 16
Party, viii, 9, 16, 43–51, 54, 55, 83, 237, 265
Paterson and Stockmeyer (polynomial evaluation method), 89–96, 99, 135, 139
Pearson correlation, 255, 260, 266
Personal health information (PHI), 43
Personally identifiable information (PII), 8, 43
Piecewise polynomial, 113

Plaintext, 14, 15, 17–19, 21, 22, 24–26, 31, 32, 38–40, 49, 55, 56, 63, 72, 75, 77–79, 102, 103, 132, 134, 135, 142, 151, 153, 171, 196, 197, 211, 212, 214, 231, 241, 243, 247–249, 252, 280, 295
Polynomial
  approximation, viii, 21–22, 57, 80, 90, 111, 115–120, 122, 127, 141, 146, 235, 246, 275, 279, 282, 286, 289, 290
Polynomial coefficients, 84, 183
Polynomial ring, 30, 31, 89, 134, 295
Post-quantum secure, 16, 295
Power basis, 117, 146
Prime, 3, 19, 130, 134
Privacy enhancing technology (PET), viii, 1, 13
Privacy-preserving federated learning (PPFL), 51
Privacy-preserving machine learning (PPML), 235–268
Privacy-preserving record linkage (PPRL), 55
Private information retrieval (PIR), 72, 73, 75, 102, 248, 249
Private key, 57
Private set intersection (PSI), 54–55
Probabilistic polynomial time (PPT), 39, 44, 294
Product depth, 27, 89–91, 93, 95, 96, 99, 101, 104, 105, 112, 120, 127–132, 134, 135, 140, 144, 145, 263
Program counter (PC), 70
Programmable bootstrapping (PBS), 80
Property inference attacks, 58
Pruning, 286–288, 295
PseudoRot, 159, 160, 162–165, 204, 205, 209–211, 213–215, 231
Pseudo-rotation, 158–161, 164, 170, 189, 201, 204–205, 209, 211, 212, 231, 252, 277, 295
Public key, 15, 27, 28, 31, 45, 47–49, 56, 59, 61–63
Public key encryption (PKE), 27, 38

Q
Quantization, 22
Quantized NNs, 20

R
Radius of convergence, 112, 114, 115, 122
Random access memory (RAM), 97, 98
Rational reconstruction, 243, 295
Reciprocal, viii, 134–136, 146, 235, 243, 266

Index

Reconstruction attacks, 58
Record linkage (RL), 55
Rectangular wide matrices, 168, 295
Rectified Linear Unit (ReLU), 57, 122, 125, 133–134, 145, 235, 273–275, 278, 284, 289
Regression models, 6, 17, 81, 243, 267, 295
Reinforcement learning, 6, 295
Relinearization, 89, 295
Relinearization key, 48
Remez algorithm, 113, 115–119, 122–123, 127, 128, 136, 141, 289, 295
Rescaling, 25, 89, 197, 295, 296
Residues number system (RNS), 19
Ring LWE (RLWE), 29–31
Rivest–shamir–adleman (RSA), 15
RNS representation, 19
Roots of a polynomial (polynomial evaluation Method), 89–96
Rot, 26, 162, 163, 170, 253
Rotatable dimension, 160, 161, 162–163, 164, 189, 203
Rotate-and-sum, 145, 153–155, 189
Rotation, 174, 186
Row-major, 17, 157, 158, 160–162, 168, 169, 171, 215, 225
Runge's phenomenon, 138

S

Same padding, 213, 296, 297
Scalar product, 103, 123
Secret key, 54
Secret sharing (SS), 50
Semantic security, 40, 70, 72, 296
Semi-honest adversaries, 44, 50, 64
Sensitive personal information (SPI), 43
Server, 38, 58, 62, 63, 81
Shared key, 45, 47–49
Shift, 21, 26, 32, 159, 162, 251, 252, 258, 278, 281, 282, 295
 left, 26
 right, 26, 251
Shortest vector problem (SVP), 29, 30
Sigmoid, 133, 235, 245, 246, 247, 248, 267, 273, 274, 289
Sign, 96, 126, 127, 131, 132, 134
Single input, single output (SISO), 210
Single instruction multiple data (SIMD), 25, 27, 69, 74, 75, 83–88, 102, 103, 132, 135, 151–172, 173–198, 201–231, 243, 245, 288
Sin/sine, 112, 113, 116, 122, 140, 142, 146

Slice, 160, 180, 208–211, 213, 214, 217, 218, 223, 227, 231
Slicing, 208–214, 217
Softmax, 57, 133, 274, 275, 280, 289
Software development kit (SDK), 18
Somewhat homomorphic encryption (SHE), 16
Sorting network, 71, 72
Square-root, 14, 56, 89, 256, 278
Standard deviation, 15, 78, 255, 256, 264, 266
Stationary (time series), 252–255, 296
Stochastic gradient descent (SGD), 236, 237, 241–243, 245, 248, 267
Streaming SIMD extensions (SSE), 84
Strides, 157, 210, 214–216, 219, 276, 277
Sum, 145, 153, 189, 197, 255
Supervised learning, 6, 7, 296
Symmetric encryption, 13
System of linear equations, 7, 30, 81, 117, 137, 139, 140, 258, 259, 261, 266, 267

T

Taylor expansion, 114, 140, 296
Taylor polynomial, 112–119, 133, 140, 141, 289, 296
Taylor series, 114, 115, 119, 122, 140, 141, 146, 296
Tensor, 156, 163–165, 174–198, 242, 247
Tensor pseudo-rotation, 158–161, 162, 164, 170, 189, 201, 204, 205, 209, 211, 212, 214, 218, 231, 252, 277, 295
Tensor rotation, 159, 174, 204, 209, 277
Tensor summation, 163–164
Threshold MP-FHE, 47
Tile shape, 174, 179
Tile tensor, 174–232
 pseudo-rotation, 204–205
 rotation, 203
 shape, 174, 179
 summation, 163–164
Time series, 64, 250–264, 266–268
Tournament, 84, 85, 132
Training, 7, 235–239, 279, 281
Training data, 6–8
Transciphering, 87, 105, 294
Transformers, 131, 274
Trigonometric functions, 137, 139, 141, 143
Trusted execution environment (TEE), 10
TTClearUnknowns, 182, 192, 193, 195, 198
TTDec, 182, 184, 188, 189
TTDup, 184, 185, 192, 193, 195
TTEnc, 178, 181
Turing complete, 88, 296
Turing machine, 6, 296

U
Unknown slots, 181–183
Unsupervised learning, 6, 296
Unused slots, 181–183
Used slots, 181–183

V
Valid padding, 213, 297
Vanishing gradients phenomenon, 297

Variance, 250, 252, 255, 258, 260, 264, 266

X
XGBoost, 61, 131

Z
Zero-knowledge proofs, 44, 50

GPSR Compliance

The European Union's (EU) General Product Safety Regulation (GPSR) is a set of rules that requires consumer products to be safe and our obligations to ensure this.

If you have any concerns about our products, you can contact us on

ProductSafety@springernature.com

In case Publisher is established outside the EU, the EU authorized representative is:

Springer Nature Customer Service Center GmbH
Europaplatz 3
69115 Heidelberg, Germany

www.ingramcontent.com/pod-product-compliance
Lightning Source LLC
Chambersburg PA
CBHW070855270426
43749CB00072B/173